THE PHYSICS OF COLLISIONLESS SHOCKS

Related Titles from AIP Conference Proceedings

774 X-Ray Diagnostics of Astrophysical Plasmas: Theory, Experiment, and Observation
Edited by R. K. Smith, July 2005, 0-7354-0259-0

719 Physics of the Outer Heliosphere: Third International IGPP Conference
Edited by V. Florinski, N. V. Pogorelov, and G. P. Zank, September 2004, 0-7354-0199-3

703 Plasmas in the Laboratory and in the Universe: New Insights and New Challenges
Edited by Giuseppe Bertin, Daniela Farina, and Roberto Pozzoli, April, 2004, 0-7354-0176-4

699 Space Technology and Applications International Forum - STAIF 2003: Conference on Thermophysics in Microgravity; Conference on Commercial/Civil Next Generation Space Transportation; 21st Symposium on Space Nuclear Power and Propulsion; Conference on Human Space Exploration; 2nd Symposium on Space Colonization; 1st Symposium on New Frontiers and Future Concepts
Edited by Mohamed S. El-Genk, February 2004, 0-7354-0171-3
CD-ROM: 0-7354-0172-1

679 Solar Wind Ten: Proceedings of the Tenth International Solar Wind Conference
Edited by Marco Velli, Roberto Bruno, and Francesco Malara, September 2003,
CD-ROM included, 0-7354-0148-9

669 Plasma Physics: 11th International Congress on Plasma Physics: ICPP2002
Edited by Ian S. Falconer, Robert L. Dewar, and Joe Khachan, June 2003,
Print: 0-7354-0133-0; CD-ROM: 0-7354-0134-9

649 Dusty Plasmas in the New Millennium: Third International Conference on the Physics of Dusty Plasmas
Edited by R. Bharuthram, M. A. Hellberg, P. K. Shukla, and F. Verheest, December 2002,
0-7354-0106-3

598 Solar and Galactic Composition: A Joint SOHO/ACE Workshop
Edited by Robert F. Wimmer-Schweingruber, December 2001,CD-ROM included,
0-7354-0042-3

471 Solar Wind Nine: Proceedings of the Ninth International Solar Wind Conference
Edited by Shadia Rifai Habbal, Ruth Esser, Joseph V. Hollweg, Philip A. Isenberg,
May 1999, 1-56396-865-7

To learn more about these titles, or the AIP Conference Proceedings Series, please visit the webpage **http://proceedings.aip.org/proceedings**

THE PHYSICS OF COLLISIONLESS SHOCKS

4th Annual IGPP International Astrophysics Conference

Palm Springs, California   26 February – 3 March 2005

EDITORS
Gang Li
Gary P. Zank
University of California, Riverside
Riverside, California

Christopher T. Russell
University of California, Los Angeles
Los Angeles, California

SPONSORING ORGANIZATIONS
IGPP, University of California, Riverside
IGPP, University of California, Los Angeles
IGPP, Los Alamos National Laboratory

Melville, New York, 2005
AIP CONFERENCE PROCEEDINGS ■ VOLUME 781

Editors:
Gang Li
Gary P. Zank

Institute of Geophysics and Planetary Physics
University of California, Riverside
1432 Geology Building
Riverside, CA 92521
USA

E-mail: gang.li@ucr.edu
zank@ucr.edu

Christopher T. Russell
Institute of Geophysics and Planetary Physics
University of California, Los Angeles
6869 Slichter Hall
Los Angeles, CA 90095
USA

E-mail: ctrussell@igpp.ucla.edu

The articles on pp. 37 – 41, 72 – 76, 165 – 169, and 185 – 190 were authored by U.S. Government employees and are not covered by the below mentioned copyright.

Authorization to photocopy items for internal or personal use, beyond the free copying permitted under the 1978 U.S. Copyright Law (see statement below), is granted by the American Institute of Physics for users registered with the Copyright Clearance Center (CCC) Transactional Reporting Service, provided that the base fee of $22.50 per copy is paid directly to CCC, 222 Rosewood Drive, Danvers, MA 01923, USA. For those organizations that have been granted a photocopy license by CCC, a separate system of payment has been arranged. The fee code for users of the Transactional Reporting Services is: 0-7354-0268-X/05/$22.50.

© 2005 American Institute of Physics

Permission is granted to quote from the AIP Conference Proceedings with the customary acknowledgment of the source. Republication of an article or portions thereof (e.g., extensive excerpts, figures, tables, etc.) in original form or in translation, as well as other types of reuse (e.g., in course packs) require formal permission from AIP and may be subject to fees. As a courtesy, the author of the original proceedings article should be informed of any request for republication/reuse. Permission may be obtained online using Rightslink. Locate the article online at http://proceedings.aip.org, then simply click on the Rightslink icon/"Permission for Reuse" link found in the article abstract. You may also address requests to: AIP Office of Rights and Permissions, Suite 1NO1, 2 Huntington Quadrangle, Melville, NY 11747-4502, USA; Fax: 516-576-2450; Tel.: 516-576-2268; E-mail: rights@aip.org.

L.C. Catalog Card No. 2005930074
ISBN 0-7354-0268-X
ISSN 0094-243X
Printed in the United States of America

CONTENTS

Preface..xi

SESSION 1
OVERVIEW

An Introduction to the Physics of Collisionless Shocks........................3
  C. T. Russell

SESSION 2
SHOCK STRUCTURES: MICROSCOPIC AND KINETIC

Electron Acceleration and Structure in the Quasi-Perpendicular
Collisionless Shock 17
  D. Burgess
On Kinetic Structure of Quasi-Perpendicular Collisionless Shocks 22
  M. Scholer and S. Matsukiyo
Global Hybrid Simulations of the Bow Shock 27
  N. Omidi, X. Blanco-Cano, and C. T. Russell
Ion Dynamics at Shocks: Ion Reflection and Beam Formation at
Quasi-Perpendicular Shocks 32
  H. Kucharek and E. Möbius
Multi-Spacecraft Observations of Interplanetary Shocks.......................37
  A. Szabo
Classical MHD Shocks: Theory and Numerical Simulation........................42
  N. V. Pogorelov
Issues for Hybrid Simulations of Collisionless Shocks 50
  D. Winske and L. Yin
A New Simulation Technique for Study of Collisionless Shocks:
Self-Adaptive Simulations 56
  H. Karimabadi, Y. Omelchenko, J. Driscoll, R. Fujimoto, K. Perumalla, and
  D. Krauss-Varban
Magnetohydrodynamics of Shocks with Reflected Particles:
Rankine-Hugoniot Relations 64
  B. Dasgupta, G. P. Zank, R. Bedros, and G. M. Webb
Electrons at Shocks..72
  K. W. Ogilvie

SESSION 3

PLANETARY BOW SHOCKS, STRUCTURE AND WAVES

The Electric Potential at the Earth's Quasi-Parallel Bow Shock:
Initial Cluster Results. 79
    R. Behlke, H. Kucharek, S. D. Bale, M. André, and E. A. Lucek
On Increasing Accuracy of Bow Shock Shape and
Position Predictions 84
    J. Merka
Field-Aligned and Gyrating Ion Beams in a Planetary Foreshock 89
    C. Mazelle, K. Meziane, M. Wilber, and D. Le Quéau
The Locations and Shapes of Jupiter's Bow Shock and Magnetopause......... 95
    R. J. Walker, S. P. Joy, M. G. Kivelson, K. Khurana, T. Ogino, and
    K. Fukazawa
Bow Shock and Upstream Waves at Jupiter and Saturn: Cassini
Magnetometer Observations 109
    C. Bertucci, N. Achilleos, C. T. Russell, M. K. Dougherty, E. J. Smith,
    M. Burton, B. T. Tsurutani, and C. Mazelle
A Review of Field-Aligned Beams Observed Upstream of the
Bow Shock 116
    K. Meziane, M. Wilber, C. Mazelle, G. K. Parks, and A. M. Hamza
Methods of Plasma Turbulence Analysis: Application to Shock Studies....... 123
    M. A. Balikhin and S. N. Walker
Observations of Turbulence near Interplanetary Travelling Shocks........... 129
    R. Kallenbach, K. Bamert, M. Hilchenbach, and C. W. Smith
Nonresonant Alfvén Waves Driven by Cosmic Rays 135
    D. Melrose
Hamiltonian Approach to Nonlinear Travelling Whistler Waves.............. 141
    G. M. Webb, J. F. McKenzie, E. Dubinin, and K. Sauer
Upstream Gyrating Ion Events: Cluster Observations and Simulations 146
    K. Sauer, M. Fränz, E. Dubinin, C. Mazelle, A. Korth, H. Rème,
    I. Dandouras, and K.-H. Glaßmeier
Ion Thermalization and Wave Excitation Downstream of Earth's
Bow Shock 151
    Y. C.-M. Liu, M. A. Lee, and H. Kucharek

SESSION 4

PARTICLE ACCELERATION AT SHOCKS

Surfing Acceleration of Ions at Relativistic, Oblique Shocks 159
    D. Üçer and V. D. Shapiro
Simulated 2D vs. 3D Shock Waves: Implications for
Particle Acceleration. 165
    F. C. Jones

Particle Acceleration at Collisionless Shocks: An Overview 170
    G. P. Zank, G. Li, G. M. Webb, J. A. le Roux, V. Florinski, X. Ao, and
    W. K. M. Rice

**The Energetic Storm Particle Event on 2003 October 24: A Test of
Diffusive Shock Acceleration Theory** 180
    D. Lario, R. B. Decker, G. C. Ho, Q. Hu, C. W. Smith, M. I. Desai, and
    A.-F. Viñas

**The Role of Quasi-Perpendicular Shocks in Solar Energetic
Particle Events** 185
    A. J. Tylka

**Energetic Particle Transport in Strong Compressive Wave Turbulence
Near Shocks** 191
    J. A. le Roux, G. P. Zank, G. Li, and G. M. Webb

**Anomalous Diffusion of Energetic Particles: Implications for Diffusive
Particle Acceleration at a Quasi-Perpendicular Shock** 196
    O. P. Verkhoglyadova and J. A. le Roux

**Simulation of SEP Acceleration and Transport at
CME-Driven Shocks** 201
    J. Kóta, W. B. Manchester, J. R. Jokipii, D. L. de Zeeuw, and
    T. I. Gombosi

Diffusive Acceleration of Ions at Interplanetary Shocks 207
    M. G. Baring and E. J. Summerlin

**The Importance of Field-Line Meandering in Particle Acceleration
at Shocks** 213
    J. Giacalone

**Energetic Particles Accelerated by Shocks in the Heliosphere: What is
the Source Material?** 219
    G. M. Mason, M. I. Desai, J. E. Mazur, and J. R. Dwyer

Solar Energetic Particle Spectral Breaks 227
    R. A. Mewaldt, C. M. S. Cohen, G. M. Mason, A. W. Labrador,
    M. L. Looper, D. E. Haggerty, C. G. Maclennan, A. C. Cummings,
    M. I. Desai, R. A. Leske, G. Li, J. E. Mazur, E. C. Stone, and
    M. E. Wiedenbeck

**Upstream Turbulence and the Particle Spectrum at
CME-Driven Shocks** 233
    G. Li, Q. Hu, and G. P. Zank

Generation of Turbulence at Shocks 240
    M. A. Lee

**Relationship of Solar Flare Accelerated Particles to Solar Energetic
Particles (SEPs) Observed in the Interplanetary Medium** 246
    R. P. Lin

Pickup Ions Upstream and Downstream of Shocks 252
    G. Gloeckler, L. A. Fisk, and L. J. Lanzerotti

SESSION 5

SHOCKS IN THE OUTER HELIOSPHERE

Observations of Energetic Ions and Electrons in the Distant
Heliosphere: 2001 - 2005.0. 261
    F. B. McDonald, E. C. Stone, L. F. Burlaga, A. C. Cummings,
    B. C. Heikkila, N. Lal, N. F. Ness, J. D. Richardson, and W. R. Webber

Search for the Heliospheric Termination Shock (TS) and
Heliosheath (HS). 267
    N. F. Ness, L. F. Burlaga, M. H. Acuña, E. C. Stone, and F. B. McDonald

Characteristics of the Termination Shock: Insights from Voyager 273
    A. C. Cummings and E. C. Stone

Voyager Observations of Interplanetary Shocks 278
    J. D. Richardson and C. Wang

Charged-Particle Acceleration at the Heliospheric Termination Shock 283
    J. R. Jokipii

A Global V-Shaped Channel Structure of the Termination Shock Due
to a Magnetic Pressure Effect, and Its Physical Connection to Bipolar
Flow Type Planetary Nebulae 289
    H. Washimi, T. Tanaka, and G. P. Zank

The Termination Shock and Beyond: MHD Modeling 294
    R. Ratkiewicz, J. Grygorczuk, and L. Ben-Jaffel

Comparison of Voyager Shocks in Solar Cycle 23 299
    J. Ashmall and J. Richardson

Initial Comparison between a 3D MHD Model and the HAFv2
Kinematic 3D Model: The October/November 2003 Events from the
Sun to 6 AU 304
    D. S. Intriligator, T. Detman, M. Dryer, C. D. Fry, W. Sun, C. Deehr, and
    J. Intriligator

SESSION 6

OTHER SHOCK RELATED PHENOMENA

3-D Hybrid Simulation of Quasi-Parallel Bow Shock and Its Effects
on the Magnetosphere 313
    Y. Lin and X. Y. Wang

3D Global Simulation of the Interaction of Interplanetary Shocks
with the Magnetosphere. 320
    C. Wang, Z. Huang, Y. Hu, and X. Guo

Spiral Shocks in Astrophysical Disks. 325
    W. K. M. Rice, G. Lodato, and P. J. Armitage

On the Fitting of Ion-Ion Drifting Plasma 331
    E. K. Kaghashvili, G. P. Zank, and B. J. Vasquez

Coronal Shock Waves Observed in Images 336
      H. S. Hudson

Proton, Electron and Ion Temperatures in Fast Shocks 342
      J. C. Raymond and K. E. Korreck

Author Index 347

PREFACE

Collisionless shocks are both a physicist's best dream and worst nightmare. The effects of a collisionless shock on a magnetized plasma (such as the solar wind) are dramatic and important. The shock affects the bulk parameters and the thermal properties of the plasma, and accelerates a small non-Maxwellian portion of plasma particles to even ultra-relativistic energies in the largest of shocks. The collisionless shock is rich in the physical processes that it nurtures. At the same time it is extremely complex. As plasma parameters change, even so subtly as a change in the magnetic field direction, the processes at work can vary drastically. Quiescent processes become turbulent. High energy particles suddenly appear. These energetic particles can damage space systems (shuttles and satellites) and lead to over-exposure in astronauts. Is it safe for human beings to venture beyond the Earth's magnetosphere to be able to set foot on Mars? The answer to this question lies in determining the efficiency and frequency of collisionless shock generation of harmful levels of relativistic particles.

Collisionless shocks are important both in the inner and the outer heliosphere. The solar wind must ultimately stop when it encounters the interstellar plasma. Since it is supersonic, it must pass through a standing shock (the termination shock) before it can be slowed by the interstellar plasma. On December 18, 2004, just before the conference was convened, the termination shock was finally crossed by spacecraft Voyager I. In its 28 year journey, Voyager I had encountered many collisionless shocks: CME-driven shocks originating from the sun, shocks associated with stream interactions, the standing bow shocks around the planets including the Earth and now finally the termination shock.

Over the last decade, tremendous progress in the understanding of the physics of collisionless shock has been made and this was the driving force for this year's IGPP International Astrophysics Conference. From February 27th to March 3rd, 2005, more than 70 scientists from all over the world in various sub-fields of collisionless shock physics gathered together in Palm Springs, California. Over five days, progress on the topics of 1) the micro-structure of collisionless shocks, 2) upstream and downstream wave activity at collisionless shocks, 3) particle acceleration at collisionless shocks and 4) collisionless shocks in outer-heliosphere, was reviewed.

This proceedings contains 54 papers. It is intended that this proceedings will serve both as a summary of our current understanding of collisionless shock physics as well as a starting point for future research, especially for young scientists in this field. The study of collisionless shocks, although already fifty years old since T. Gold first suggested their existence in 1955, still remains one of the most active research areas in space physics, plasma physics and astrophysics.

The conference organizers wish to express their thanks to all participants and especially to those who reviewed manuscripts in this volume. Finally, we would like to thank Ms. Adele Corona for her help in organizing and managing the conference.

G. Li
G. P. Zank
C. T. Russell

SESSION 1
OVERVIEW

An Introduction to the Physics of Collisionless Shocks

C. T. Russell

Institute of Geophysics and Planetary Physics
and
Department of Earth and Space Sciences
University of California, Los Angeles, CA 90095-1567, USA.

Abstract. Collisionless shocks are important in astrophysical, heliospheric and magnetospheric settings. They deflect flows around obstacles; they heat the plasma, and they alter the properties of the flow as it intersects those obstacles. The physical processes occurring at collisionless shocks depend on the Mach number (strength) and beta (magnetic to thermal pressure) of the shocks and the direction of the magnetic field relative to the shock normal. Herein we review how the shock has been modeled in numerical simulations, the basic physical processes at work, including dissipation and thermalization, the electric potential drop at the shock, and the formation of the electron and ion foreshocks.

INTRODUCTION

Systems of collisional particles evolve with time to maxwellian distributions with very few particles at high energies. However, in collisionless plasmas, frequently a small minority of the particles gain a disproportionate share of the energy and become non-maxwellian. Collisionless shocks enable the acceleration of a small fraction of the particles to very high energies. This occurs not only at the bow shock of the Earth at moderately low energies, but also in astrophysical shocks at highly relativistic energies. The elementary-particle nuclear reactions created when those ultra-relativistic particles encounter the Earth's atmosphere, once provided the only means to study elementary particles, and detectors carried into "space" on high altitude balloons were an important tool in nuclear physics.

Today we know much more about the role of shocks in collisionless plasmas. Especially important is their role in the processes of space weather, in the control of the dynamics and energization of the terrestrial magnetosphere, and the energization of the radiation belts in which our spacecraft operate. When interplanetary shocks intercept the terrestrial magnetosphere, they lead to the rapid compression of the magnetosphere. This rapid compression can both heat the plasma and accelerate particles that surf on the shock wave. Even after a shock has passed the Earth, its effects are still felt through the changed properties of the solar wind plasma through which it passed. The solar wind plasma is heated and its bulk velocity increased by the shock. The density and magnetic field strengths are enhanced and the general level of turbulence increased. These all can in turn modify the properties of the magnetospheric plasma and magnetic field that are confined by the solar wind flow. Furthermore, every planet has a standing bow shock that decelerates and deflects the supersonically flowing solar wind around the planetary obstacle. When the flowing plasma is diverted, its properties are changed, and it is these altered properties that control other plasma processes such as reconnection between the solar wind and magnetospheric plasma.

SOME HISTORY

A long-studied phenomenon of magnetospheric dynamics is the sudden impulse, in which the magnetic field as measured on the surface of the Earth is rapidly enhanced. Over 150 years ago it was identified as a disturbance launched from the Sun. In 1955 T. Gold, who also gave us the term magnetosphere, postulated that the sudden impulse must be caused by a collisionless shock because the pressure front was so thin. This postulate, taken as predicting the existence of collisionless shocks in magnetized plasmas, is certainly correct. A pressure wave in the solar wind would be much broader if the pressure wave were subsonic so that the bulk velocity on either side of the wave front changed less than the velocity of the pressure wave. Taken as a postulate on the cause of sudden impulses it is not completely correct for not all sudden impulses are created by collisionless shocks. Pressure-balanced tangential discontinuities are also quite thin. These structures do not propagate but are carried past the magnetosphere rapidly. It is the dynamic pressure measured in the Earth's frame that is important in determining the compressional state of the magnetosphere, and that causes the sudden impulse.

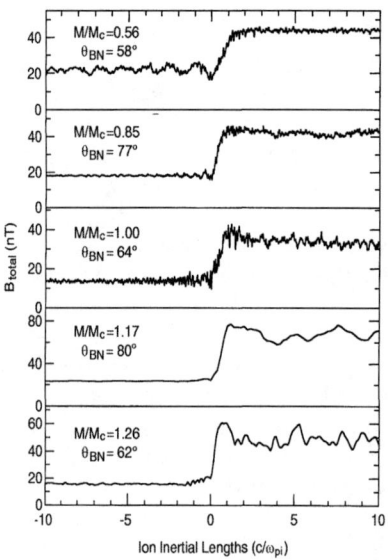

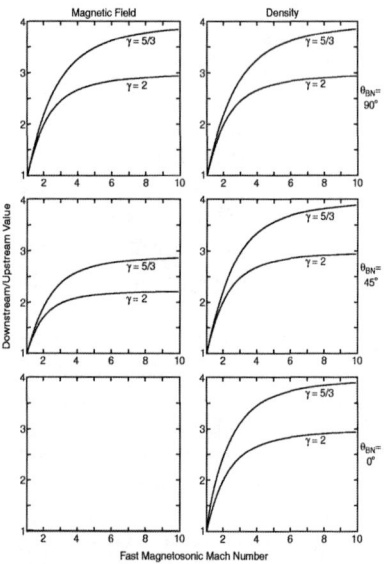

FIGURE 1. Magnetic field profile of low Mach number, low beta shocks observed by the dual co-orbiting satellites ISEE-1 and ISEE-2. M_c is the critical Mach number where the downstream flow velocity along the normal equals the downstream sound speed.

FIGURE 2. Jump in the magnetic field and density across MHD shocks of varying angle between the interplanetary magnetic field and the shock normal (θ_{BN}) plotted versus increasing magnetosonic Mach number as calculated from the Rankine-Hugoniot conditions for $\beta=1$. For these conditions and $\theta_{BN}=0°$ the field does not change across the shock.

The concept of a standing shock in front of the magnetosphere took seven more years to be proposed but surprisingly when it was proposed, it was done so simultaneously in two articles in the same issue of the Journal of Geophysical Research in September 1962 [1,2]. By this time interplanetary spacecraft were being built, and soon high-altitude measurements were being returned. The measurements needed to resolve the bow shock discontinuity had to await the launch of the first Orbiting Geophysical Observatory [3] but the mapping of the location of the bow shock could be done with measurements of lower cadence on the Interplanetary Monitoring Platforms and the permanence of this feature of the interaction was quickly confirmed e.g. [4].

High time resolution measurements are important for studying the bow shock, not only because it is thin but also because it is in motion. This motion was a serious impediment to comparison with theory until the launch of the co-orbiting spacecraft ISEE-1 and ISEE-2. The two ISEEs had an adjustable separation that allowed the shock velocity relative to the spacecraft to be deduced from the timing of the appearance of the shocks at the two locations [5]. Figure 1 shows five crossings of the bow shock by ISEE-1 and 2 under a variety of solar wind conditions, albeit all at low beta and low Mach numbers when the magnetic field was at a large angle to the shock normal. As can be seen the shock is thin, comparable or less than an ion inertial length. The four spacecraft Cluster mission is now extending this work [6].

FLUID TREATMENTS

The jump conditions across the shock can be readily derived from the Rankine-Hugoniot equations of magnetohydrodynamics. These equations are based on the conservation of mass, momentum and energy and Maxwell's equations across a thin discontinuity together with the usual assumptions of magnetohydrodynamics. These equations allow the downstream conditions to be derived from the upstream conditions as a function of the upstream Mach number, the magnetic field direction relative to the flow, the beta of the plasma and the polytropic index that governs the compressibility of the plasma. Figure 2 shows, as a function of Mach number, the jumps in the magnetic field and density for three directions of the upstream magnetic field relative to the shock normal (θ_{BN}) and two values of the polytropic index, γ. The maximum jumps in the magnetic field and density are 3 for $\gamma=2$ and 4 for $\gamma=5/3$. The temperature is not bounded in this way but increases with Mach number. The downstream thermal pressure is, of course, bounded since the source of the downstream thermal pressure is the upstream dynamic pressure. Figure 2 also illustrates that the density is only weakly dependent on the angle of the magnetic field to the shock normal but the magnetic field downstream is very sensitive to this direction.

While the Rankine-Hugoniot equations provide a local solution of the jump in the plasma conditions across the shock, we are often interested in the global interaction problem including the spatially evolving conditions in the subsonic (and supersonic) regions behind the shock. Furthermore, while the Rankine-Hugoniot equations give solutions, they provide no insight into the mechanism for dissipation. Thus we have many reasons for going beyond the Rankine-Hugoniot equations despite their great utility.

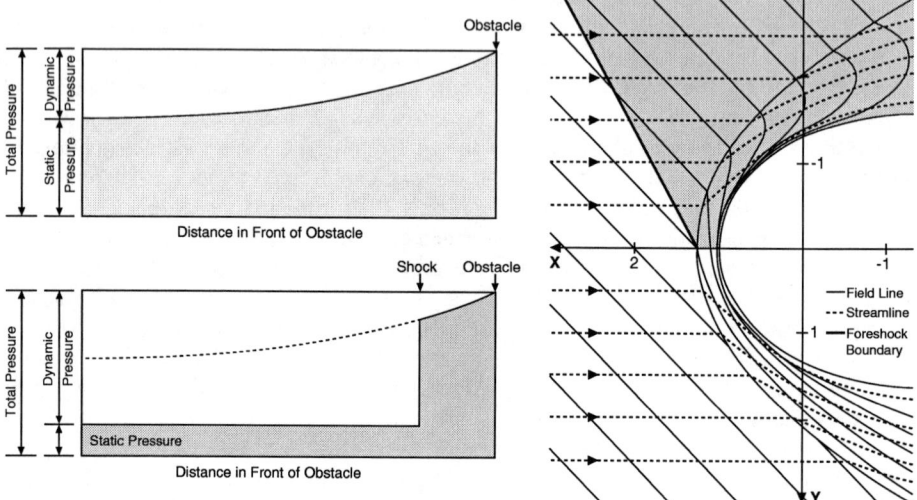

FIGURE 3. Schematic illustration of the deflection of flow by an obstacle under subsonic conditions (top) and how the formation of a shock in a supersonic flow produces the subsonic gradient in pressure needed to deflect the flow.

FIGURE 4. Flow and magnetic field lines in the gasdynamic solution of the solar wind-magnetosphere interaction.

Figure 3 gives us a one-dimensional introduction to the physics of the global interaction of a flowing fluid or gas with an obstacle to the flow. In a subsonic interaction (top) a pressure gradient forms in the flow and the flow is slowed and diverted around the obstacle. In the supersonic case (bottom) the flow does not have sufficient thermal pressure to set up the gradient force needed to slow and deflect the flow without the formation of a shock. The shock forms at the location where the compressed flow behind the shock can flow between the shock and the obstacle. Obviously this depends on the compression of the plasma that is in turn controlled by the Mach number and gamma. As the Mach number approaches unity, the shock moves to infinity. As the obstacle becomes sharper (less blunt), the shock moves toward the obstacle. As the obstacle becomes more absorbing, the shock moves toward the obstacle. If the object absorbs all the flow, as the Earth's moon essentially does, then no shock is present.

The classic solution of the global interaction with the Earth's magnetosphere is the convected-field gas-dynamic treatment by [7] shown in Figure 3. In this treatment there are no magnetic forces and the obstacle is fixed in size. There is only one wave affecting the flow, the compressional wave with an isotropic velocity. We see the deflection of the flow and the draping of the magnetic field. The solution gives generally useful values at high Mach numbers but is not accurate at low Mach numbers where magnetic forces are very important. The model also is incorrect at the subsolar magnetopause where magnetic forces that are present in the magnetic plasma of the magnetosheath enable the plasma to flow around the magnetopause, causing a depletion layer rather than the pile-up predicted by the convected-field gas-dynamic model. Also not produced by the model is the foreshock, shaded in Figure 3, where kinetic effects are important.

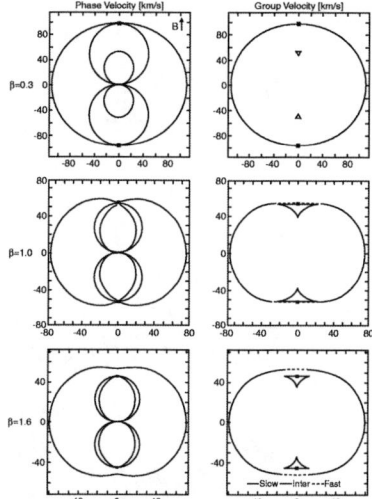

FIGURE 5. Phase and group velocities of the three MHD waves (fast, intermediate and slow) as a function of the angle of the velocity and the magnetic field for three plasma betas.

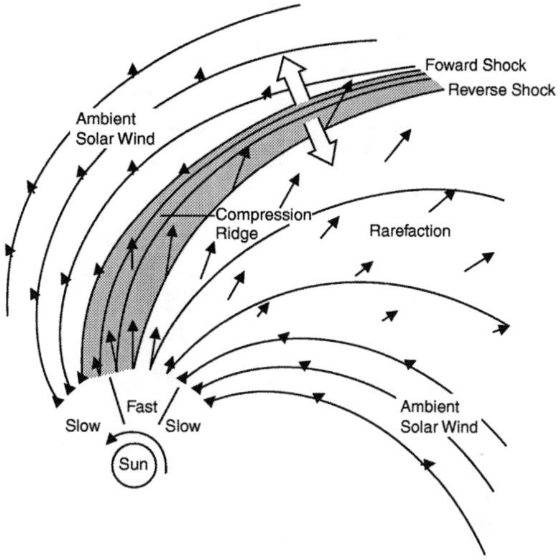

FIGURE 6. Schematic of the interaction of a fast and a slow solar wind stream as a function of heliocentric distance. As the speed of the fast mode compressional wave drops with distance from the Sun, the change in velocity across the interaction region becomes greater than the fast mode speed and shocks can form.

In a magnetized plasma there are three waves that affect the flow and are necessary for the complete description of the interaction: the fast, the intermediate and the slow modes. The phase and group velocities of these waves are shown as a function of the direction of the velocity to the magnetic field for three values of beta in Figure 5. The intermediate wave's energy is closely guided by the magnetic field and the slow wave is almost equally as well guided, even though they both can propagate with phase fronts over a wide range of angles to the field. The fast compressional wave that can propagate at any angle is the work horse of the interaction. Nevertheless they all have some role to play, and, when the magnetic forces are allowed to act in a global simulation, the plasma depletion layer is created as observed [8]. These global MHD simulations also produce a variable-size obstacle that is important to understanding the dynamic magnetosphere. However useful this approach is, it still has its limitations. Even Hall-MHD does not produce the correct wave velocities in a moderate-beta plasma and we must use a kinetic approach to obtain a correct global solution. Moreover, the MHD approach like gas dynamics does not produce a foreshock, the sine qua non of the solar wind interaction with a planetary magnetosphere.

Shocks (and foreshocks) occur in other contexts too. The Sun launches coronal mass ejections into the solar wind and they move faster than the solar wind ahead of them. Here the obstacle is moving but this does not change the overall physical picture and the interaction much resembles that shown in Figure 4. Shocks can also exist in the solar wind in a steady state due to the spatially varying properties of the solar wind across the solar surface. These shocks have a sufficiently different geometry and evolution that it is

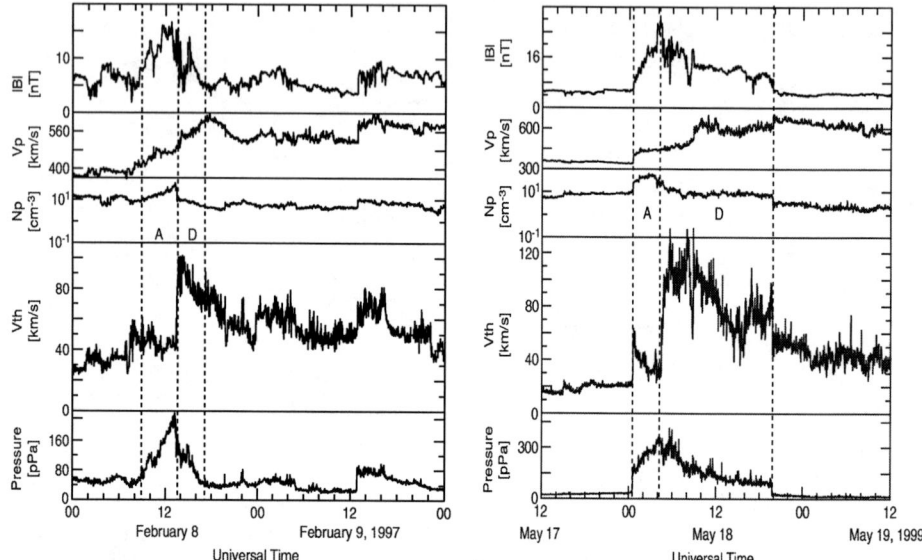

FIGURE 7. Interplanetary magnetic field strength, solar wind speed, number density ion thermal speed and total pressure (magnetic plus thermal) for two stream interactions, one that produced no shocks and one that produced two shocks.

worth mentioning them. Figure 6 illustrates the interaction that leads to these other shocks. Here a fast stream has overtaken a slow stream in the solar wind. This collision can occur because the magnetic structure of the Sun, that controls the solar wind velocity, is not symmetric with respect to the Sun's rotation axis. Thus, as the Sun rotates, the fast stream can overtake the slow stream. Initially the collision just produces a pressure ridge between the two streams slowing down the fast stream and deflecting it and speeding up the slow stream and deflecting it. As the streams move outward the velocity of the fast compressional drops, and the change in velocity across the discontinuity becomes larger than the fast mode velocity and a shock can form. Eventually the change in velocity is great enough for two shocks, forward and reverse to form. We note that the pressure ridge is generally quite simple in the total pressure, magnetic plus thermal, but in any one constituent (magnetic field, density, temperature) the profile might be very complex. Figure 7 illustrates this for two stream interactions, one without shocks and one with two shocks. Some might at first identify the May 18, 1999 event on the right as the interplanetary manifestation of a CME (an ICME) but the increasing solar wind speed through the event and the occurrence of maximum pressure at a discontinuity in both temperature and density distinguish this from ICMEs.

SIMULATIONS OF ION KINETICS

In order to understand the dissipation and thermalization at the bow shock one has to determine how the individual particles behave. In addition, particles can move back into the solar wind and affect the fluid parameters of the flow. The former problem can be treated with local (planar shock) solutions and these local particle simulations have been used for many years to understand mainly the ion kinetics. Treating both ions and electrons in the same simulation remains difficult because of the large ratio (1836) of masses. Fully kinetic treatments are even more difficult when we attempt to understand the spatial variation in the processes occurring in the global interaction between the solar wind flow and a planetary magnetosphere. In this case we have to perform a global kinetic simulation with realistic geometry. To achieve a credible (stationary) solution is currently impossible following the kinetics of both protons and electrons but a hybrid approach using kinetic ions and fluid electrons is possible, and both 2D and 3D global interactions have been run.

One set of hybrid simulations is particularly instructive in understanding the formation of the shock and magnetosheath (and other features of the interaction) because of the systematic exploration of parameter space [9]. In this study the authors explored obstacles of varying size from below an ion inertial length to many times the ion inertial length. For the smallest obstacles showing any disturbance in the flowing plasma only a whistler wave is seen. At larger scale sizes a compressional wake is found. Above the ion inertial length the obstacle produces a compressional wave with a trailing wake. Above about 20 ion inertial lengths the shock forms with a regular magnetosheath. Most importantly the global hybrid simulation reproduces a realistic foreshock region in which the interaction of the upstream waves with the quasi-parallel shock (shock reformation) is reproduced as well as the correlation of wave types and particle distribution functions. Figure 8 shows a section of one of these simulations showing the upstream waves and quasi-parallel shock as color-coded in Bz, the component of the field into the page, thus emphasizing transverse waves. Two cuts are illustrated on the figure, one C1 along the region of field-aligned beams where the wave perturbations are transverse to the field and a second, C2, where the waves are compressional and associated with ring-beam (gyrating) ions. These latter waves appear to be essential to the formation of the quasi-parallel shock [10,11].

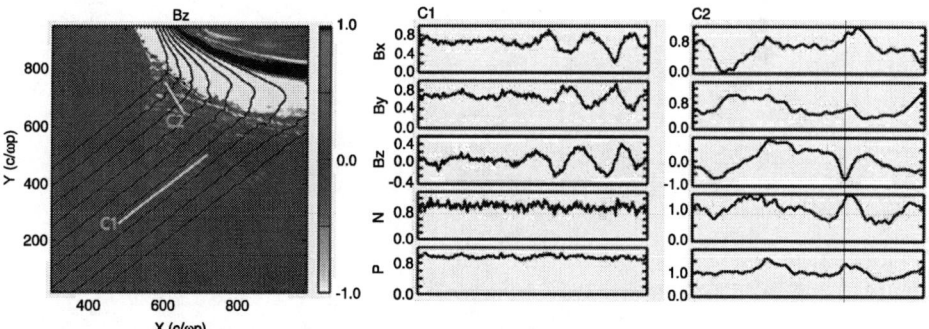

FIGURE 8. The Bz component of the magnetic field (component into the page) over a portion of the grid in a 2D hybrid simulation of the solar wind interaction with a magnetic obstacle. Cut C1 shows non-compressive waves upstream from the quasiparallel shock. Cut C2 shows compressive waves forming near the shock and at the leading edge of the foreshock region.

SHOCK DISSIPATION

The behavior of the ions and electrons at the shock is sufficiently complex that it is difficult to provide an adequate review in a paper of this length but it is important to at least have a qualitative understanding of the behavior of the particles at the shock. Figure 9 attempts to explain how Mach number affects the dissipation process. In the top panel (2 traces) the energy of the incoming flow is high enough and the thermal spread low enough that the whole distribution passes through the electric potential drop of the shock and no particles are reflected. The downstream distribution function is heated and slowed by the compression, and the waves present are sufficient to produce the conditions required by the Rankine-Hugoniot equations. In the bottom 3 traces we show a high Mach number shock with a larger potential barrier that reflects some of the ions back into the solar wind. In a quasi perpendicular shock (in which the angle between the upstream magnetic field and the shock normal is greater than 39°) these ions turn around and drift back into the shock and enter the magnetosheath where they are eventually thermalized as illustrated in the bottom panel. It should be emphasized that both the magnetic field and the electric field participate in the reflection process.

One might expect that the electric potential that slows the ions would accelerate the electrons an equivalent amount but electrons receive little heating at the shock [12]. The reason that electrons can avoid being accelerated is tied to the magnetic structure of the

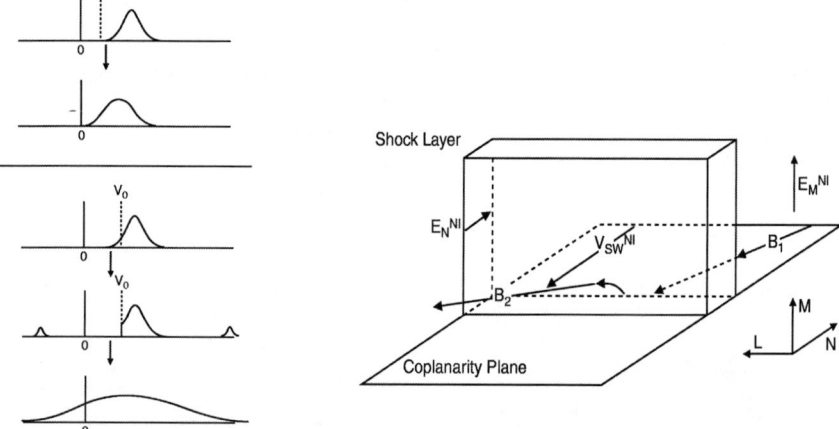

FIGURE 9. Schematic illustration of the variation of shock dissipation with Mach number. In the top half of the panel the solar wind approaches a weak shock with a low electric potential drop so that the entire ion distribution can pass through without reflection. In the lower half a strong shock has a potential drop that turns around part of the distribution so that it returns to the solar wind only to gyrate around the magnetic field and cross the shock at higher speed. The downstream ions are heated in crossing the shock and thermalized with increasing distance down-stream.

FIGURE 10. Schematic illustration of how the electrons move along the magnetic field and are able to avoid the full effect of the potential drop across the shock by moving across interplanetary electric potentials.

shock, at least at low Mach numbers where the heating paradox occurs. Electrons follow the magnetic field as shown in Figure 10 while ions move straight across the shock parallel to the normal. The magnetic field rotates as it crosses the shock in just the direction to move the electrons across solar wind equipotentials to compensate for much of the electric potential drop across the shock normal [13]. This phenomenon is caused by the difference in masses of electrons and ions. As a result the electrons that are generally warmer than ions in the solar wind are the cooler species in the magnetosheath and the plasma sheet by about a factor of 7.

ELECTRON AND ION UPSTREAM WAVES

Electrons and ions are observed streaming along magnetic field lines into the solar wind in front of all planetary shocks. In discussing the dissipation at the shock we noted above that some portion of the ions are often reflected from the shock. Over much of the shock these particles drift back into the magnetosheath but when the angle between the magnetic field and the shock normal becomes less than about 39° the ions can travel upstream along the magnetic field and escape being convected into the magnetosheath. This is the basis for drawing the starting point for the ion foreshock in Figure 11. The direction of the ion foreshock boundary depends on the parallel velocity of the ions along the magnetic field and the drift of ions in the solar wind electric field. Electrons are accelerated back into the solar wind at a much higher velocity and therefore have a foreshock boundary much closer to the tangent field line and they can be accelerated right up to the tangent point (albeit with infinitesimal flux). This beam produces Langmuir oscillations at the plasma frequency of the solar wind.

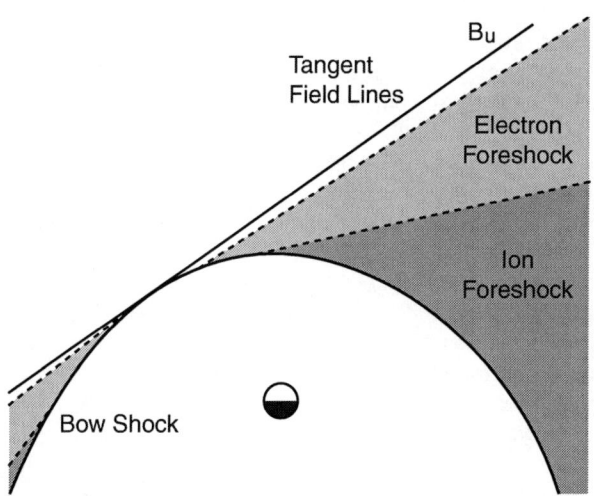

FIGURE 11. Schematic illustration of the location of the electron and ion foreshocks in relationship to the magnetic field line that is tangent to the bow shock.

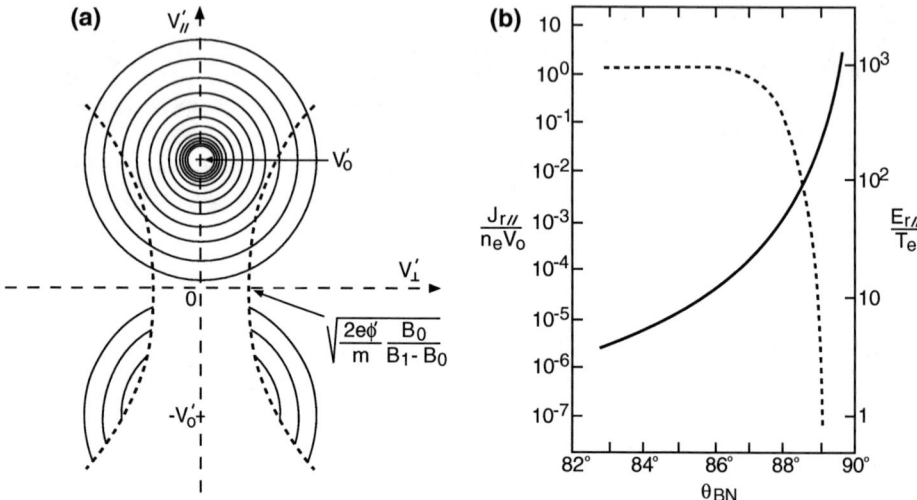

FIGURE 12. Schematic illustration of the production of an electron beam at the bow shock by the fast Fermi acceleration mechanism, shown in the deHoffman-Teller reference frame. The dashed lines in the left-hand panel show the portion of the electron distribution that cannot pass through the ramp because of the magnetic mirror effect countering the potential drop. The closer the magnetic field line is to tangency the further is the distribution above the axis and the smaller is the portion of the distribution reflected and the higher energy of the reflected ions. B_o, B_1, are upstream and downstream magnetic fields; V'_o the deHoffmann-Teller speed, ϕ, the shock potential; n_e and T_o, the upstream electron density and temperature; $J_{r\parallel}$ and $E_{r\parallel}$, the flux and energy per charge of the reflected beam along the magnetic field, and θ_{BN} the shock normal angle [14].

It is instructive to briefly review the reflection of the electrons, often called fast-Fermi acceleration, to gain insight into the nature of this electron beam. We follow the model of [14]. Figure 12 shows the electron angular distribution on the left-hand side in the deHoffman-Teller frame in which the solar wind flow is parallel to the shock normal. The closer is the magnetic field to being tangent to the shock normal the higher is this transformation velocity so that the center of the electron distribution moves upward in this diagram. The dashed line shows the part of the distribution that cannot cross the shock because of the combined action of the magnetic and electric fields. As the deHoffman-Teller velocity becomes faster and faster the reflected portion of the distribution moves to higher and higher energy and the flux reflected becomes less and less. This is made more quantitative in the panel to the left. The result is a Langmuir wave generation region whose extent is controlled by the curvature of the shock.

The waves associated with ions are even more complex as they propagate and steepen while they simultaneously alter the particle distributions that produce them. This close coupling and mutual interaction makes it difficult to sort out cause and effect and underlines the need to use both observation and global simulation to determine the physics of this region.

CONCLUDING REMARKS

At best this paper has been a cursory review of some of the physics of collisionless shocks. Much has been omitted, especially in the area of high Mach number shocks where the magnetic field becomes turbulent and large overshoots appear in the magnetic field just downstream from the shock. The bow shocks of the outer planets are particularly strong and exhibit significant overshoots. We have given wave steepening in the upstream region short shrift and not discussed shocklets and discrete wave pockets (whistlers) that arise when the waves have room to evolve before they are swept into the magnetosheath. We have not addressed waves generated at the shock and propagating into the solar wind. Some of these form standing waves that may be considered to be part of the shock structure. Others, such as the so called 1-Hz waves, can propagate far into the solar wind. We have also concentrated solely on shocks associated with the fast mode.

Collisionless shocks are important for many reasons, including their role in energization of charged particles, but also because of their effect on the bulk properties of the flow. They are important intellectually as they are a point of common interest for astrophysicists, plasma physicists and space physicists. They also show the importance of a strong link between observation and simulation and the importance of global simulation.

Those interested in further discussions of the collisionless shock are referred to two AGU Geophysical Monographs [15,16] and several thematic issues of Advances in Space Research [17,18,19].

ACKNOWLEDGMENTS

The author has benefited from the collaborations with many colleagues in his studies of the shock and foreshock. These include E. W. Greenstadt, C. F. Kennel, J. T. Gosling, M. E. Thomsen, X. Blanco-Cano, N. Omidi, G. Le, J. Newbury, M. Farris and M. M. Mellott. This work was supported by the National Science Foundation under research grant ATM 04-02213.

REFERENCES

1. W .I. Axford, "The interaction between the solar wind and the Earth's magnetosphere", *J. Geophys. Res.*, **67**, 1962, 3791-3796.
2. P. J. Kellogg, "Flow of plasma around the Earth", *J. Geophys. Res.*, **67**, 1962, 3805-3811.
3. R. E. Holzer, M. G. McLeod, and E. J. Smith, "Preliminary results from OGO-1 search coil magnetometer: Boundary positions and magnetic noise spectrum", *J. Geophys. Res.*, **71**, 1966, 1481-1486.
4. D. H. Fairfield, "Average and unusual locations of the Earth's magnetopause and bow shock", *J. Geophys. Res.*, **76**, 1971, 6700-6716.
5. C. T. Russell, and E. W. Greenstadt, "Initial ISEE magnetometer results: Shock observation", *Space Sci. Rev.*, **23**, 1979, 3-37.
6. C. P. Escoubet, C. T. Russell, and R. Schmidt, eds., 1997, 658pp, Kluwer Academic Publishers, Dordrecht, The Netherlands.
7. J. R. Spreiter, A. L. Summers, and A. Y. Alksne, "Hydromagnetic flow around the magnetosphere", *Planet. Space Sci.*, **14**, 1966, 223-253.

8. Y. L. Wang, J. Raeder, C. T. Russell, T. D. Phan, and M. Manapat, "Plasma depletion layer: Event studies with a global model", *J. Geophys. Res.*, **108**, 2003, 1010, doi:10.1029/2002JA009281.
9. X. Blanco-Cano, N. Omidi, and C. T. Russell, "How to make a magnetosphere", *Astronomy and Geophysics*, **45**, 2004, p3.14 - 3.17.
10. N. Omidi, X. Blanco-Cano, and C. T. Russell, Macro-structure of collisionless bow shocks: 1. Scale length, *J. Geophys. Res.*, 2005, submitted.
11. X. Blanco-Cano, N. Omidi, and C. T. Russell, Global hybrid simulations: Nature of ULF waves in the foreshock, *J. Geophys. Res.*, 2005, submitted.
12. M. F. Thomsen, J. T. Gosling, S. J. Bame, and M. M. Mellott, "Ion and electron heating at collisionless shock near critical shock number", *J. Geophys. Res.*, **90**, 1985, 137-149.
13. C. C. Goodrich, and J. D. Scudder, "The adiabatic energy change of plasma electrons and the frame dependence of the cross-shock potential at collisionless magnetosonic shock waves", *J. Geophys. Res.*, **89**, 1984, 6654-6662.
14. M. M. Leroy, and A. Mangeney, "A theory of energization of solar wind electrons by the Earth's bow shock", *Annales Geophysics*, **2**, 1984, 449-456.
15. R. G. Stone and B. T. Tsurutani (Eds.), Collisionless Shocks in the Heliosphere: A Tutorial Review, *AGU Geophys. Monograph,* **34**, 1985, 114pp, American Geophysical Union, Washington, DC.
16. B. T. Tsurutani, and R. G. Stone (Eds.) Collisionless Shocks in the Heliosphere: Reviews of Current Research, *AGU Geophys. Monograph,* **35**, 1985, 301pp, American Geophysical Union, Washington, DC.
17. C. T. Russell, ed., Multipoint Magnetospheric Measurements, *Adv. Space Res.*, **8**(9), 1988, 464pp, Pergamon Press, Oxford.
18. C. T. Russell, ed., The Magnetosheath, *Adv. Space Res.*, **14**, 1994, 135pp, Pergamon Press, Oxford.
19. C. T. Russell, ed., Physics of Collisionless Shocks, *Adv. Space Res.*, **15**(8/9), 1995, 544pp, Pergamon Press, Oxford.

SESSION 2
SHOCK STRUCTURES:
MICROSCOPIC AND KINETIC

Electron Acceleration and Structure in the Quasi-perpendicular Collisionless Shock

D. Burgess

Astronomy Unit, Queen Mary, University of London

Abstract. Electron acceleration at quasi-perpendicular shocks is a key problem in collisionless shock physics, in the context of the Earth's bow shock and other astrophysical situations. Fast Fermi acceleration, or reflection by adiabatic mirroring is a robust mechanism, but predicts that the highest energies are produced over a very small shock angle range, close to perpendicular where the reflected flux is decreasingly small. Pitch angle scattering has been shown to be effective in broadening the parameter range where this process is important. Using 2D hybrid simulations and electron test particle simulations, we show that ripples and oscillations of the shock surface are efficient scatters of suprathermal electrons. The results indicate that power law energy distributions can be obtained for both upstream and downstream energetic electrons, over a reasonably wide range of shock angles.

Keywords: Collisionless shock, particle acceleration, plasma simulation
PACS: 96.50.Fm, 96.50.Pw, 52.35.Tc

INTRODUCTION

Electron acceleration at quasi-perpendicular shocks is crucial for both radio emission in planetary foreshocks and as a possible seed population mechanism at astrophysical shocks. In terms of observations in the terrestrial foreshock, there are a small number of key observations that any modelling or theory should explain. Anderson et al. [1] using ISEE data showed a crossing of the upstream edge of the electron foreshock which illustrated the presence of velocity dispersion. Electrons with energies greater than 16 keV showed a flux increase before that of lower energies. Moving deeper into the electron foreshock, the fluxes in the > 16 keV and 5.3 keV channels decreased, whereas fluxes at 1.5 keV rose and thereafter were present throughout the period of connection with the bow shock. Later work [2] showed evidence for loss-cone depletion at low (just above thermal) energies. This is interpreted as indicating that the more energetic electrons are only accelerated near where the tangent field lines touch the bow shock surface. It was estimated that this corresponded to an acceleration region on the bow shock of width about $\sim 10^4$ km, and where the shock normal angle θ_{Bn} is in the range 1°–5° from perpendicular.

An alternative view was put forward by Gosling et al. [3] who studied the variation of the electron distribution through a set of shock crossings. They found that the most energetic electrons are most commonly found immediately at, or downstream of, quasi-perpendicular shocks, but rarely associated with quasi-parallel shocks. Furthermore, the suprathermal flux takes the form of a power law tail (with exponent 3–4), which emerges smoothly from the thermal distribution. The field aligned component which escapes upstream is observed in the ramp, and appears to be at similar phase space density levels as the downstream population. It was argued that simple magnetic mirroring could not explain these observations, and that the upstream electron beams could be explained by leakage from the heated and energized downstream distribution.

Given the magnetic field jump at the quasi-perpendicular shock, and the evidence in favour of reflection of incoming solar wind electrons, it is natural to consider the process of magnetic mirroring. The first analytical studies [4, 5] calculated the reflected distribution function and moments as functions of the shock parameters, and the shock normal angle in particular. The model assumed time-steady, one-dimensional fields, and a planar shock surface. The process of adiabatic conserving reflection (fast Fermi acceleration) is best studied in the de Hoffman-Teller Frame (HTF) in which the shock is at rest and the upstream flow is aligned with the upstream magnetic field direction. In this frame there is zero convection electric field, and particles thus conserve energy (provided that the fields remain time-steady). The reflection process in the HTF may conserve energy, but in the normal incidence frame (NIF) there is an energy gain which basically depends on the HTF transformation speed V_{HT}. From the geometry of the shock fields $V_{HT} = V_{NIF} \tan \theta_{Bn}$, so that V_{HT} increases dramatically as θ_{Bn} approaches 90°. However, for an incident solar wind distribution, as θ_{Bn} increases, the portion which will mirror comes increasingly from the outer parts of the distribution. Thus the reflected density decreases as θ_{Bn} approaches 90°, and furthermore, it will begin to depend strongly on the exact details of the trans-thermal part of the distribution. For this reason studies have used either a kappa or core+halo solar wind distribution.

The cross-shock potential within the shock layer also plays a role in the reflection process, since electrons which might otherwise reflect, but with low enough energy are pulled through into the downstream region by the potential. In order to model this effect self-consistently a one-dimensional hybrid simulation was used to provide the time and spatially varying fields, in which test particle electrons were followed [6]. Since exact trajectories were computed this work also relaxed the assumption of conservation of the magnetic moment. Nevertheless, most of the results of the simple adiabatic theory were confirmed. These simulation studies did reveal one implication of the theory: the reflection process could take a fairly long time for some particles, so that motion along the magnetic field meant a large distance scale for the reflection. This led to a study of the effect of shock curvature based on fields from 1D hybrid simulations [7].

Vandas [8] modelled the effects of shock curvature and finite shock width, and, after comparisons with observations, concluded that simple adiabatic reflection, although qualitatively successful, could not explain the anisotropies and energy spectra observed both upstream and downstream of the shock. He concluded that some other process, such as pitch angle scattering must be operating. The effect of pitch angle scattering within adiabatic reflection was studied by Krauss-Varban [9] using some ad hoc assumptions about the scattering process. It was found that pitch angle scattering could be very effective at redistributing energized electrons, leading to significant fluxes over a much broader range of shock angles.

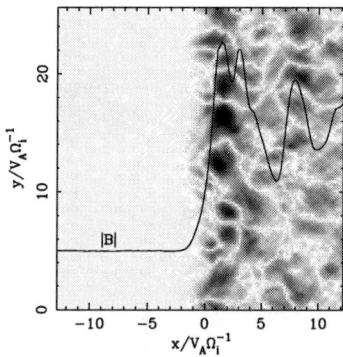

FIGURE 1. Image map of shock normal component of magnetic field for a shock with $\theta_{Bn} = 88°$, $M_A = 6$; the magnetic field profile averaged over the y direction is superimposed (from Lowe and Burgess [10]).

SIMULATION OF ELECTRON ACCELERATION

We report here work done to extend earlier modelling to include the effects of structure within the shock ramp along the magnetic field direction. Two-dimensional hybrid simulations have been carried out with the upstream magnetic field in the simulation plane as in earlier simulations [11]. The results show considerable structuring along the shock front, which is illustrated in Fig. 1. It should be noted that when the upstream magnetic field direction is principally out of the simulation plane, then the shock structure is very similar to the results of 1D simulations. Further details of the structuring in the shock have been presented by Lowe and Burgess [10], who showed that the structuring can be interpreted as ripples in the shock surface. These ripples propagate with a range of wave vectors, but with a constant phase speed, which was found to be equal to the Alfvén speed at the position of the shock overshoot. Other field components are present with predominantly shorter wavelengths which have wave vectors oblique to the shock surface. These waves and structuring are likely to be related to whistler wave packets in the foot and ramp, perhaps generated by nonlinear steepening [12].

The rippling of the shock, i.e., nonstationary structuring along the magnetic field lines, has a major effect on the dynamics of suprathermal electrons at the shock. This is investigated by following test particle electron trajectories in the fields calculated by the hybrid simulation. All timesteps are used and interpolation is carried out in space (bicubic) and in time (linear). The time variation within the hybrid simulation and the high speed of electrons along the magnetic field leads, in some cases, to rapid variations of the fields as felt by the particles. In order to preserve accuracy it is thus necessary to use adaptive time-stepping. Preliminary results of this model have been presented [13], where it was pointed out that the motion of electrons along the field line, and the temporal evolution of the field magnitude along the field line, could lead to particles being trapped within the shock structure, and being scattered.

The key results from these simulations can be illustrated by Fig. 2, which shows the final distributions obtained by initially releasing an isotropic velocity shell of electrons

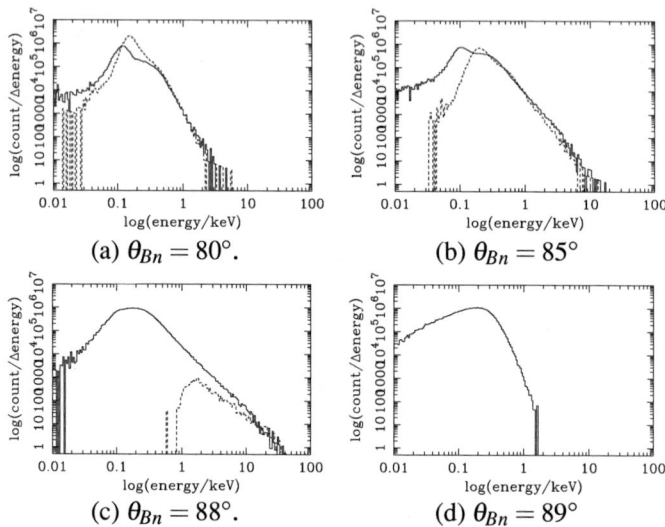

FIGURE 2. Differential energy spectra (arbitrary units) showing upstream (dashed) and downstream (solid) final electron energy distributions for a fixed Mach number, and for a range of values of θ_{Bn}. The initial distribution is an isotropic shell of energy 100 eV (omitting particles that would not interact with the shock).

of energy 100eV (omitting electrons that would not interact with the shock) upstream. Similar results are obtained for initial energies up to at least 10keV. The results show power law behavior of both upstream and downstream particles. Furthermore, contrary to the predictions of simple adiabatic reflection theory, the flux of reflected particles is relatively high over a range of θ_{Bn} away from 90°. Finally, the level of both upstream and downstream distributions is similar in the power law region. This would mean that the upstream distribution will appear to emerge from the energized electron distribution within the shock ramp, in line with observations.

Following individual trajectories shows that scattering within the ramp is a key feature of the most energetic particles. However, there is still a signature of adiabatic reflection, since some particles are reflected in the foot of the shock, before entering the region where the structuring is most effective. An illustration of the type of trajectory which is associated with the largest energy gain is shown in Fig. 3. This shows a particle, of initial energy 500 eV, which senses the full shock ramp, and undergoes strong pitch angle changes both on its path into the ramp, and then as it transits back out through the ramp before it finally escapes the shock with an energy of about 6 keV.

Conclusions

Modelling electron motion in the structured, time-varying fields of the shock ramp has highlighted the importance of pitch angle scattering in the development of the electron

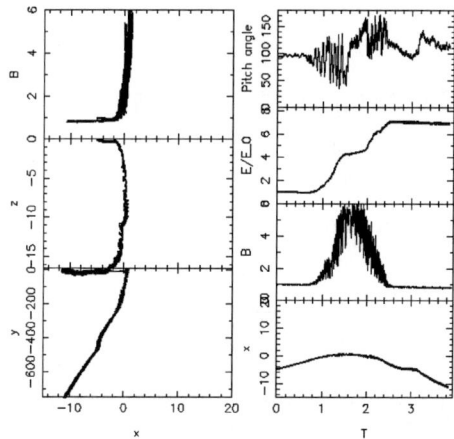

FIGURE 3. Trajectory of electron with initial energy of 500eV interacting with a shock with $\theta_{Bn} = 88°$, $M_A = 6$. The trajectory history in both space and time is shown.

distribution and the acceleration of electrons. It has been demonstrated that appreciable scattering, leading to power law energy distributions, can result from structuring on ion scale lengths, as seen in hybrid simulations. The effect of smaller scale structuring has yet to be evaluated, but it can only augment the processes illustrated here.

ACKNOWLEDGMENTS

We acknowledge the assistance of R.E. Lowe and O. Moullard. Parts of this work have been supported from research grants from PPARC (UK).

REFERENCES

1. K. A. Anderson, R. P. Lin, F. Martel, C. S. Lin, G. K. Parks, and H. Rème, *Geophys. Res. Lett.*, **6**, 401–404 (1979).
2. R. J. Fitzenreiter, J. D. Scudder, and A. J. Klimas, *J. Geophys. Res.*, **95**, 4155–4173 (1990).
3. J. T. Gosling, M. F. Thomsen, S. J. Bame, and C. T. Russell, *J. Geophys. Res.*, **94**, 10011–10025 (1989).
4. M. M. Leroy, and A. Mangeney, *Annales Geophys.*, **2**, 449–456 (1984).
5. C. S. Wu, *J. Geophys. Res.*, **89**, 8857–8862 (1984).
6. D. Krauss-Varban, D. Burgess, and C. S. Wu, *J. Geophys. Res.*, **94**, 15089–15098 (1989).
7. D. Krauss-Varban, and D. Burgess, *J. Geophys. Res.*, **96**, 143–154 (1991).
8. M. Vandas, *J. Geophys. Res*, **106**, 1859–1872 (2001).
9. D. Krauss-Varban, *J. Geophys. Res*, **99**, 2537–2551 (1994).
10. R. E. Lowe, and D. Burgess, *Annales Geophysicae*, **21**, 671–679 (2003).
11. D. Winske, and K. B. Quest, *J. Geophys. Res*, **93**, 9681–9693 (1988).
12. V. Krasnosselskikh, B. Lembège, P. Savoini, and V. V. Lobzin, *Phys. Plasmas*, **9**, 1192 (2002).
13. R. E. Lowe, and D. Burgess, *Geophys. Res. Lett.*, **27**, 3249–3252 (2000).

On Kinetic Structure of Quasi-Perpendicular Collisionless Shocks

Manfred Scholer* and Shuichi Matsukiyo[†]

*Max-Planck-Institut für extraterrestrische Physik, Garching, Germany
[†]Earth System Science and Technology, Kyushu University, Fukuoka, Japan

Abstract.
We discuss one-dimensional electromagnetic full particle simulations of quasi-perpendicular collisionless shocks with physical mass ratio in the context of recent Cluster bow shock measurements. The density ramp scale in units of the ion inertial length obtained from simulations of high ion beta shocks ($\beta_i \approx 1$) does not increase with Mach number, contrary to what has been reported from Earth's bow shock measurements. Several possibilities are offered which could explain this discrepancy. In low beta shocks the modified two-stream instability is important in the foot of the shock and can lead to large density and electric field spikes. It is suggested that recent Cluster observations of short scale electric field spikes at the quasi-perpendicular Earth's bow shock can be explained in terms of the modified two-stream instability.

DENSITY RAMP SCALES IN QUASI-PERPENDICULAR SHOCKS: HIGH BETA SIMULATIONS

In the last few years the observations by the Cluster four spacecraft mission at Earth's bow shock have provided new information on shock structure and shock related processes. One of the important questions in collisionless shock physics is to which degree the steepening of the magnetosonic wave leading to the shock is balanced by either dissipation or dispersion. The spatial scales associated with dispersion is the electron or ion inertial scale length, whereas dissipation due to gyroviscosity associated with gradients of the ion stress tensor introduces the gyroscale of the ions in the shock. Bale et al. [1] have shown that the largest density transition scale of the quasi-perpendicular shock is, independent of shock (magnetosonic) Mach number, the convected ion gyroradius. Determining shock normals and velocities by the timing technique they fitted shock profiles with a hyberbolic tangent function and derived the transition scale L for a large number of quasi-perpendicular shocks. The transition scale L in units of the ion inertial lenght clearly increases monotonically with the shock fast magnetosonic Mach number M_f, while the transition scale is almost constant with magnetosonic Mach when evaluated in units of the convected gyroradius, where the magnetic field is determined by the average downstream magnetic field. Bale et al. [1] argue that this demonstrates the importance of gyroviscosity over dispersion in shock dissipation.

We have performed a number of one-dimensional full particle simulations of quasi-perpendicular shocks in order to compare the ramp scales obtained from the simulations with the Bale et al. [1] Cluster result. Details of the collisionless shock simulations

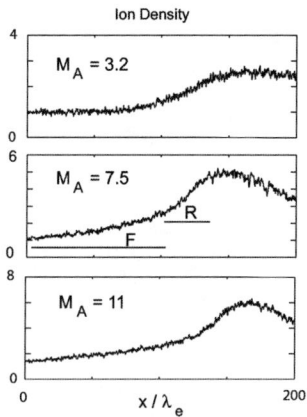

FIGURE 1. The ion density profile as obtained from PIC simulations of almost perpendicular shocks with different Alfvén Mach numbers. For the $M_A = 7.5$ shock (middle panel) both foot (F) and ramp (R) regions are indicated.

can be found in [2]. Under Mach number we understand in the following the Alfvén Mach number (Alfvén velocity v_A), distances are given in units of the electron inertial length $\lambda_e = c/\omega_{pe}$ and time is in units of the inverse ion gyrofreqency Ω_{ci}. A second important parameter in PIC simulations is the ratio of electron plasma frequency to electron gyrofrequency, ω_{pe}/Ω_e, which can also be written as $(c/v_A)(m_e/m_i)^{1/2}$. In our PIC simulations shown here this ratio is between 2 and 4, while in reality at Earth's orbit this ratio is more than one order of magnitude larger. Increasing ω_{pe}/Ω_e is computationally extremely intensive. We will show in the following that the structure of a quasi-perpendicular shock depends strongly on the ion beta. Figure 1 shows ion density profiles for various shock strengths obtained from one-dimensional full particle shock simulations. The shock normal magnetic field angle Θ_{Bn} was assumed to be 87° and the ion and electron beta is $\beta_i = \beta_e = 1$. For these simulations the physical ion to electron mass ratio has been assumed (here 1840) and the ratio of the ratio of electron plasma to gyro-frequency is assumed to be $\mu = \omega_{pe}/\Omega_e = 4$. It can immediately be seen that the ramp scale does not increase with inertial length, contrary to what has been reported in [2]. Only a more and more extended foot appears at higher Mach number.

Figure 2 shows the ramp scale in units of the ion inertial length as a function of Alfvén Mach number for 5 shock simulations. All parameters were identical, and only the Mach number has been changed. The ramp scale L has been determined by fitting straight lines to the ramp and the foot. L was then defined as the difference in distance between the intersection of the two lines at the base of the ramp and the position of the overshoot. As can be seen from Figure 2 the ramp scale in units of the ion inertial length obtained from the simulations is constant or decreases slightly with Mach number. When evaluating the convected gyroradius (in units of the ion inertial length) with the overshoot magnetic field strength it is about constant with Mach number. Thus the ramp scale is also independent of Mach number when evaluated in units of the convected gyroradius. But

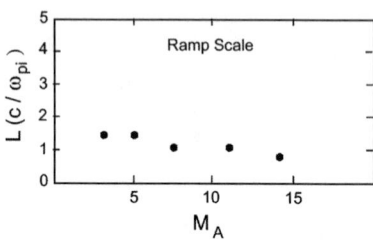

FIGURE 2. Ramp scale in units of ion inertial length for 5 different simulations versus Alfvén Mach number. All other parameters are kept constant.

the convected gyroradius has to be evaluated with the overshoot magnetic field and not by the averaged field downstream of the overshoot.

Another important question is how well the magnetic field/density profile represents the cross shock potential profile. Figure 3 shows from top to bottom the main magnetic field component, the cross shock potential, and the ion density profile through a high Mach number shock. It can be seen that the shock potential increases already considerably in the foot and reaches a maximum before the magnetic field maximum. Close inspection shows that the electric field in the shock normal direction leading to the potential has a large contribution in the foot due to specularly reflected ions: these ions are turned around in the magnetic field and obtain large velocities parallel to the shock. Their $\mathbf{v} \times \mathbf{B}$ term contributes to the electric field in the shock normal direction.

The differences between the Cluster observations and the simulations deserve further detailed investigation. Several possibilities could lead to this discrepancy. (1) When fitting a hyperbolic tangent to the high Mach number simulation profiles we obtain a ramp scale which is different from the true ramp scale: the fit essentially results in a scale which is given by the shock foot. (2) The simulations are one-dimensional and thus can not reproduce instabilities in the current direction in the shock ramp. The shock current may excite the lower hybrid drift instability in the ramp leading to a widening of the ramp. (3) We have only investigated almost perpendicular shocks $\Theta_{Bn} = 87°$ while Bale et al. [1] analyzed quasi-perpendicular shocks with a variety of shock normal - magnetic field angles. At more oblique shocks the ramp may widen and the ramp scales obtained from simulations could be closer to the Bale et al. result.

ELECTRIC FIELD SPIKES IN THE SHOCK FOOT AND RAMP: LOW BETA SIMULATIONS

As discussed above for the case of high beta shocks an important aspect in collisionless shock physics is the spatial scale over which changes in the electric field occur and its relation to the scale size over which changes in the magnetic field occur. Walker et al. [3] reported small scale high amplitude spike-like features in the electric field during a number of quasi-perpendicular bow shock crossings. Several features can be clearly identified in these observations: the magnetic field profile is different at different

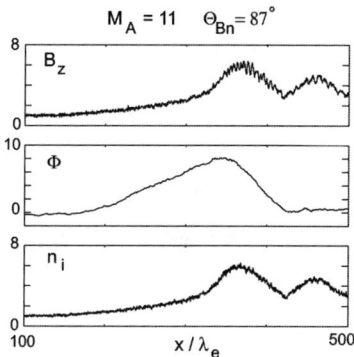

FIGURE 3. Magnetic field, cross shock potential, and ion density through a high Mach number shock.

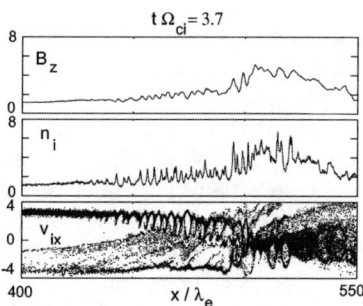

FIGURE 4. Magnetic field B_z component, ion density, and v_{ix} phase space during a particular time in a low beta shock simulation (from [6]).

spacecraft and large disturbances in the electric field begin in the magnetic foot region and continue through the overshoot/undershoot region. From observations at a number of shocks Walker et al. concluded that the spikes, which appear in pairs, have scale sizes of the order of 1-5 electron inertial lengths (c/ω_{pe}), a value much smaller than that of the magnetic ramp scale. We have indeed observed very similar electric field spikes in low beta ($\beta_i < 0.4$) shock simulations and suggest that these spikes can be attributed to the modified two-stream instability.

It has been suggested on the basis of PIC simulations that quasi-perpendicular shocks can reform on the time scale of the inverse ion gyrofrequency (e.g, [2], [4]). At low (ion) beta perpendicular shocks reformation is due to the accumulation of specularly reflected ions at the upstream edge of the foot. In realistic mass ratio simulations of slightly oblique shocks the modified two-stream instability (MTSI) can get excited: in not exactly perpendicular shock simulations wave vectors with a component parallel to the magnetic field are allowed for, so that the beam mode can interact with the whistler mode [5]. It has been shown that the beam ions responsible for the MTSI in the foot are the incoming solar wind ions. Because of the presence of specularly reflected ions

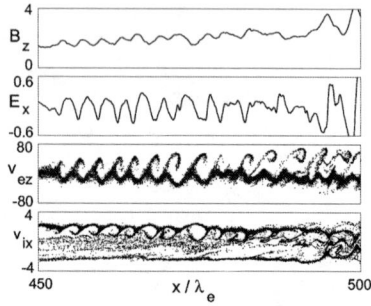

FIGURE 5. Magnetic field B_z component, electric field component in the shock normal direction, electron v_{ez} and ion v_{ix} phase space distribution in the foot and ramp of a low beta shock (from [6]).

the bulk velocity of the total ion population in the foot decreases. Then the electron bulk velocity should also decrease in the foot when requiring zero electrical current in the shock normal direction. This leads to a velocity difference between incoming solar wind electrons and ions. Figure 4 from [6] shows from top to bottom the main magnetic field component, the ion density, and ion v_{ix} phase space at a certain time in a a low beta ($\beta_i = 0.05$) shock simulation. One can see large density spikes and vortices in the incoming solar wind ion distribution in the foot region. Figure 5 shows the magnetic field B_z component, the electric field in the shock normal direction, electron phase space in the direction parallel to the magnetic field, and ion phase space in the shock normal direction in the foot in higher spatial resolution. Interesting are the large electric field spikes during the development of the MTSI. We propose here that the electric field spikes seen by Walker et al. [3] are possibly due to the MTSI and are not electrostatic in nature. The vortices in the ion phase space density of the incoming ions grow and eventually lead to a mixing between incoming ions and reflected ions. After thermalization a new ramp appears at the upstream edge of the former foot which suggests that the MTSI is largely responsible for the shock reformation process.

REFERENCES

1. S. D. Bale, F. S. Mozer, and T. S. Horbury, *Phys. Rev. Lett.* **91**, 265004 (2003)
2. M. Scholer, I. Shinohara, and S. Matsukiyo, *J. Geophys. Res.* **108**, doi:10.1029/2002JA009515 (2003)
3. S. N. Walker, H. St. C. K. Alleyne, M. Balikhin, M. Andre, and T. S. Horbury, *Ann. Geophys.* **22**, 2291 (2004)
4. T. Hada, M. Onishi, B. Lembege, and P. Savoini, *J. Geophys. Res.* **108**, doi:10.1029/2003JA009339 (2003)
5. S. Matsukiyo, and M. Scholer, *J. Geophys. Res.* **108**, doi:10.1029/2003JA010080 (2003)
6. M. Scholer, and S. Matsukiyo, *Ann. Geophys.* **22**, 2345 (2004)

Global Hybrid Simulations of the Bow Shock

N. Omidi[1], X. Blanco-Cano[2], C. T. Russell[3]

1. Solana Scientific Inc. (omidi@solanasci.com); 2. Instituto de Geofisica, UNAM; 3. IGPP, UCLA

Abstract. This paper summarizes recent results from global hybrid (kinetic ions, fluid electrons) simulations of bow shocks or waves associated with solar wind interaction with magnetic dipoles of various strength. By virtue of resolving ion temporal and spatial scales, global hybrid simulations account for collissionless dissipational processes at and upstream of the shock and their effects on the macrostructure of the bow shock, ion foreshock and the magnetosheath. The results demonstrate that as the level of magnetization increases and the dipole becomes a more effective obstacle, the quasi-perpendicular part of the bow shock forms first and that formation of quasi-parallel part of the bow shock is tied to the generation of oblique magnetosonic waves which steepen to form shocklets in the upstream region.

Keywords: Bow Shock, Foreshock, kinetic simulations
PACS: 94.30.Va

INTRODUCTION

Over the past three decades, our knowledge of collisionless shocks in general and planetary bow shocks, in particular, has increased greatly (see e.g. review articles in [1,2]). Multiple spacecraft such as ISEE and AMPTE, coupled with local, electromagnetic hybrid (kinetic/PIC ions, fluid electrons) and full particle simulations have led to a much better understanding of the shock micro-structure and dissipation processes under a variety of Mach numbers, shock normal angles and plasma beta (ratio of kinetic to magnetic pressure). In particular, the shock is divided into sub- and super-critical, as well as, quasi-parallel and perpendicular classes based on Mach number and shock normal angle respectively. Examination of the observed amount of heating in electrons and ions across the shock has established that ions are the dominant species in the dissipational processes and determination of the temporal and spatial scales of the shock. The only exception to this general conclusion could be the super-critical, perpendicular regime where electron dynamics and scales may be important (see e.g. article by Scholer in this book). The next important set of questions regarding the physics of the bow shock and its influences on the magnetosphere are related to the macrostructure of the shock under a variety of solar wind conditions. In recent years, we have used global (2-D in space, 3-D in fields and currents) hybrid simulations of solar wind interaction with magnetic dipoles of various strengths to gain a better understanding of the resulting bow shocks/waves, as well as, magnetospheres (e.g. [3,4,5,6,7]).

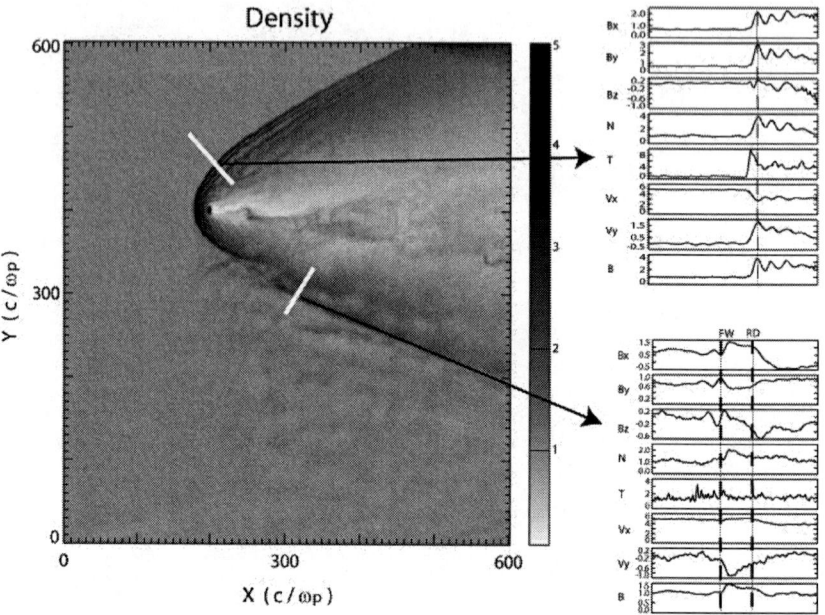

FIGURE 1. Plot of density for Dp=2.1 illustrating bow shock/wave. The right panels show cuts across the quasi-parallel and perpendicular parts. All quantities normalized to solar wind values.

In this paper, due to lack of space, we only briefly summarize the results and conclusions of these studies and refer the interested readers to them for more details and comprehensive list of references.

RESULTS

The weakest type of interaction between the solar wind and a magnetic dipole corresponds to the generation of a whistler wake which does not affect the flow of the solar wind. As the dipole strength increases, fast and slow magnetosonic wakes are also generated which result in flow diversion and formation of a plasma tail. Formation of a bow shock/wave does not take place until the dipole strength is large enough for the plasma to become stagnant upstream of the dipole (i.e. formation of some form of magnetopause). We define the parameter Dp as the distance (normalized to ion skin depth), upstream of the dipole, where dynamic pressure of solar wind and magnetic pressure are in balance. The appearance of a bow shock/wave occurs at Dp ~ 1. The left panel in Fig. 1 shows the density in the simulation box for Dp = 2.1, illustrating an example of such a bow shock/wave. The top and bottom panels on the right hand side of the figure show variations of plasma and field values across the upper and lower portions of the bow shock/wave respectively.

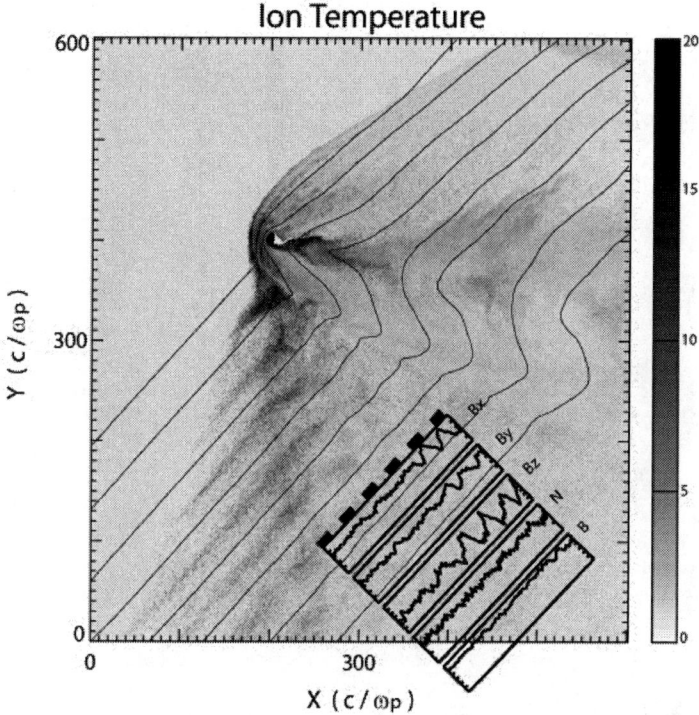

FIGURE 2. Ion temperature plot showing ion foreshock for Dp=2.1. Insert shows example of sinusoidal, ULF waves generated in the foreshock along the dashed line.

In this run, the interplanetary magnetic field (IMF) makes a 45° angle with the X-axis and thus the upper (lower) portion of the bow shock corresponds to quasi-perpendicular (parallel) geometry. The top right-hand panel in Fig. 1, shows changes in plasma and fields consistent with those associated with quasi-perpendicular, fast magnetosonic shocks. On the other hand, the changes seen in the lower panel are not consistent with a shock crossing and have been interpreted as a combination of a fast magnetosonic wave followed by a rotational discontinuity. The former results in flow diversion while the latter accommodates the change in the direction of magnetic field from the solar wind to the draped fields in the tail. Thus, the quasi-perpendicular part of the bow shock has formed while the quasi-parallel has not and overall constitutes a bow shock/wave. Even though the quasi-parallel section of the bow shock has not formed, ion reflection near the nose of the bow shock and leakage from the quasi-parallel magnetosheath lead to a population of backstreaming ions whose interaction with the solar wind results in generation of ULF waves. This ion foreshock region is illustrated in Fig. 2 which shows ion temperature and magnetic field lines in the simulation box. The insert shows the profile of parallel propagating, non-compressional, sinusoidal, ULF waves generated in the foreshock region. These waves are generated through the right-hand resonant ion-beam instability [8] and result in scattering of the field aligned beam into a broader distribution in pitch angle without affecting the solar wind in any significant way. They posses all the properties of 30 second, sinusoidal, ULF waves observed in the earth's foreshock [9].

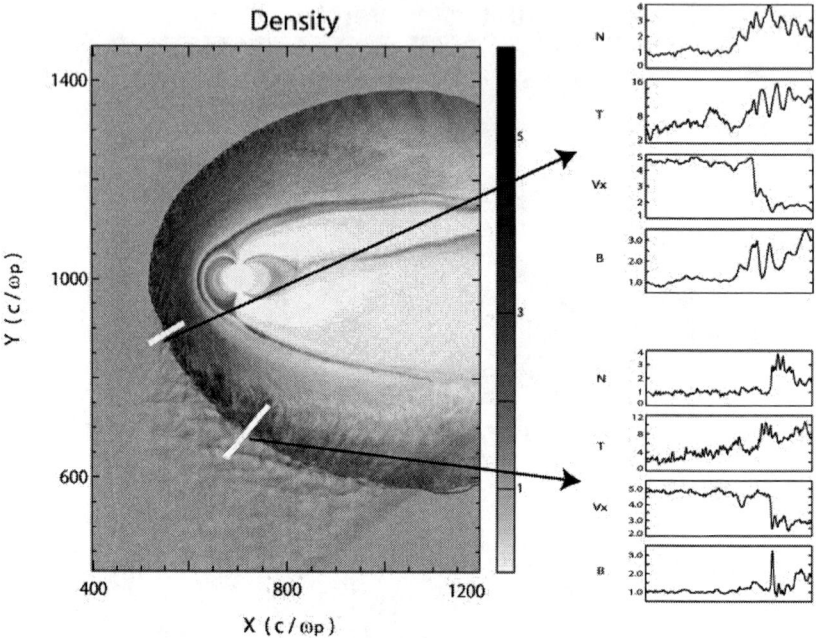

FIGURE 3. Same as in Fig.1, except Dp=64 and quasi-parallel shock is formed.

It was shown by [4] that when Dp ~ 20, the resulting magnetosphere has a terrestrial structure and the bow shock is well developed. Figure 3 shows the results of simulations for Dp = 64, which corresponds to Mercury's level of magnetization, in a format similar to that in Fig. 1. The IMF in this run has the same orientation as the Dp =2.1 case, and the upper (lower) portion of the bow shock corresponds to quasi-perpendicular (parallel) geometries. The right hand panels show 2 crossings of the quasi-parallel shock with one (top) closer to the nose and the other (bottom) further in the flank. As can be seen, in both cases the transition from solar wind to the magnetosheath takes place in a number of steps which is the result of local heating and deceleration of the solar wind by the upstream generated waves. The shock profile is highly turbulent and time varying due to convection of the upstream generated ULF waves into the shock and its cyclic reformation (e.g. [10]). Examination of the ULF waves generated in this run and the ion distribution functions in the foreshock by [5,7] has revealed the presence of both parallel and obliquely propagating waves with the former being generated by field aligned ion beams and the latter by gyrating ions closer to the quasi-parallel shock. The right hand panels in Fig. 4 show examples of both waves, with the oblique ones (top) having steepened to form shocklets (e.g. [9]) and the parallel ones (bottom) having a sinusoidal wave form as in the Dp = 2.1 case. That these two types of waves are generated through different instabilities is consistent with the suggestion by [11] who used linear theory and local, 2-D hybrid simulations and observations of ion beams in the foreshock [12]. The results also demonstrate that generation of shocklets is essential for the formation of the quasi-parallel shocks.

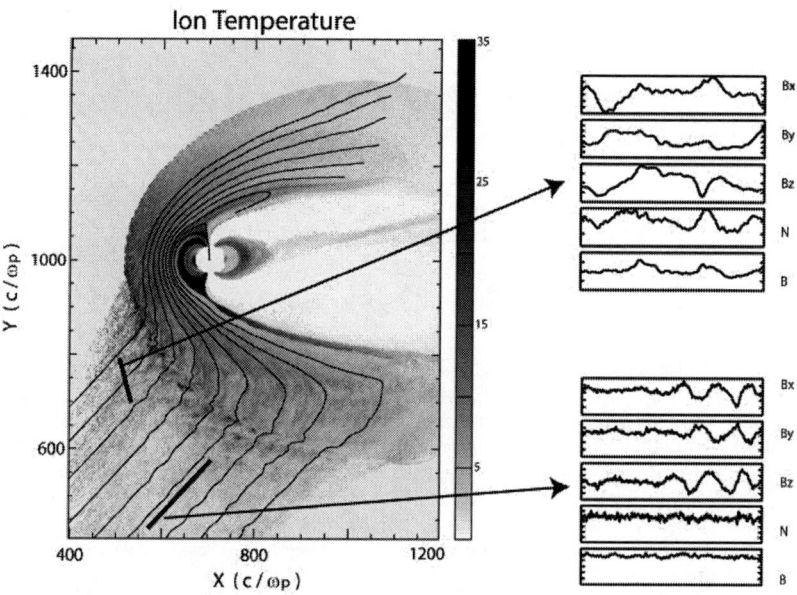

FIGURE 4. Same as in Fig. 2 except Dp=64 and both parallel and oblique ULF waves are generated.

ACKNOWLEDGMENTS

This work was supported by NASA grants NAG5-12946 and NAG5-11754 at UCSD and NSF grant ATM 04-02213 at UCLA.

REFERENCES

1. Tsurutani, B. and R. Stone, eds., *Collisionless Shocks in the Heliosphere*, AGU Monographs 30,31, Washington DC, 1985.
2. Russell, C. T., *Physics of Collisionless Shocks*, Editor, Advances in Space Research, Pergamon, 1995.
3. Omidi, N., X. Blanco-Cano, C. T. Russell, H. Karimabadi and M. Acuna, "Hybrid simulations of solar wind interaction with magnetized asteroids: General characteristics, *J. Geophys. Res.*, 107, 1487, 2002.
4. Omidi, N., X. Blanco-Cano, C. T. Russell, and H. Karimabadi, "Dipolar magnetospheres and their characterization as a function of magnetic moment", *Adv. Space Res.*, 33, Issue 11, 1996, 2004.
5. Omidi, N., X. Blanco-Cano, C. T. Russell, "Macro-Structure of collisionless bow shocks: 1. Scale lengths, *J. Geophys. Res.*, Submitted, 2005.
6. Blanco-Cano, X., N. Omidi and C. T. Russell Hybrid simulations of solar wind interaction with magnetized asteroids: Comparison with Galileo observations near Gaspra and Ida, *J. Geophys. Res.*, 108, 11-1, 2003.
7. Blanco-Cano, X., N. Omidi and C.T. Russell, Macro-Structure of Collisionless Bow Shocks: 2. Wave properties, *J. Geophys. Res.*, submitted, 2005.
8. Gary, S.P., J.T. Gosling and D. Forslund, The electromagnetic ion beam instability upstream of the Earth's bow shock, *J. Geophys. Res*, 86, 6691, 1981.
9. Russell, C.T. and M. Hoppe, Upstream waves and particles, *Space Sci Revs*, 34, 155, 1983.
10. Burgess, D.,Cyclic behavior at collisionless quasi-parallel shocks, *Geophys. Res. Lett.*, 16, 345, 1989.
11. Omidi, N., H. Karimabadi, D. Krauss-Varban and K. Killen, Generation and nonlinear evolution of oblique magnetosonic waves: Application to foreshock and comets, in *Solar System Plasmas in Space and Time*, Geophysical Monograph 84, p. 71, 1994.
12. Fuselier, S. A., *Physics of Collisionless Shocks*, ed. C. T. Russell, Advances in Space Research, Pergamon, 43, 1995.

Ion Dynamics at Shocks: Ion Reflection and Beam Formation at Quasi-perpendicular Shocks

Harald Kucharek and Eberhard Möbius

University of New Hampshire Space Science Center Durham, NH, 03824, USA

Abstract. The physics of collisionless shocks is controlled by the ion dynamics. The generation of gyrating ions by reflection as well as the formation of field-aligned ion beams are essential parts of this dynamic. On the one hand reflection is most likely the first interaction of ions with the shock before they undergo the downstream thermalization process. On the other hand field-aligned ion beams, predominately found at the quasi-perpendicular bow shock, propagate into the distant foreshock region and may create wave activity. We revisit ion reflection, the source and basic production mechanism of field-aligned ion beams, by using multi-spacecraft measurements and contrast these observations with existing theories. Finally, we propose an alternative production mechanism.

Keywords: Earth's bow shock, ion reflection, ion acceleration multi-spacecraft measurements
PACS: 96.50.Fn, 95.30.Qd,50.20.-j

INTRODUCTION

Early space missions explored the foreshock region of the Earth's bow shock and its variety of ion distributions. At the upstream edge of the ion foreshock energetic (> 10keV) field aligned ion beams (FAB) of low densities are present. These beams originate from a fraction of the incoming solar wind accelerated by shock drift acceleration at the quasi-perpendicular portion of the Earth's bow shock. In addition to shock drift acceleration, leakage of ions out of the magnetosheath is thought to contribute to this distribution, in particular for low energy beams. These lower-energy FABs excite low frequency monochromatic waves. Ions are trapped in these waves, which leads to the formation of gyro-phase bunched ion distributions. Efficient wave particle scattering is thought to be the basic mechanism that scatters a gyro-phase bunched distribution into an intermediate distribution. Upstream of the quasi-parallel regime of the Earth's bow shock diffusive particle distributions are found. These distributions consist of 150keV-200keV ions and they are nearly isotropic. These distributions are accompanied by large amplitude magnetic fluctuations.

Although significant progress has been achieved in understanding the global dynamics of the ion distributions in the foreshock region, some underlying production mechanisms are still not fully understood, and models may sometimes even seem contradictory to observations. The origin and the basic production mechanism of field-aligned ion beams are such an examples.

GYRATING IONS AND BEAM DISTRIBUTIONS

Reflected gyrating ion distributions, which extend to about one gyro radius into the upstream region, have been carefully distinguished from the field-aligned beam distribution (e.g. Gosling and Robson, 1985; Thomsen, 1985) and discussed as two different distributions. While the specular reflection of the gyrating distribution has been explained in a straight-forward manner as a reflection in the shock potential, for specific phases of the incoming ions assisted by gyro-motion in the compressed interplanetary magnetic field (IMF) downstream of the shock, the generation mechanism and the source of the FABs are not so readily understood. Although the kinematics and energetics of the beams can be derived correctly in terms of a perfect reflection of the incoming ions under energy conservation along the upstream IMF in the de-Hofman-Teller (dHT) frame (Sonnerup, 1969; Paschmann *et al.*, 1980), the microphysics of their generation at the shock is still under debate. Recent Cluster results show that the ion beam distribution is closely related to the gyrating ion distribution formed by specularly reflected ions. Möbius *et al.*, (2001) found clear evidence that the gyrating and beam distributions are intimately connected. In fact the beam distribution that escapes from the shock along the magnetic field lines emerges from the low pitch-angle wing of the specularly reflected ion distribution in the shock ramp under flux conservation.

SOURCE AND GENERATION OF FIELD-ALIGNED BEAMS

A number of models to produce FAB's have been proposed: For example Sonnerup (1969) demonstrated that solar wind protons could easily be energized to a rather energetic ion beam if the bow shock managed to turn the incoming ions around in such a way that they left the shock reasonably well field-aligned after reflection. He assumed that the particle energy was preserved in the dHT-frame and the motion remained field-aligned after reflection, but he did not specify a reflection process. In observations with ISEE Paschmann *et al.* (1980) actually found that the peak energy of ion beams as a function of the magnetic field orientation relative to the solar wind and to the shock normal agreed well with the prediction of this model. This scenario is also referred to as "adiabatic reflection" because of the apparent conservation of the magnetic moment μ. However, in observational studies of the reflection process by Paschmann *et al.* (1982) and in simulations Leroy *et al.* (1981, 1982) showed that μ is far from constant during ion reflection at the quasi-perpendicular bow shock.

An alternate source for FAB's could be leakage of magnetosheath ions that have been heated downstream of the shock. In an idealized model Edmiston *et al.* (1982) proposed that plasma is heated and thermalized in a thin layer at the shock. They calculated how ions from a hot Maxwellian distribution in this layer could return upstream. Schwartz *et al.* (1983) proposed a modified version of this model. They suggested that magnetosheath particles are accelerated by the shock potential mainly along the shock normal and that its component parallel to the magnetic field constitutes the resulting guiding center motion back upstream.

Tanaka *et al.* (1983) proposed a more self-consistent non-local model based on observations by Paschmann *et al.* (1982), simulations by Leroy *et al.* (1981), and the work by Edmiston *et al.* (1982). As solar wind encounters the quasi-perpendicular section of the bow shock, part of the incoming solar wind distribution is specularly reflected and creates a gyrating ion distribution that is swept downstream. Its high perpendicular temperature is the source of free energy for electromagnetic ion cyclotron (EMIC) waves downstream of the shock. Subsequent efficient pitch angle scattering produces particles with a high enough velocity parallel to the magnetic field so that they can escape upstream. Tanaka *et al.* (1983) pointed out that this model is consistent with a large fraction of the beams observed by Paschmann *et al.* (1980), but fails to explain the most energetic ion beams. Furthermore, these models could not be distinguished on the basis of the range of parallel velocities of the FABs observed in the Earth's foreshock region by ISEE-2.

Cluster studies by Kucharek *et al.* (2004) suggest a resolution of question where the beams originate. They analyzed several quasi-perpendicular shock crossing with CIS and followed the spatial and temporal evolution of the reflected and transmitted ion populations across the shock. Figure 1 shows a composite plot during the crossing at 18:48 UT on March 31, 2001, from downstream to upstream, including a snapshot in the shock ramp. The middle panel shows the magnetic field as a function of time, and in the lower/upper panel the ion distributions, parallel and perpendicular to the mean interplanetary magnetic field, orientation indicated by arrows, are shown for three different locations: downstream, at the ramp, and upstream of the bow shock. The dark blue shaded areas in the magnetic field profile indicate the integration times for the ion distributions. Downstream, the shape of the ion distribution is more elongated perpendicular to the magnetic field. The phase space is filled with ions up to a parallel velocity of 1000 km/s. In the shock ramp gyrating ions appear, whose phase space density extends in parallel velocity, substantially exceeding the limit of $v \approx 1000$ km/s. Upstream of the shock (right hand distribution), this part of the distribution decouples from the core and forms a collimated beam along the mean interplanetary magnetic field. Note that the beam occupies a portion of the phase space that is empty downstream. Indeed, the upstream distribution is recorded six minutes later (~2 Earth radii upstream of the bow shock) and we are not connected with the same field line as the distribution at the shock ramp. However, simultaneous multi-spacecraft observations with cluster at these different regions of the bow shock show the same picture (see.Kucharek *et al.*, 2004). If we then assume that there are no waves or other upstream magnetic field structures the field line would intersects the shock at about the same Θ_{Bn}. From quantitative studies of ion energies in this event it appears as if the basic escape condition is violated and the conditions are far from reflection under conservation of energy. At the very high shock normal angle of this event (74.5°) the beam ions should not be able to escape upstream. However, perpendicular shocks are dynamic and even small-scale structures can cause deviations from the average Θ_{Bn} both can modify the critical conditions so that ions can escape upstream.

In the lower right of this figure we present a possible mechanism that is based on the idea of direct reflection and subsequent scattering. Ions will escape upstream if after a final encounter they have sufficient parallel guiding center velocity to prevent their return to the shock surface. Ions that are finally located in the fat dark portion of this circle marked "escape" have persistently a positive normal velocity and will escape. These ions have nearly the maximal beam speed as deduced using adiabatic reflection (Sonnerup, 1969), although the new picture hints at potential microscopic processes. The Sonnerup model would predict a narrow (-point like) distribution. In the new model however a larger portion of the circle will result in escaping particles, and it is uncertain what the center of the total population would be, although simulations (Burgess, 1989) provide some indication.

Whether it is justified to describe this scenario as pitch angle diffusion scattering is debatable. Scattering due to fluctuations and irregularities in the shock fields (e.g., within the foot, ramp, and/or overshoot regions) almost certainly does not preserve kinetic energy in the dHT-frame, as the circle in the figure suggests. Such fluctuations propagate at relatively small speeds relative to the bulk plasma flow. Using a one- dimensional self- consistent hybrid simulations and test particle techniques Burgess et al., 1987 investigated the formation of FAB's by direct reflection at oblique collisionless shocks. They examined the beam distribution and reproduce these basic characteristics

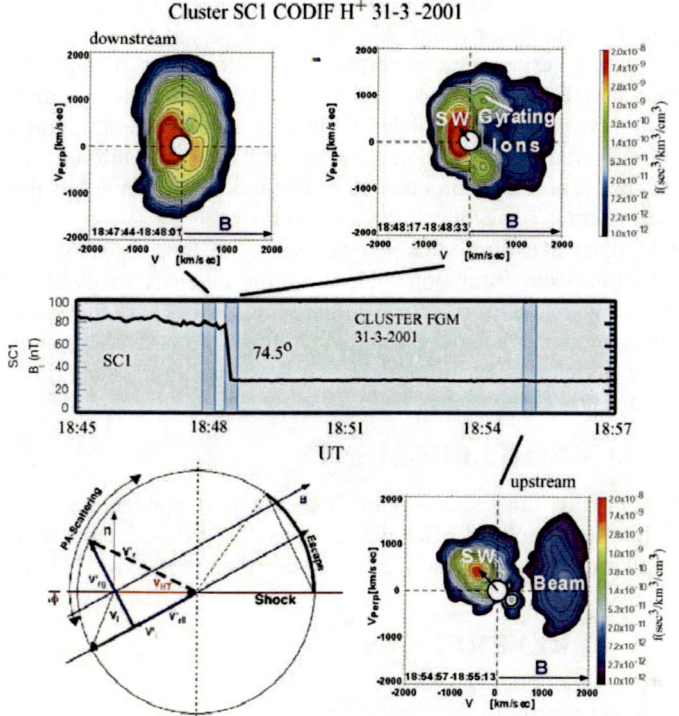

FIGURE 1. Ion distributions measured by Cluster during a shock crossing at March 31, 2001, and a schematic drawing of an alternative productions mechanism (lower left). (Adapted and modified from Kucharek *et al*, (2004))

such as peaked energy spectrum, velocity anisotropy, and a gyro-phase bunched component. They also pointed out that with increasing Θ_{BN} reflected ions originate from the wings of the incident solar wind distribution implying that field-aligned ion beams could form at Θ_{BN} above 60°.

CROSS SHOCK POTENTIAL

Ions have large gyro radii and penetrate deeper into the shock than electrons Thus their different turning points create two separated sheets of opposite electric charge with an electric field pointing upstream perpendicular to the shock. The field accelerates electrons through the shock and decelerates ions. The resulting potential created is called cross-shock potential. If the total energy of the ion exceeds the potential drop it penetrates the shock, otherwise it is reflected. Reflected ions gyrate about the upstream magnetic field with large gyro radii and move in the direction of the motional electric field gaining energy from it. They may have multiple encounters with the shock and will finally be convected downstream.

SUMMARY

Recent observations with Cluster have shown that the distribution of the beam ions originates from gyrating ions in the shock ramp. Also Cluster observed FABs at very high shock normal angles, with a very surprising result. The beam ions originate from the gyrating ion distribution and they do not leak from downstream. In the case studied the upstream escape conditions seem to be violated. However, small-scale structures can cause deviations of the average Θ_{Bn}. Both can modify the critical conditions so that ions can escape upstream. Therefore, processes right in the shock ramp and the dynamics of the shock itself seem to be responsible for these ion beams.

There are a number of open questions. For instance what are the critical shock parameters that control the ion beam formation? How are the intensity of gyrating ions and FAB's is correlated with the cross-shock potential, the shock structure, the magnetic field, and the upstream solar wind conditions at high and low Mach number supercritical shocks? These issues and the importance of wave particle interactions inside the shock front are currently being investigated.

ACKNOWLEDGMENTS

We would like to thank many unnamed individuals that help to analyze the data. Furthermore we would like to thank M. Scholer and the ISSI team (energetic foreshock ions) for helpful discussions. This work has been supported by NASA contract NAS5-30613.

REFERENCES

1. Burgess, D., *Geophys. Res. Let.*, **16**, 163–166, 1989.
2. Burgess, D. *Ann. Geophys.*, **5**, 133-145, 1987.
3. Edmiston, J.P., C.F. Kennel, D. Eichler, *Geophys. Res. Let.*, **9**, 531-534, 1982.
4. Gosling, J. and A. Robson In: B. T. Tsurutani and R. G. Stone (eds.): *Collisionless Shocks in the Heliosphere*, Geophys. Monogr. Ser. vol 35.Washington, D.C.: AGU, pp. 141–152. 1985.
5. Leroy, M. M., C. C. Goodrich, D. Winske, C. S. Wu, K. Papadopoulos: GRL. **8**, 1269–1272., 1981.
6. Leroy, M. M., D. Winske, C. C. Goodrich, C. S. Wu, and K. Papadopoulos, *J. Geophys. Res.*, **87**, 5081,1982.
7. Möbius,E. H. Kucharek, C. Mouikis, E. Georgescu, L. M. Kistler, M. A. Popecki, M. Scholer J. M. Bosqued, H. Rème, +17 co-authors, *Ann. Geophys.*, **19**, 1411, 2001.
8. Paschmann, G. N. Sckopke, S. J. Bame, and J.T. Gosling, *Geophys. Res. Let.*, **9**, 881, 1982.
9. Paschmann G. N. Sckopke I. Papamastorakis, J.R. Asbridge, S.J. Bame, and J.T. Gosling, *J. Geophys. Res.*, **85**, 4689, 1980.
10. Kucharek, H., E. Möbius, M. Scholer, C. Mouikis, L.M. Kistler, T. Horbury, A. Balogh, H. Réme, and J.M. Bosqued, *Ann. Geophys.*, **22**, 2301, 2004.
11. Sonnerup, B. U. O., *J. Geophys. Res.* **74**, 1301, 1969.
12. Schwartz, S., M. Thomsen, and J. Gosling, *J. Geophys. Res.* **88**, 2039–2047, 1983.
13. Tanaka, M., C. Goodrich, D.Winske, and K. Papadopoulos., *J. Geophys. Res.* **88**, 3046, 1983.
14. Thomsen, M.F., J.T. Gosling, S.J. Bame, W.C. Feldman, G. Paschmann, N. Sckopke, *Geophys. Res. Lett.*, **10**, 1207-1210, 1983.

Multi-Spacecraft Observations of Interplanetary Shocks

Adam Szabo

Laboratory for Solar and Space Physics, NASA Goddard Space Flight Center, Greenbelt, MD 20771, USA

Abstract. This paper presents the analysis of a number of interplanetary shocks observed by near-Earth solar wind monitors. Single and four-spacecraft shock normal determination methods are compared and it is demonstrated that the four-spacecraft timing method is not generally more accurate than the single-spacecraft Rankine-Hugoniot fit. Moreover, the single-spacecraft methods, unlike the four-spacecraft ones, allow the determination of possible shock surface curvatures. Examples are presented where significant shock curvature are observed that can lead to 5-10 minutes forecast error for bow shock arrival times.

Keywords: Interplanetary Shocks, Shock Fitting Techniques.
PACS: 96.50 Fm, 95.75 Pq

INTRODUCTION

There is a strong correlation between interplanetary (IP) shocks impinging on the magnetosphere and geomagnetic disturbances [e.g., 1, 2]. IP shocks, as all solar wind pressure events, tend to disrupt the magnetopause surface leading to magnetopause transient events [3] that, in turn, can initiate magnetic reconnection resulting in substorm onset [e.g., 4]. IP shocks have also been connected to sudden commencements and aurora brightening [e.g., 5].

In order to better understand the geo-effectiveness of incoming interplanetary shocks, it is necessary to investigate their 3D geometry on the magnetospheric scale-length. Since multi-spacecraft observations are necessary for the determination of the three-dimensional geometry or shape of shocks [e.g., 6], very few studies have attempted to address this question. Generally it is assumed that the incoming IP shocks are planar on the scale-size of the magnetosphere. Indeed, in a recent study *Russell et al.* [7] analyzed a single IP shock with four solar wind satellites and found that three of them were consistent with the planarity assumption. However, deviation from planarity has been reported before [8, 9, 10, 11]. Specifically, *Szabo et al.* [9] has found, analyzing a few ACE, WIND and IMP 8 observed IP shocks, that with larger inter-spacecraft separation, perpendicular to the solar wind flow direction, larger angular deviation between individual shock normals was possible.

This paper presents multi-spacecraft observations and the various techniques to analyze a number of IP shocks establishing the range of shock surface curvatures on magnetospheric scale lengths, hence the uncertainties in the shock arrival times into the magnetosphere.

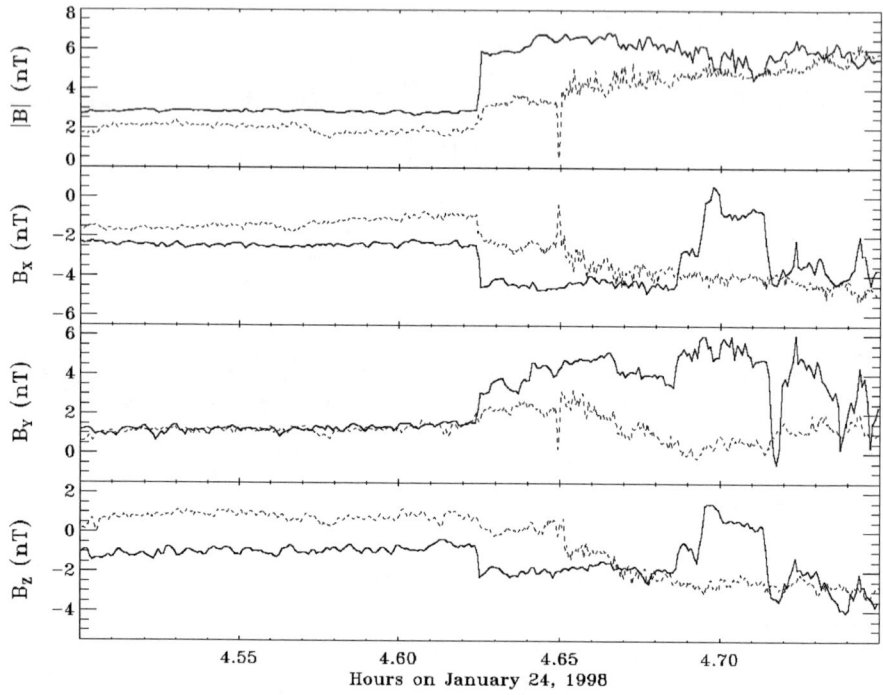

FIGURE 1. WIND (solid line) and ACE (dashed line) magnetic field observations of the January 24, 1988 interplanetary shock. The ACE data has been time shifted by 10.4 minutes.

SINGLE SPACECRAFT SHOCK FITTING TECHNIQUES

While for some IP shocks the multi-spacecraft observed magnetofluid parameters are strikingly similar, significant differences often arise. As an example, Fig. 1 shows vector magnetic field observations by ACE and WIND of the January 24, 1998 IP shock. The ACE data is time shifted by 10.4 minutes to align the shock observations and facilitate comparison. At this time, both spacecraft were near L1 and within 50 R_E of each other. This example, and there are many such cases, illustrates that significant variations in shock characteristics are common on magnetospheric scale lengths. And no matter what shock fitting technique is employed, different shock normal directions and speeds are likely to result.

Several techniques have been developed to determine the geometrical and physical characteristics of shocks in the magnetohydrodynamic (MHD) regime. All of them rely on some subset of conservation and Maxwell's equations and assume that the microphysical properties inside the shock layer can be ignored or avoided. The pre-averaged methods of shock normal and speed determination (e.g., magnetic coplanarity or velocity coplanarity) though easy to implement and use, suffer from the limitation that they use only a small subset (one or two out of nine equations)

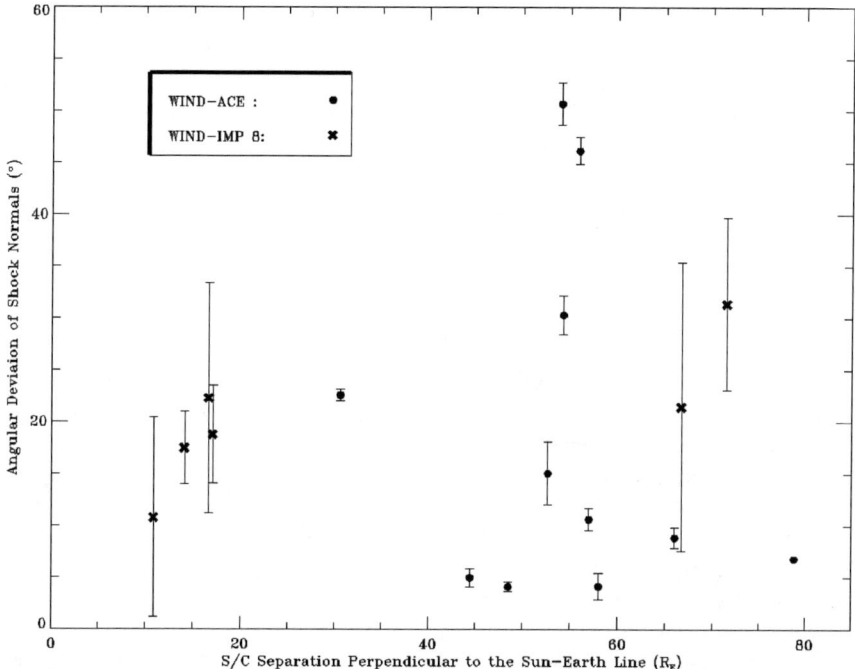

FIGURE 2. WIND/ACE and WIND/IMP 8 observations of the deviations between the locally determined shock surface normals for the same events as a function of spacecraft separation perpendicular to the Sun-Earth line.

of the complete MHD Rankine-Hugoniot jump conditions, and further assume that the average upstream and downstream magnetofluid parameters are the asymptotic MHD values. Reducing the measured data set on each side of the shock to a single averaged point, before the actual analysis is performed, makes these procedures prone to errors introduced by waves and systematic ambient variations in the solar wind data. Since the different pre-averaged methods use different subsets of the Rankine-Hugoniot jump conditions with different associated sensitivities and assumptions, they often give disparate solutions for the same set of measurements.

Iterative schemes, such as the *Vinas and Scudder* Rankine-Hugoniot (RH) method [12], try to resolve this problem by solving directly a larger set of the jump condition equations for self-consistent asymptotic magnetofluid parameters on both sides of the shock that are in closest agreement with the observations. While this RH method is fast and optimal, it does not use the energy and normal momentum flux conservation equations, equations that require knowledge of fluid temperature or pressure. *Szabo* [13] extended this method to include the last two jump condition equations. This technique has also the advantage that besides using the complete set of jump conditions, it uses full error propagation allowing the determination of statistically significant uncertainty regions for the shock normal and speed.

Using the above-described Vinas-Scudder-Szabo RH method, a number of IP shocks have been analyzed that were observed by two different spacecraft (either

WIND and ACE or WIND and IMP 8). Figure 2 shows the angular deviation between the shock normals determined from the individual spacecraft measurements for the same shock event as a function of inter-spacecraft separation perpendicular to the Sun-Earth line. Solid circles mark shocks observed by WIND and ACE, while crosses refer to WIND and IMP 8 observations. Considerable deviation from planarity is apparent with the envelope increasing with separation distance. However, by no means all shocks appear to have significant curvatures on the scale length of the magnetosphere.

THE FOUR SPACECRAFT METHOD

Once multi-spacecraft observations of the same IP shock are available, the time and position of the shock passage alone can be used to estimate the surface normal and speed [e.g., 8; 14]. However, a minimum of four spacecraft observations, that are not coplanar, are necessary and it is intrinsically assumed that these shocks are planar. Moreover, multi-spacecraft methods are very sensitive to observational uncertainties, especially for closely spaced spacecraft such as the L1 constellation. To match the accuracy of single spacecraft methods the time of the shock encounter needs to be known to a fraction of a second and the corresponding positions to better than 1 R_E. These requirements are at the very limit of the capabilities of current solar wind monitors. Nevertheless, multi-spacecraft methods are useful as a comparison tool to establish a baseline from which possible deviations are searched.

To illustrate the sensitivity of the four spacecraft method on the observed timing of the shock passage, the case of August 10, 1998 shock is analyzed in detail. This event is unusual in that five spacecraft observed the passage of the shock outside of the Earth's bow shock: ACE, WIND, IMP 8, Geotail and Interball Tail. Since only four is required for the determination of the shock normal direction and speed it is possible to establish the stability of the solution as the five possible combination of four-spacecraft observations are evaluated. The thus obtained shock normals and speeds are listed in the first five rows of Table 1 with the name of the spacecraft not used for the calculations indicated. The uncertainties are based on the time resolution of the available highest cadence data, in this case, the magnetic field observations. The angular variations between the five shock normals obtained ranges between 3 and 11 degrees. For comparison, Table 1 shows the Rankine-Hugoniot (RH) normals computed for the WIND, ACE and IMP 8 measurements. The RH uncertainties are due to measurement uncertainties propagated through the fitting process and to the

TABLE 1. Comparison of fitting techniques for the August 10, 1998 shock (GSE coordinates)

Fit Method	Shock Normal Direction	Shock Speed (km/s)
4-S/C w/o ACE	-0.57 ± 0.02, -0.32 ± 0.04, 0.76 ± 0.03	326 ± 15
4-S/C w/o WIND	-0.59 ± 0.03, -0.35 ± 0.05, 0.73 ± 0.04	348 ± 18
4-S/C w/o IMP 8	-0.53 ± 0.02, -0.31 ± 0.04, 0.78 ± 0.03	307 ± 15
4-S/C w/o Geotail	-0.48 ± 0.03, -0.25 ± 0.05, 0.84 ± 0.04	265 ± 18
4-S/C w/o Interball	-0.51 ± 0.02, -0.25 ± 0.04, 0.82 ± 0.03	290 ± 15
RH - WIND	-0.57 ± 0.06, -0.55 ± 0.08, 0.62 ± 0.08	346 ± 32
RH - ACE	-0.45 ± 0.07, -0.19 ± 0.11, 0.87 ± 0.04	304 ± 31
RH - IMP 8	-0.51 ± 0.07, -0.31 ± 0.09, 0.80 ± 0.05	347 ± 33

requirement that a self-consistent solution be found. Even though this particular shock is relatively weak, increasing the RH uncertainties well above typical levels, the stability of the 4-spacecraft method does not appear to be significantly higher. Moreover, the differences between the RH normals are larger than the computed uncertainties implying that the shock is not planar. Hence the solution of the RH method is generally to be preferred over the 4-spacecraft method.

SUMMARY

In this paper, it has been shown that while a number of interplanetary shocks appear to be very close to planar geometry on magnetospheric scale lengths, significant surface curvature is not unusual. It has been established that solutions of four-spacecraft timing methods are not generally superior in accuracy to the single-spacecraft Rankine-Hugoniot fits. Moreover, the single-spacecraft RH method allows the determination of the surface curvature of the interplanetary shocks. The observed curvatures can result in 5-10 minutes errors in the predicted arrival times at the Earth's bow shock, using simple ballistic propagation of a planar shock surface, making magnetosheath and magnetospheric shock propagation studies difficult.

REFERENCES

1. Tsurutani, B. T., W. D. Gonzalez, A. L. C. Gonzalez, F. Tang, J. K. Arballo and M. Okada, *J. Geophys. Res.,* **100**, 21,717-21,733, 1995.
2. Gonzales, W. D. and B. T. Tsurutani, The interplanetary causes of magnetic storms: "A review", in *Magnetic Storms*, eds. B. T. Tsurutani, W. D. Gonzales, Y. Kamide and J. K. Arballo, AGU Monograph 98, Washington, D.C., 1998.
3. Sibeck, D. G. and P.T. Newell, *J. Geophys. Res.,* **100**, 21,773-21,778, 1995.
4. Wu, C. C., *J. Geophys. Res.,* **105**, 7533-7543, 2000.
5. Tsurutani, B. T., X.-Y. Zhou, J. K. Arballo, W. D. Gonzales, G. S. Lakhina, V. M. Vasyliunas, J. S. Pickett, T. Araki, H. Yang, G. Rostoker, T. J. Hughrs, R. P. Lepping and D. Berdichevsky, *J. Atmos Sol.-Terr. Phys.,* **63**, 513-522, 2001.
6. Thomsen, M. F., *Adv. Space Res.,* **8**, (9)157, 1988.
7. Russell, C. T., Y. L. Wang, J. Reader, R. L. Tokar, C. W. Smith, K. W. Ogilvie, A. J. Lazarus, R. P. Lepping, A. Szabo, H. Kawano, T. Mukai, S. Savin, Y. I. Yermolaev, X.-Y. Zhou and B. T. Tsurutani, *J. Geophys. Res.,* **105**, 25,143, 2000.
8. Russell, C. T., J. T. Gosling, R. D. Zwickl and E. J. Smith, *J. Geophys. Res.,* **88**, 9941, 1983.
9. Szabo, A., R. P. Lepping, J. Merka, C. W. Smith, and R. M. Skoug, "The evolution of interplanetary shocks driven by magnetic clouds", in *Proceedings of "Solar Encounter: The First Solar Orbiter Workshop"*, pp. 383-387, eds. E. Marsch, V. Martinez Pillet, B. Fleck and R. Marsden, ESA SP-493, Puerto de la Cruz, Tenerife, Spain, 2001.
10. Szabo, A., C. W. Smith, and R. M. Skoug, "The transition of interplanetary shocks through the magnetosheath", in *Solar Wind 10*, AIP Conference Proceedings, Melville, New York, 2003.
11. Szabo, A., "Interplanetary Discontinuities and Shocks in the Earth's Magnetosheath", in *Multiscale Processes in the Earth's Magnetosphere*, eds. J-A Sauvaud and Z. Nemecek, Kluwer, 2004.
12. Vinas, A. F. and J. D. Scudder, *J. Geophys. Res.,* **91**, 39, 1986.
13. Szabo, A., *J. Geophys. Res.,* **99**, 14,737, 1994.
14. Schwartz, S. J., "Shock and discontinuity normals, Mach numbers, and related parameters", in *Analysis Methods for Multi-Spacecraft Data*, pp., 249-270, eds., G. Paschmann and P. W. Daly, ESA Publication Division, Noordwijk, The Netherlands, 1998.

Classical MHD shocks: theory and numerical simulation

Nikolai V. Pogorelov

Institute of Geophysics and Planetary Physics, University of California, Riverside, CA 92521

Abstract. Recent results are surveyed in the investigation of the behavior of shocks in ideal magnetohydrodynamics (MHD) and corresponding structures in dissipative/resistive plasma flows. In contrast to evolutionary shocks, a solution of the problem of the nonevolutionary shock interaction with small perturbations is either nonunique or does not exist. The peculiarity of non-ideal MHD is in that some nonevolutionary shocks have dissipative structures. Since this structure is always non-plane, it can reveal itself in problems where transverse perturbations do not exist due to symmetries restrictions. We discuss the numerical behavior of nonevolutionary shocks and argue that they necessarily disappear once the problem is solved in a genuinely three-dimensional statement.

Keywords: Numerical simulation; magnetohydrodynamics; multi-fluid plasma; interplanetary and interstellar magnetic fields; solar wind plasma
PACS: 02.60.Cb, 52.30.Cv, 52.30.Ex, 52.65.Kj, 96.50.Bh, 96.50.Ci, 96.50.Fm

INTRODUCTION

Magenetohydrodynamic (MHD) discontinuities are subdivided according to the presence (or absence) of the mass flux through them [11]. With no flux, they are represented by contact or tangential discontinuities, depending on whether the normal component of the magnetic field is not or is equal to zero, respectively. If the mass flux is non-zero and density is continuous, the mass flux is equal to $m = B_n(\rho/4\pi)^{1/2}$. Such discontinuities are called Alfvénic. Their only effect on the upstream flow is in rotation of the velocity and magnetic field vectors without changing their magnitudes. All other discontinuities are called shocks. Not all shocks that are solutions to Rankine–Hugoniot relations are admissible in ideal magnetohydrodynamics. It is well-known that any small plane perturbation of the state vector in the hyperbolic system can be represented as a sum of linear simple waves moving with corresponding characteristic velocities λ_i. Each simple wave is characterized by its amplitude. If we consider a shock with velocity W, each wave is either incoming or outgoing, depending on the sign of $\lambda_i - W$. A shock is called evolutionary if the problem of its interaction with small perturbations has a unique solution. The necessary conditions for that were given by [1, 40]: the number of outgoing waves, whose amplitudes should be determined, must be equal to $n-1$, where n is the number of shock (Hugoniot) relations. An additional relation is necessary to find a perturbed shock velocity. For strictly hyperbolic, convex MHD systems these conditions reduce to Lax inequalities [28]. If the number of unknowns is greater than the number of relations, the amplitudes cannot be resolved uniquely and therefore must depend on one ore more arbitrary functions of time. Such discontinuities either do not exist or there are physical reasons to impose additional relations of them, which are indepen-

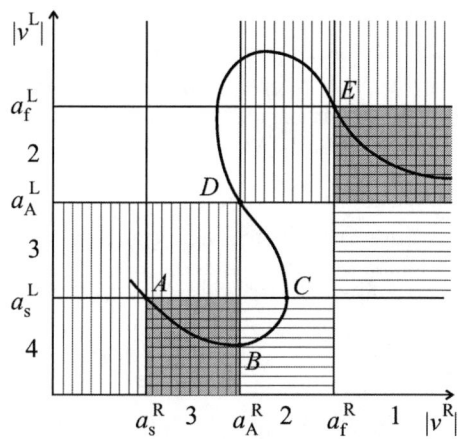

FIGURE 1. MHD adiabatic shock curve

dent of the conservation laws [26]. Otherwise, if the number of unknowns is less than the number of relations, there is no solution in the linear approximation, and a shock splits into evolutionary discontinuities. Here I summarize theoretical results regarding the nonevolutionary shock behavior in ideal MHD and related dissipative structures in non-ideal plasma. Clarification of these results may be useful for interpreting numerical simulations, since numerical dissipation/resistivity can be orders of magnitude higher than physical in space plasmas.

MHD SHOCK EVOLUTIONARY DIAGRAM AND CLASSIFICATION OF SHOCKS

There are two important peculiarities of MHD shocks: (i) relations for the wave amplitudes in the problem of the shock interaction with small perturbations split into two groups (plane-polarized and Alfvén, or rotational, waves; see [27]); (ii) a Jouget point (a point where the shock velocity coincides with one of the characteristic velocities) on the adiabatic shock (Hugoniot) curve has a vertical tangent [22, 24].

Let us assign superscripts R and L to the quantities ahead of and behind the shock, respectively. The Hugoniot curve can be considered as a curve in the functional space with the initial point at $\mathbf{U}^R$. If we know the state vector $\mathbf{U}$ ahead of the shock and the shock velocity W, we can determine $\mathbf{U}^L$. In this sense the Hugoniot curve is a function of only one parameter. That is, the velocity behind the shock v_n^L can be represented as $v_n^L = v_n^L(\sigma)$, where σ is a parameter on the Hugoniot curve, say, the arc length calculated from the initial point $\mathbf{U}^R$ in the functional space. In Fig. 1, we show a Hugoniot curve in the (v_n^R, v_n^L) space. Here $v_n(\sigma)$ is plotted in the real scale along the horizontal axis and schematically along the vertical axis, by retaining only the inequalities between the flow velocity v and fast a_f, slow a_s, and Alfvén a_A velocities on the both sides of the shock and the continuity condition for the curve itself. Figure 1 reflects the fact that in ideal MHD only fast ($|v^R| > a_f^R$, $|v^L| < a_f^L$) and slow ($a_s^R < |v^R| < a_A^R$, $|v^L| < a_s^L$) shocks are evolutionary. They are shown by shaded rectangles. Vertical and horizontal hatching refer to shocks evolutionary only to rotational and plane (compressive) perturbations, respectively. Points A and E correspond to weak evolutionary shocks. Point D corresponds to a rotational discontinuity. If we enumerate the horizontal and vertical stripes, as it is done in Fig. 1, all shock transitions will also be enumerated as $i \to j$. Thus, $1 \to 2$ and $3 \to 4$ shocks are evolutionary. As shown in [20], only transitions with increasing rectangle number are entropy increasing. Shocks exhibiting $2 \to 3$, $1 \to 3$, $2 \to 4$, and $1 \to 4$ transitions are nonevolutionary with respect to Alfvén perturbations. The solution does not exist, since there is only one outgoing transverse wave. Shocks $2 \to 3$ and $1 \to 4$ are also nonevolutionary with respect to perturbations in magnetosonic properties. The solution is nonunique for the former and does not exist for the latter. In contrast to evolutionary shocks, when crossing a nonevolutionary shock, the magnetic field vector always changes the orientation of its tangential component to the opposite. A theory of 2D MHD shocks have been developed in [36]. Alfvén waves disappear in two dimensions while $1 \to 3$ and $2 \to 4$ shocks become evolutionary. They ensure rotation of the magnetic field, which in three dimensions is performed only by Alfvén waves. However, $1 \to 4$ (overcompressive) and $2 \to 3$ (undercompressive) shocks remain nonevolutionary.

MHD RIEMANN PROBLEM

The Riemann problem for the time evolution of an arbitrary MHD discontinuity was first solved in [21]. It has also been proven that this solution can be uniquely represented by a combination of evolutionary discontinuities and rarefaction waves. The latter statement is obviously valid also for initial data representing pre- and post-shock states of any nonevolutionary discontinuity. An MHD piston problem was solved in [3]. Since the exact solution of the MHD Riemann problem is too multi-variant to be used in regular computations, approximate solutions of the MHD Riemann problem have also been obtained by a number of authors [26]. It is worth noting that an approximate solution [10], where rarefaction waves were modelled by corresponding rarefaction shocks (obeying jump conditions and involving decreases in entropy), took special care of selecting only evolutionary shocks. In contrast to that, paper [44] investigated uniqueness conditions for the MHD Riemann problem disregarding the fact that this solution is always unique if only evolutionary waves are involved.

MHD SHOCKS AS LIMITS OF DISSIPATIVE STRUCTURES

It has been shown in [20] that both evolutionary and nonevolutionary shocks can have dissipative structures in the form of a non-deforming travelling wave. If magnetic viscosity is much less than molecular viscosity, nonevolutionary shocks have no structure [25]. Papers [41, 42, 43] were the first to have analyzed the effect of finite resistivity on transient processes in nonevolutionary shock waves affected by small perturbations. It is therefore remarkable that those papers are not cited by a number of authors who investigate shock wave structures and rediscover analogous results. We say that the shock is parallel if the magnetic field is parallel to the shock normal both ahead of and behind the shock. If magnetic field vector is parallel to it only ahead of (behind) the shock, the latter is called a switch-on (switch-off) shock. The results obtained by L. Todd can be summarized as follows.

- Nonevolutionary (trans-Alfvénic) shocks with transitions $1 \to 3$, $1 \to 4$, and $2 \to 4$ can have steady structure. This structure is nonunique, but can be fixed by the choice of one parameter, e. g., the integral $I_z = \int_{-\infty}^{\infty} B_z dx$ of the out-of-plane component of the magnetic field vector over the shock structure. Note that this integral is constant once the z-components of the velocity and magnetic field vanish at $x = \pm\infty$, that is, for planar shocks.
- $2 \to 3$ shocks have no steady structure, and they should therefore be called nonstationary.
- Structures of nonevolutionary shocks can contain gas dynamic subshocks.
- When subjected to small perturbations, shock profiles of nonevolutionary shocks linearly grow with time. Thus, nonlinear effects must eventually dominate and disrupt the shock *in some way*. $2 \to 3$ trans-Alfvénic shocks are wave generators, whereas other nonevolutionary shocks are flux-collectors.

Using a simplified system describing MHD flows when the angle between the x-axis and the magnetic field vector **B** is small, important results have been obtained in [23, 46, 47, 48] about unsteady processes occurring inside nonevolutionary shock structures. Similar analysis on the basis of the full MHD system was performed in [31, 32, 33]. According to [23], a nonevolutionary $2 \to 3$ shock has a structure represented by one of the two symmetric integral curves, while a nonevolutionary $1 \to 3$ shock can be associated with a one-parametric (I_z being chosen as a parameter) set of integral curves. For $1 \to 3$ shocks $|I_z| \leq I_z^*$, where I_z^* is the value of I_z for a $2 \to 3$ shock with the same velocity. If a $1 \to 3$ shock structure starts interacting with out-of-plane perturbations, it will accumulate I_z until it becomes a $2 \to 3$ shock emitting a fast shock. Further interaction will asymptotically shift a $2 \to 3$ shock to a rotational discontinuity (see Fig. 1) with rotation of the magnetic field by π. This scenario supports the statement that nonevolutionary shocks are rather stable with respect to small perturbations. However, the smallness of the perturbation cannot guarantee stability of the nonevolutionary shock structure if we supply sufficiently large $v_m^{-1} I_z$ into it [26]. This implies that the described scenario occurs in a similar way if we decrease the magnetic viscosity coefficient v_m. Since numerical dissipation is frequently much higher than physical in space plasmas, nonevolutionary shock structures can manifest themselves even in cases where transverse perturbations are available. This behavior is totally spurious, and every attempt should be undertaken to avoid it [4, 13, 31].

A very important and detailed mathematical analysis of existence, stability, and bifurcation of dissipative profiles of trans-Alfvénic shock waves has been performed recently in [15, 16, 17, 18]. For a physicist, it is important to realize, however, that most of the analyzed unsteady phenomena refer to resistive (laboratory) plasma. They occur nearly instantaneously in space, where both molecular and magnetic Reynolds numbers are very large. E. g., it has been shown in [13] that for a typical interplanetary shock with

$a_A = 40$ km/s and the width $l = 5 \times 10^4$ km, the time of existence for an trans-Alfvénic shock is $1200/\delta\varphi$ s, where $\delta\varphi$ is a small rotation of the magnetic field caused by a small-amplitude Alfvén perturbation. Since the flow time at 9 AU is about 3×10^7 s, this amplitude should be extremely small.

Paper [4] investigated the process of eliminating a nonevolutionary compound wave from the solution of the co-planar Riemann problem suggested in [9]. It was shown that since the Hugoniot curve has a vertical tangent at point C in Fig. 1, a solution can be constructed where a trans-Alfvénic shock represented by this point is followed by a slow rarefaction wave one of whose sides moves with the shock velocity $W = \lambda_s^L = v_n^L - a_s^L$. This compound wave lies in the part of the Hugoniot curve that is nonevolutionary in three dimensions. For this reason, if we add out-of-plane perturbations, the $2 \to 4$ shock will start accumulating I_z until it splits into a $2 \to 3$ shock and a slow shock. Further supply of I_z will be transferring the shock properties from point C to point D, which corresponds to Alfvén discontinuities. Since a slow shock–rarefaction wave junction occurs now not at the Jouget point C, the waves will interact until a pure slow shock is obtained [4, 26]. These processes last shorter for higher grid resolution and/or for lower dissipation numerical schemes.

NONEVOLUTIONARY SHOCKS IN THE PRESENCE OF SYMMETRIES

The properties of nonevolutionary (intermediate, trans-Alfvénic) shocks described here make it natural to expect them to appear in cases where transverse perturbations are excluded either by the physical statement of the problem or by the reduction of its dimension via introducing a symmetry axis or a symmetry plane. For example, it has long been known (see [27]) that fast parallel (or gas dynamic) MHD shocks, for $a_A^R > c_e^R$ (c_e is the speed of sound), are evolutionary only up to a certain limiting magnetic field strength. Fast switch-on shocks become evolutionary for stronger fields. Thus, if we consider an axisymmetric, magnetic field aligned plasma flow over an obstacle, a bow shock must be parallel in its forward point. If we increase the magnetic field ahead of the shock, sooner or later this shock will become inadmissible. On the other hand, switch-on shocks are prohibited by the axial symmetry of the flow. Similar situation may occur if we assume the presence of a symmetry plane. As a result, we obtain a solution with a complicated combination of shocks, some of them being nonevolutionary [12]. Nonevolutionary shocks of that kind disappear in the solar wind interaction with the interstellar medium if the problem is solved in the genuinely three-dimensional formulation [37, 38, 39].

It is rather easy to destroy an analogous nonevolutionary shock structure that appears in the plane, magnetic field aligned flow over an infinitely conducting cylinder. It only suffices to make this flow non-plane by adding an infinitesimal component of the magnetic field parallel to the cylinder axis [26]. As found by Yee & Sjögreen (2004; private communication), the plane solution is not steady if the space resolution is sufficiently high. The analysis of the shock types [12] for this configuration (see Fig 3a, where the Mach and Alfvén numbers are equal to $M_\infty = v_\infty/c_{e\infty} = \sqrt{13.5}$ and $A_\infty = v_\infty/a_{A\infty} = 1.5$, respectively) shows that some segments of the shock configuration behind the bow shock

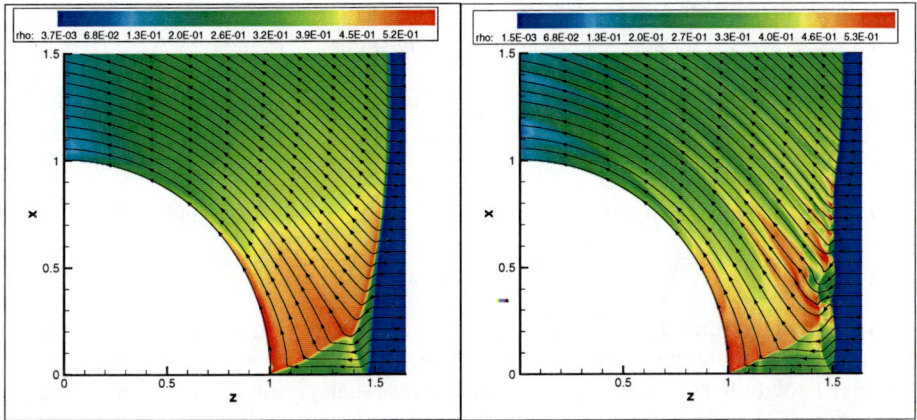

FIGURE 2. Density distribution and streamline (magnetic field line) for the plasma flow over an infinitely conduction cylinder with the Mach and Alfvén numbers equal to $\sqrt{13.5}$ and 1.5, respectively: (a) a stationary solution and (b) an unsteady solution

are of the $2 \to 3$ type and therefore should be nonevolutionary with respect to perturbations in magnetosonic quantities. For this reason, the plane stationary solution obtained with the TVD Lax–Friedrichs scheme [4] becomes unsteady (see Fig. 2b) if a Roe-type scheme [26] is used on a sufficiently fine mesh.

STABILITY OF MHD SHOCKS

It is worth mentioning that the problem of evolutionarity differs from the problem of stability of MHD shocks in some respects. Nonlinear stability of shocks should be understood as existence (at least for short times) of a solution to the MHD system involving a shock dividing the regions of continuity. Alongside with the nonlinear stability problem one can consider a linearized stability problem (LSP). In the latter, one should analyze well-posedness of the mixed-type hyperbolic problem for small perturbations of the left- and right-hand sides of the shock. For stability, perturbations that vanish at large distances from the shock should not exhibit exponentially growing modes. Initial conditions for perturbations at $t = 0$ are given. Boundary conditions are expressed by the Rankine–Hugoniot shock relations at the shock front, which is initially at $x_1 = 0$ and becomes perturbed according to the formula $x_1 = \delta f_s(t, x_2, x_3)$. Thus, the perturbation of the shock front δf_s is also among the sought quantities. If we consider a one-dimensional LSP, the shock profile becomes plane and depends only on t. It is obvious that a one-dimensional LSP is always well-posed for evolutionary shocks. In that sense one might call plane evolutionary shocks linearly stable with respect to small plane perturbations. However, it is widely accepted to preserve the term evolutionarity for this case. It has been proved in [19] that fast parallel and perpendicular MHD shocks in polytropic media are linearly (weakly) stable for $1 < \gamma < 3$. The restriction $\gamma < 3$ has been removed in [2] for perpendicular and in [7] for parallel shocks. It was shown in [6] that all admissible MHD shocks are weakly stable for weak magnetic fields ($B^2/4\pi\gamma p \ll 1$). Fast parallel shocks are uniformly stable [7] if and only if

$$F\left(M + \sqrt{M^2 + \frac{2}{\gamma-1}}\right) > 0,$$

where M is the sonic Mach number behind of the shock, $q = a_A^L/c_e^L$, and

$$F(z) = (zM-1)z^4 + q^2\{(zM-1)(z^2-2)z^2 - q^2(z^2-1)^2\}.$$

Analysis of these formulae together with the evolutionary conditions shows that the region of uniform stability is a subregion of the evolutionary region. A rather small portion of evolutionary shocks with non-small q (note that q is always less than M, while $\sqrt{\gamma-1/2\gamma} < M < 1$) is called neutrally stable. Shocks belonging to this region may appear nonlinearly unstable. Slow shocks are also unstable for strong magnetic fields [8] but the range of their instability is typically wider than that for fast shocks. Two-dimensional instability domains for slow MHD shocks in ideal gas with $\gamma = 5/3$ were found in [29] and for three dimensions in [14]. Note also that the latter paper showed that 3D perturbations did not enlarge the 2D stability domain for fast shocks. A complete stability analysis of fast MHD shocks in monoatomic adiabatic plasma has been made in [45], where both the regions of weak and uniform stability of the LSP were found. We note that the domain of uniform stability is a subdomain of weak stability, where the uniform Kreiss–Lopatinskii condition is satisfied. The rest of the weak stability region corresponds to so-called neutral stability, where small perturbations of upstream and downstream quantities, as well as of a planar shock front, are not changing in time. The importance of the notion of uniform stability is in that uniformly stable shocks are nonlinearly stable [8, 30, 34, 35]. In particular, for uniformly stable MHD shocks a short-time existence theorem is valid. Of interest for plasma physicists is the fact that both slow and fast (evolutionary) shocks always have a domain of 2D instability for sufficiently strong magnetic fields behind the shock.

ACKNOWLEDGMENTS

This research was supported by NASA grants NAG5-12903 and NNG05GD45G, and NSF award ATM-0428880. The author is grateful to A. A. Barmin, A. G. Kulikovskii, and G. P. Zank for numerous discussions. Special thanks to Y. L. Trakhinin for his comments on uniform stability of MHD shocks.

REFERENCES

1. Akhiezer, A.I., Lyubarskii, G.Y., & Polovin, R.V. (1959), Soviet Physics – JETP, 8, 507–511
2. Anile, A. M., & Russo, G. (1989), in *Nonlinear Wave Motion*, A. Jeffrey (ed.), Pitman Monographs and Surveys in Pure and Applied Mathematics, 43, 11–21, John Wiley, New York
3. Barmin, A.A., Gogosov, V.V. (1960), Soviet Physics Doklady, 134, No. 5, 1041
4. Barmin, A. A., Kulikovskii, A. G., & Pogorelov, N. V. (1996), J. Comput. Phys., 126, 77
5. Blokhin, A. M. (1994), *Strong Discontinuities in Magnetohydrodynamics*, Nova Science, New York
6. Blokhin, A. M., & Druzhinin, I. Y. (1989), Siberian Math. J., 30, 511–524.

7. Blokhin, A. M., & Trakhinin, Y. L. (1999), Eur. J. Mech. B/Fluids, 18, 197–211
8. Blokhin, A. M., & Trakhinin, Y. L. (2003), *Stability of Strong Discontinuities in Magnetohydrodynamics and Electrodynamics*, Nova Science, New York
9. Brio, M., Wu, C.C. (1988), J. Comput. Phys., 75, 400
10. Dai, W., Woodward, P. R. (1994), J. Comput. Phys., 111, 354
11. De Hoffmann, F., & Teller, E. (1950), Phys. Rev., 80, 692–703
12. De Sterck, H., & Poedts, S. (1999), Astron. Astrophys., 343, 641
13. Falle, S. A. E. G., Komissarov, S. S. (2001), J. Plasma Phys., 65, 29
14. Filippova, O. L. (1991), Fluid Dyn., 26, 897–904
15. Freistühler, H., Fries, C., Rohde, C. (2000), Prepr. 60, Max-Planck-Istitut für Mathematik in den Naturwissenschaften, Leipzig
16. Freistühler, H., Rohde, C. (2001), Math. Comput., 71, 1021
17. Freistühler, H., Rohde, C. (2002), Prepr. 4, Max-Planck-Istitut für Mathematik in den Naturwissenschaften, Leipzig
18. Fries, C. (2000), Arch. Rational Mech. Anal., 152, 141
19. Gardner, G. S., & Kruskal, M. D. (1964), Phys. Fluids, 7, 770–706
20. Germain, P. (1959) *Contribution a la théorie des ondes de choc en magnétodynamique des fluides*, Office National d'Etudes et de Recherches Aeronautiques, Publ. 97, Paris
21. Gogosov, V. V. (1961), J. Appl. Math. Mech., 25, 148
22. Hanyga, A. (1976) Polish Acad. Sci. Publications of Geophysics, A-1 (98), Panstvowe wydavnitstvo naukowe, Warszawa
23. Kennel, C. F., Blandford, R. D., & Wu, C. C. (1990), Phys. Fluids B, 2, 253
24. Kulikovskii, A.G. (1979) Fluid Dynamics, 14, No. 2, 317
25. Kulikovskii, A.G., Lyubimov, G.A. (1965) *Magnetohydrodynamics*, Addison Wesley, Reading, MA
26. Kulikovskii, A. G., Pogorelov, N. V., & Semenov, A. Yu., (2001) *Mathematical Aspects of Numerical Solution of Hyperbolic Systems*, 560 pp., Chapman & Hall / CRC, Boca Raton, Florida
27. Landau, L. D., Lifshitz, E. M. (1984), *Electrodynamics of Continuous Media*, Pergamon, Oxford
28. Lax, P. D. (1957), Comm. Pure Appl. Math., 10, 537
29. Lessen, M., & Deshpande, M. V. (1967), J. Plasma Phys., 1, 463–472
30. Majda, A. (1983), *The Existence of Multidimensional Shock Fronts*, Mem. Amer. Math. Soc., 43, AMS, Providence
31. Markovskii, S.A. (1998), Phys. Plasmas, 5, 2596
32. Markovskii, S. A., & Skorokhodov, S. L. (2000a), Phys. Plasmas, 7, 158
33. Markovskii, S. A., & Skorokhodov, S. L. (2000b), J. Geophys. Res., 105(A6), 12705
34. Métivier, G. (2001), in *Advances in the Shock Wave Theory*, 25–103, Progr. Nonlinear Differential Equatuins Appl., 47, Birkhäuser, Boston
35. Métivier, G., & Zumbrun, K. (2005), J. Diff. Equations, 211, 61–134
36. Myong, R. S., Roe, P. L. (1997), J. Comput. Phys., 147, 545
37. Pogorelov, N. V., Matsuda, T. (2000), Astron. Astrophys., 354, 697
38. Pogorelov, N. V., Matsuda, T. (2004), J. Geophys. Res., 109, A02110, doi:10.1029/2003JA009998
39. Pogorelov, N. V., Zank, G. P., & Ogino, T. (2004), Astrophys. J., 614, 1007
40. Syrovatskii, S.I. (1959), Soviet Physics–JETP, 8, 1024–1028
41. Todd, L. (1964), J. Fluid Mech., 18, 321
42. Todd, L. (1965), J. Fluid Mech., 21, 193
43. Todd, L. (1966), J. Fluid Mech., 24, 597
44. Torrilhon, M. (2003), J. Comput. Phys., 192, 73
45. Trakhinin, Y. (2003), Commun. Math. Phys., 236, 65–92
46. Wu, C. C. (1990), J. Geopys. Res., 95, 987
47. Wu, C. C., & Kennel, C. F. (1992), Phys. Rev. Lett., 68, 56
48. Wu, C. C., & Kennel, C. F. (1993), Phys. Fluids B, 5, 2877

Issues for Hybrid Simulations of Collisionless Shocks

D. Winske and L. Yin

Division of Applied Physics, Los Alamos National Laboratory, Los Alamos, NM 87545 USA

Abstract. We discuss ways in which well-known hybrid simulation methods can be extended to address ongoing issues related to the structure and properties of collisionless shocks. Focusing primarily on slow-mode shocks, we discuss how the basic physics model for the electrons can be enhanced and how kinetic models can be embedded in more global calculations.

INTRODUCTION

Hybrid codes, in which ions are treated kinetically using particle-in-cell methods and electrons are modeled as an inertia-less fluid, have been used successfully for many years to model the structure of fast-mode collisionless shocks in the heliosphere, including the Earth's bow shock, bow shocks of other planets, interplanetary shocks, cometary shocks, and the termination shock, slow-mode shocks in the magnetotail as well as intermediate shocks and other discontinuities (see recent reviews of shocks [1] and hybrid code methods [2]). This success rests on the fact that the major dissipation mechanisms for these types of shocks reside on ion length and time scales that are well resolved in hybrid simulations. There are of course, additional dissipation processes that occur on various scales, which are resolved by full-particle simulations methods [3]. As we will show here, some of this additional physics can be included in hybrid codes by improving the electron fluid model while remaining on ion scales.

In the hybrid method, the electromagnetic field equations are solved on a rectilinear grid. The electric $\mathbf{E}$ and magnetic $\mathbf{B}$ fields are interpolated to the position of each ion to compute the force which is then used to advance its velocity and position by a time step Δt, which is a small fraction of the ion gyro-period. The positions and velocities of the ions are then recollected on the grid to update the ion density and current. The magnetic field is advanced in time from Faraday's law and the electric field is computed from the electron momentum equation, using the fact that the electron mass $m_e = 0$, the electrons are quasi-neutral so that on this spatial scale the electron and ion densities are equal ($n_e = n_i$), and the electron current is related to the ion current and curl $\mathbf{B}$ by Ampere's law, i.e.,

$$\mathbf{E} = -\frac{\mathbf{V}_i \times \mathbf{B}}{c} - \frac{\nabla p_e}{en_i} - \frac{\mathbf{B} \times (\nabla \times \mathbf{B})}{4\pi e n_i}, \tag{1}$$

where $\mathbf{V}_i$ is the ion flow velocity, c is the speed of light, e is the electronic charge and p_e is the scalar electron pressure, and resistive effects are ignored. The electron pressure is determined by an equation of state, e.g., an adiabatic law that is typically used in shock simulations.

This model has been successfully employed to model slow-mode shocks that occur in the magnetotail. Slow shocks are characterized by an inflow speed that is less than the local Alfven speed, a jump in the plasma density at the shock, a corresponding drop in the plasma flow speed normal to the shock, a decrease in the magnetic field transverse to the shock and a corresponding increase in the transverse flow speed. In other words, unlike a fast shock, at a slow-mode shock the transverse magnetic field is reduced, while the plasma flow is increased and diverted away from the shock normal direction. The formation of such slow shocks can be studied in two spatial dimensions by inducing reconnection to occur locally where two plasmas flowing in from the tail lobes interact to form a steady-state, Petschek-like configuration [4]. The problem can be reduced to one spatial dimension by having the inflowing plasma interact with a reflecting wall, where the transverse magnetic field is set to zero. For modest angles of the upstream magnetic field relative to the shock normal ($\theta_{Bn} \sim 55° - 75°$), well-defined shocks are usually formed [5], characterized by some ions streaming back from the shock into the upstream along the magnetic field direction and a trailing magnetic wavetrain that is predicted by theory [6], in general agreement with observations [7].

ELECTRON PHYSICS MODEL

In the case of very oblique flows ($\theta_{Bn} \sim 84°$), however, similar hybrid simulations indicate that a shock is not generated, presumably because the ion dissipation is not sufficient to allow the shock to form. Yin et al. [8] have argued, following Hesse and Winske [9] for magnetic reconnection in the absence of a strong guide field, that the electron model in (1) needs to be generalized, by replacing ∇p_e with $\nabla \bullet \mathbf{P}_e$, where $\mathbf{P}_e$ is the electron pressure tensor. $\mathbf{P}_e$ is advanced in time using,

$$\frac{\partial \mathbf{P}_e}{\partial t} = -\mathbf{v}_e \cdot \nabla \mathbf{P}_e - \mathbf{P}_e \nabla \cdot \mathbf{v}_e - \mathbf{P}_e \cdot \nabla \mathbf{v}_e - (\mathbf{P}_e \cdot \nabla \mathbf{v}_e)^T$$

$$-\upsilon(\mathbf{P}_e - p\mathbf{1}) + \frac{e}{m_e c}(\mathbf{B} \times \mathbf{P}_e - \mathbf{P}_e \times \mathbf{B}), \qquad (2)$$

where $\mathbf{V}_e$ is the electron flow velocity, $\mathbf{1}$ is the unit tensor, $(\ )^T$ denotes the transpose of the tensor, $p = (1/3)\mathrm{Tr}(\mathbf{P}_e)$ and υ is parameter that characterizes the rate at which the pressure tensor isotropizes (due to microphysics that is not specified in the calculations). We note that in this equation the electron mass specifically appears; this term evidently involves fast electron gyro-motion, which can be included implicitly on ion time scales. The spatial resolution implied in (2) remains on scales comparable to the ion inertial length (c/ω_i). We have also ignored the divergence of the heat flux in (2), although with some analytical reduction it can also be included [10]. Hybrid simulations of the aforementioned case with oblique flow at $\theta_{Bn} = 84°$ using this extended electron physics model demonstrate that the slow wave will steepen into a shock with a trailing wavetrain whose damping rate depends on the parameter υ [8].

To further demonstrate that these off-diagonal terms in the electron pressure tensor are correctly modeled by (2), Yin et al. [11] have also carried out a full particle simulation in 1-D of the $\theta_{Bn} = 84°$ slow shock and evaluated the components of $\mathbf{P}_e$ that are self-consistently generated in this calculation. The full particle simulation also indicates how an anisotropy in the diagonal electron pressure terms develops [12], due to parallel electron heating in the shock transition region from kinetic Alfven waves, i.e., ion cyclotron waves that propagate nearly perpendicular to the magnetic field, with a large electric field parallel to the local magnetic field. Figure presents results from the simulation showing electron and ion phase space (velocity component parallel to the local magnetic field direction versus spatial coordinate x) at $\Omega_i t = 1000$. The shock ramp is located at $510 < x\omega_i/c < 540$. The oscillations in the width of the electron distribution are due to resonant interactions with the kinetic Alfven waves that lead to the hotter distribution downstream ($x > 550\ c/\omega_i$). The counterstreaming ions in the ramp generate the waves. The backstreaming ion flow is disrupted upstream ($x \sim 470\ c/\omega_i$), which is similar to that seen in low ion beta, quasi-perpendicular fast shocks when they are unsteady and periodically reform [1]. Work is underway to characterize the parallel electron heating and incorporate its consequences in a generalization of (2).

EMBEDDING

Embedding allows a more detailed physics characterization of local regions enclosed in more global environments, whose effects extend beyond their limited spatial extent. For example, one may wish to include kinetic physics occurring in narrow magnetospheric boundary layers in global solar-wind MHD calculations. More specifically, one may want to model the region near the reconnection site in the magnetotail more accurately than by just including an ad hoc, enhanced localized resistivity. Such a resistivity will lead to reconnection, of course, but the global manifestations of this simplified model for the reconnection process may be different from what one obtains through a full kinetic model that is restricted to a small region near the reconnection site.

The enhanced electron pressure tensor model discussed above has been used in hybrid simulations of reconnection [9,13] and compared with results of full particle simulations [14] as well as Hall-MHD simulations [13,15] that use the same electric field (2) to demonstrate that the model correctly reproduces the reconnection rate and the magnitude and spatial extent of the off-diagonal pressure terms. The validated Hall MHD model with the electron pressure tensor model has then been embedded in an MHD calculation of driven reconnection in a realistic magnetotail equilibrium model [13,16] and compared with a full Hall-MHD simulation of the same tail configuration.

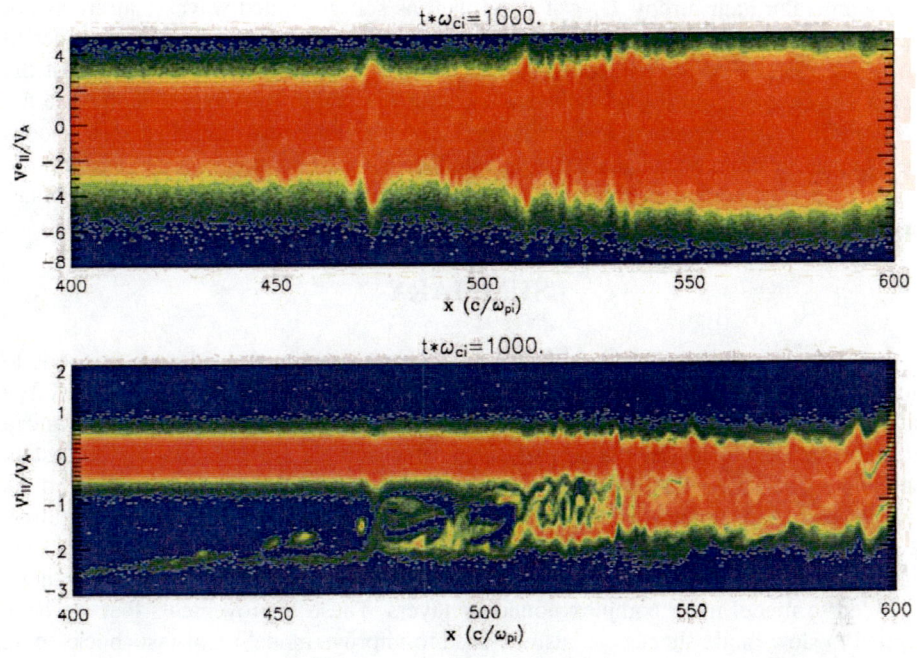

FIGURE 1. Results of full particle simulation of $\theta_{Bn} = 84°$; $v_{\parallel}$ - x phase space for the electrons (top) and ions (bottom) indicates electron heating in the shock ramp due to waves from ion counterstreaming.

The two calculations are essentially indistinguishable, indicating that embedding in this manner provides a viable method to model the reconnection process without imposing an artificial resistivity. The embedded calculation includes a small overlap region with the MHD calculation where waves from the inner region are damped away. It should be noted, however, that the Hall-MHD model does not include all of the kinetic physics. Hybrid calculations of reconnection indicate that there are additional effects due to finite ion gyro-radius dynamics near the X-point. These can be added through the introduction of test ions that are used to calculate off-diagonal ion pressure terms, which can be included in the Hall-MHD treatment as well [17].

Embedding is relatively easy for reconnection simulations because the ion flows are not large and few waves are generated in the vicinity of the diffusion region. Embedding of fast shocks in an MHD simulation may be more challenging. As is well known [1], super-critical, quasi-perpendicular fast shocks produce reflected ions that propagate into the downstream, contribute to a large ion temperature anisotropy ($T_{i\perp} > T_{i\parallel}$) and generate large amplitude ion cyclotron and mirror waves in the downstream magnetosheath. Similarly, quasi-parallel shocks have a large upstream region containing ions backstreaming from the shock (that can in some sense be characterized by a temperature anisotropy $T_{i\perp} < T_{i\|}$), as well as self-generated waves that are swept back into the shock. An embedded hybrid simulation would require novel techniques to determine the size of the embedded region, ways to damp out the waves (at least the longer wavenumber portion of the wave spectrum) as well as methods to isotropize the ion pressure. An important feature is the energetic ion population that such shocks produce, so the embedded calculation must either be large enough in space and time to include the acceleration process self-consistently or provide a model for the shock-generated energetic ion spectrum.

SUMMARY

In summary, we have outlined issues related to extending hybrid code methods to more complex problems related to the modeling of collisionless shocks in space. We have discussed how the electron physics model can be generalized, why an advanced model is needed to model the formation of very oblique slow shocks in the distant magnetotail, and ways that the model can be further refined. We have also addressed the issue of embedding more detailed hybrid and Hall-MHD simulation models in global, MHD calculations. This method has been successfully applied to accurately model the diffusion region in steady-state magnetic reconnection in the near tail and can be extended to model more complex boundary layers. These improvements that we have applied to slow-mode shocks can also be used to improve modeling of fast shocks, e.g., to include the electron temperature anisotropy that develops in the shock ramp which leads to the generation of whistler waves and perhaps contributes to shock unsteadiness [1]. In addition, embedding could be used to include the propagation of interplanetary shocks and resulting ion acceleration in global MHD simulations of the solar wind or to model more detailed physics of the termination shock in coupled global MHD-interstellar neutrals calculations.

ACKNOWLEDGMENTS

This work was funded in part by the NASA Grant No. NNH04AB221.

REFERENCES

1. B. Lembege, J. Giacalone, M. Scholer, T. Hada, M. Hoshino, V. Krasnoselskikh, H. Kucharek, P. Savoini and T. Terasawa, *Space Sci. Rev.* **110**, 161-226 (2004).
2. D. Winske, L. Yin, N. Omidi, H. Karimabadi and K. Quest, "Hybrid Simulation Codes: Past, Present and Future" in *Space Plasma Simulation,* edited by J. Büchner, C. T. Dum and M. Scholer, Springer-Verlag, Berlin, 2003, pp.136-165.
3. P. L. Pritchett, "Particle-in-Cell Simulations of Plasmas—A Tutorial" in *Space Plasma Simulation,* edited by J. Büchner, C. T. Dum and M. Scholer, Springer-Verlag, Berlin, 2003, pp.1-24.
4. D. Krauss-Varban and N. Omidi, *Geophys. Res. Lett.* **22**, 3271-3274 (1995).
5. N. Omidi and D. Winske, *J. Geophys. Res.* **97**, 14801-14821 (1992).

6. F. V. Coroniti, *Nucl. Fusion* **11**, 261-283 (1971).
7. Y. Saito, T. Mukai, T. Terasawa, A. Nishida, S. Machida, M. Hirahara, K. Maezawa, S. Kokubun and T. Yamamoto, *J. Geophys. Res.* **100**, 23567-23581 (1995).
8. L. Yin, D. Winske, W. Daughton and F. V. Coroniti, *Geophys. Res. Lett.* **31**, L09803 (2004).
9. M. Hesse and D. Winske, *J. Geophys. Res.* **99**, 11177-11192 (1994).
10. M. Hesse, M. Kuznetsova and J. Birn, *Phys. Plasmas* **11** 5387-5397 (2004).
11. L. Yin, D. Winske, W. Daughton and F. V. Coroniti, *J. Geophys. Res.* **110**, in press (2005).
12. W. Daughton, D. Winske, L. Yin and S. P. Gary, *J. Geophys. Res.* **106**, 25031-25039 (2001).
13. L. Yin and D. Winske, *Phys. Plasmas* **10**, 1595-1604 (2003).
14. M. Hesse, K. Schindler, J. Birn and M. Kuznetsova, *Phys. Plasmas* **6**, 1781-1795 (1999).
15. L. Yin, D. Winske, S. P. Gary and J. Birn, *J. Geophys. Res.* **106**, 10761-10775 (2001).
16. L. Yin and D. Winske, *J. Geophys. Res.* **107**, No. A12, 1485 (2002).
17. L. Yin, D. Winske, S. P. Gary and J. Birn, *Geophys. Res. Lett.* **28**, 2173-2176 (2001).

A New Simulation Technique for Study of Collisionless Shocks: Self-Adaptive Simulations

H. Karimabadi[1], Y. Omelchenko[1], J. Driscoll[1], R. Fujimoto[2], K. Perumalla[2], D. Krauss-Varban[1]

[1]SciberQuest, Inc., Solana Beach, CA, 92075, USA
[2]Georgia Institute of Technology, Atlanta, GA, 30332, USA

Abstract. The traditional technique for simulating physical systems modeled by partial differential equations is by means of time-stepping methodology where the state of the system is updated at regular discrete time intervals. This method has inherent inefficiencies. In contrast to this methodology, we have developed a new asynchronous type of simulation based on a discrete-event-driven (as opposed to time-driven) approach, where the simulation state is updated on a "need-to-be-done-only" basis. Here we report on this new technique, show an example of particle acceleration in a fast magnetosonic shockwave, and briefly discuss additional issues that we are addressing concerning algorithm development and parallel execution.

Keywords: Self-adaptive, simulations, particle-in-cell, hybrid, shock waves, particle acceleration
PACS: 52.65.Rr, 52.65.Ww, 96.50.-e, 96.50.Fm, 96.50.Pw

INTRODUCTION

The strongly disparate temporal and spatial scales commonly occurring in many complex physical systems pose a significant computational challenge and necessitate a leap in simulation technology. Although the Adaptive Mesh Refinement (AMR) methodology is a powerful technique for addressing large variations in spatial scales, the time update of variables is still done through traditional time-stepping methodology (referred to here as standard time-stepping or STS). This has inherent inefficiencies and suffers from the usual Courant-Friedrichs-Levy (CFL) limitations. We have developed a novel simulation technique by abandoning time-stepping and replacing it with an event-driven method[1-5]. Event-driven simulations have their origins in operations research and management science, and more recently have found application in war games and telecommunications but have not been applied to complex physical systems. We have combined traditional mesh discretization techniques with a novel discrete-event methodology and developed several plasma and fluid simulation codes[1-5]. In our discrete-event simulation (DES) approach[3-4], the traditional measure of time advance, Δt is replaced by the meaningful physical information unit, Δf. In effect, this technique introduces an individual adaptive time line for every computational entity, enabling truly *asynchronous* time integration of the system state variables. As a result, at any given time a DES model has to process only changes to its global state that exceed the minimum information unit, Δf. This eliminates unnecessary computation in the inactive regions. Several parallel applications of this technology have been built, ranging from an electrostatic particle code[1,5] to an electromagnetic hybrid (electron fluid, particle ions) code[2,4] to a diffusion-convection equation solver[3], and have demonstrated superior metrics:

Faster: By eliminating unnecessary computations, speed-ups as large as a factor of 300 were achieved in one-dimension, with further enhancements being expected in 2D and 3D.

More Accurate: By updating the system state based on local Δf's rather than the global timestep, the user has more effective control over the desired numerical accuracy.

Stable: The DES codes run successfully in regimes where standard codes are subject to explosive numerical instabilities.

DES ALGORITHMS

As shown in Figure 1, a discrete-event simulation (DES) system can be broken into two components: (1) the models and (2) the parallel simulation executive that manages events and the progression of simulated time. Development of next generation plasma codes requires innovations in both components.

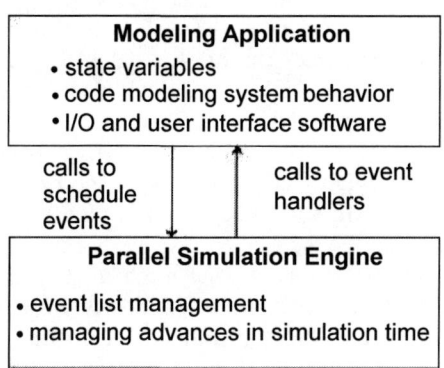

FIGURE 1. The two main components of a discrete event simulation.

Field equations are discretized in space in the conservation form. Each computational mesh cell is assigned discrete states associated with the temporal evolution of local field and particle quantities. Transitions from one temporal state to another are called "events". Time integration of each field component is delayed by a time interval that is computed based on the magnitude of its predicted rate of change. Particles are scheduled for advance in each cell based on their current velocities, local field magnitude and cell size. Each computation cell keeps a registry of increments to its original state (the state used for the prediction) caused by the neighboring cells and reschedules events (time advances) to earlier times if the cell state is significantly altered during the predicted time delay. The DES code programming architecture is drastically different from conventional (time-driven) codes. In particular, each mesh cell has a means of polling its neighbors and fetching global simulation information using its local data handlers. It is also "aware" of its role in establishing communication with remote (distributed) parts of the system or applying proper boundary conditions. A nontrivial problem is to preserve fluxes across mesh cell interfaces. In explicit time-driven codes adjacent cells are always advanced with fluxes taken at the same time level. DES cells schedule themselves asynchronously and therefore special care is taken to ensure that field quantities in cells with common interfaces are always integrated in time with identical fluxes across the common boundaries[3,4]. The key distinguishing elements of our DES algorithm are:

1. Each particle is advanced based on its own (time varying) time step.

2. Fields are updated *locally* and *only* when there is a change in that cell.

3. The solution is monitored and evolved based on the required accuracy rather than based on a pre-selected time-step size.

We have developed a library of C++ classes (SciDES) designed to provide a set of discrete-event software tools for implementing finite difference and particle-in-cell methods for the solution of coupled partial differential equations and equations of particle motion. SciDES standardizes fundamental data structures and algorithms for programming distributed time-dependent scientific models on block-structured computational domains and formalizes the most essential aspects of the distributed physics-based DES models in the form of a pseudo-distributed architecture. This pursues several goals. First, the SciDES API separates the computational physics algorithms from the communication issues by abstracting them into well defined concepts (C++ classes) and providing all the necessary "go-between" implementation details. Second, it fosters more efficient cooperation of computational physicists with computer scientists working on the distributed discrete-event engine algorithms since it allows substitution of pseudo-distributed plug-in modules by their Message Passing Interface (MPI) counterparts in a plug-and-play fashion without breaking the physics core of the code. In addition, the ability to run virtual distributed simulations on a single CPU enables testing various physical mechanisms that provide important insight into predictive properties of physics-based parallel discrete-event simulations.

An example of our SciDES architecture is shown in Figure 2. The class MPDES abstracts the virtual multi-processor DES environment. In this diagram dark dashed arrows represent ownership (the "has a" relationship) and light dashed arrows mark class instantiation from template classes.

PARALLELIZATION

The parallelization of asynchronous (event-driven) continuous PIC models presents a number of challenges. As in conventional (time-driven) simulations, it is

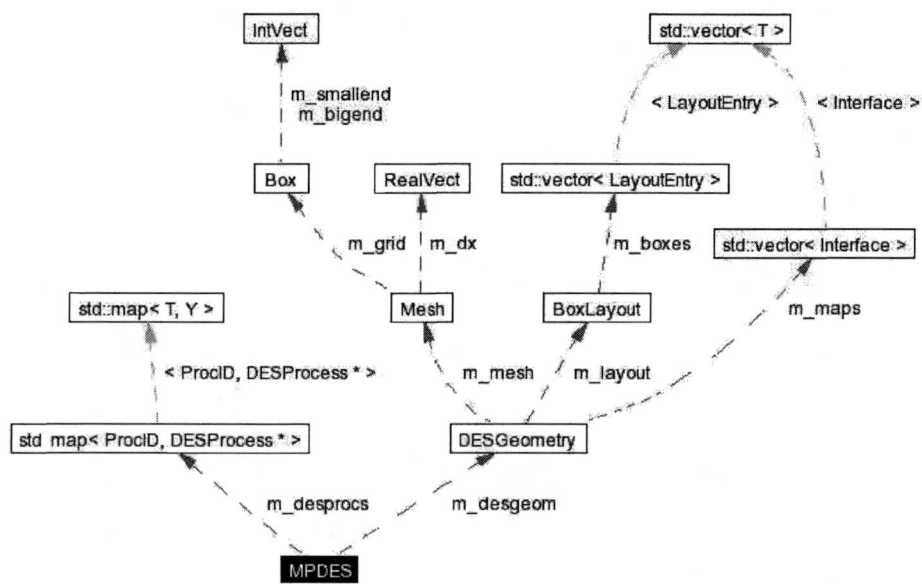

FIGURE 2. The *MPDES* class collaboration diagram. The MPDES object encapsulates the global simulation geometry properties and defines the table of virtual DES processes.

realized by decomposing the global computation domain into subdomains. In each subdomain, the individual cells and particles are aggregated into containers, which are mapped to distributed parallel processors in a way that achieves maximum load balancing efficiency. The parallel execution of conventional (time-driven) simulations is achieved by copying field information from the inner lattice cells to the ghost cells of the neighboring subdomains and exchanging out-of-bounds particles between the processors at the end of each update cycle. In contrast, in parallel asynchronous PIC simulations both particle and field events are not synchronized by the global clock (i.e. they do not take place at the same time levels throughout the simulation domain), but occur at arbitrary time intervals. This may introduce synchronization problems if some processors get ahead (in simulated time) of other processors (the "optimistic" approach, described later). As a result, a processor may receive an event message from a neighbor with a simulation time stamp that is in the receiver's past, thus causing a causality error. On the other hand, parts of a distributed discrete-event simulation can be forced to execute synchronously with remote tasks corresponding to the neighboring subdomains (the "conservative" approach). If so, the parallel speed-up critically depends on the underlying domain decomposition technique and additional predictive ("look-ahead") properties of the simulation in question. Regardless of the approach taken, it is important to note that DES computations offer substantial efficiencies compared to conventional explicit time-driven simulations due to the reduction in the amount of computation to be performed.

The following are some of the important issues that must be addressed in parallel discrete event simulations of continuous systems:

Synchronization: This is by far the paramount issue to be carefully resolved for achieving the best parallel execution performance. Broadly there are two approaches commonly used - conservative and optimistic.

Conservative: This approach always ensures *safe* timestamp-ordered processing. However, runtime performance is critically dependent on *a priori* determination of an application property called *lookahead*, which is roughly dependent on the degree to which the computation can predict future computations without global information. In the conservative approach, events that are beyond the next lookahead window are blocked until the window advances sufficiently far to cover those events. Typically the lookahead property is very hard to extract in complex applications, as it tends to be implicitly defined in the source code interdependencies. The appeal of this approach however is that it is one of the easiest schemes to implement if the lookahead can be specified by the application.

Optimistic: This approach avoids blocking by optimistically processing events beyond the lookahead window. When some events are later detected to have been processed in an incorrect order, the system invokes compensation code such as state restoration or reverse computation. Since blocking is not used, the lookahead value is not as important, and could even be specified to be zero. While this approach eliminates the problem of lookahead extraction, it has a different challenge – namely, support for compensation code.

Combination: Sometimes it might help to have some parts of the application execute optimistically ahead (e.g., parts for which lookahead is low are hard to extract), while other parts execute conservatively (e.g., parts for which lookahead is large, or for which compensation code is difficult to generate). In such cases, a combination of conservative and optimistic synchronization techniques can be appropriate.

Load Balancing: As with any parallel/distributed application, the best performance is obtained when the load is evenly balanced across all resources. In parallel simulation in particular, load imbalance can have a very adverse effect. This is because typically the slowest processor can hold back the progress of simulation (virtual) time, which in turn slows down even those processors that are relatively lightly loaded. In optimistic methods, a poor load distribution can lead to an excessive amount of rolled back computation.

Automated/Adaptive: Automated schemes are preferable for load-balancing at runtime. These schemes vary with the particular synchronization approach used.

Support Primitives: In order to permit automated/adaptive load balancing by the system, it is important to provide appropriate primitives to the application, so that application-level entities can be easily moved across processors by the system in a transparent manner as needed.

Modeling and Runtime Interface: To be able to decouple the implementation details of the parallel simulation executive from the application/models, it is best to define the model-simulation interface in an implementation-independent fashion. Not only this helps avoid reimplementation of models whenever the engine programming structure is modified, but it also permits experimentation with multiple synchronization and load-balancing approaches for the same application. Additionally, it enables engine-level optimizations to remain transparent to the application, so that the application-developer is not burdened or sidetracked with such issues during model development.

With the preceding issues in mind, we are carefully developing appropriate interfaces and implementations of our parallel execution engine. A brief description of our approach follows:

- The synchronization issue is being resolved by providing a transparent interface that does not mandate one synchronization approach over another. The underlying implementation is also being developed such that different model entities can chose different synchronization (conservative or optimistic execution style), as is most appropriate for them.
- The load balancing issue is being addressed by the use of an "indirect messaging" interface layer that decouples application entities from their processor mapping.
- The modeling and runtime interface is also kept abstract and flexible, so that radically alternative implementations can be implemented underneath the interface.

EXAMPLE

We now demonstrate the power of this new methodology through an example. Figures 3a-c show the spatial profile of a fast magnetosonic shockwave as obtained from our new DES-based hybrid code[4]. Figures 3d and 3e show the field and average particle update rates, respectively. It is clear that the resulting waves span a wide range in spatial and temporal scales. The dashed lines in Fig. 3d-e indicate a timestep of $\Omega t = 0.005$. In the time-stepping simulation (STS), one chooses a fixed time increment whereas in DES the field update rate is automatically adjusted by the algorithm and each particle is advanced based on its own time-varying timestep. As figure 3d demonstrates, choosing a fixed uniform timestep would over-resolve the frequency in some regions and under-resolve it in other regions. Aside from computational inefficiency, this leads to less accurate solutions compared to DES as demonstrated in Figure 4. Figure 4a shows that in the DES model ions entering the near-front upstream region are trapped and pre-decelerated (enhanced counts at low energies) by the well-resolved turbulence generated in the vicinity of the shock front (Fig. 3b). On the contrary, the ion distribution function in STS is found insensitive to the upstream location of the reference upstream position because of the inability of STS to separate high-frequency wave activity from thermal noise.

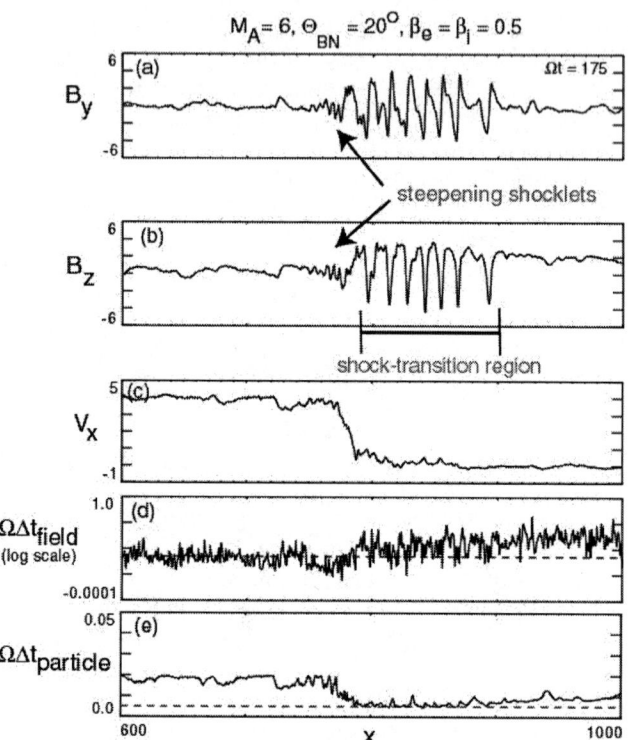

FIGURE 3. DES Simulation of a fast magnetosonic shockwave and the associated particle acceleration. DES self-adapts by reducing local field update timesteps in order to resolve high-frequency turbulence preceding the shock front (d). STS refers to the standard time-stepped simulation.

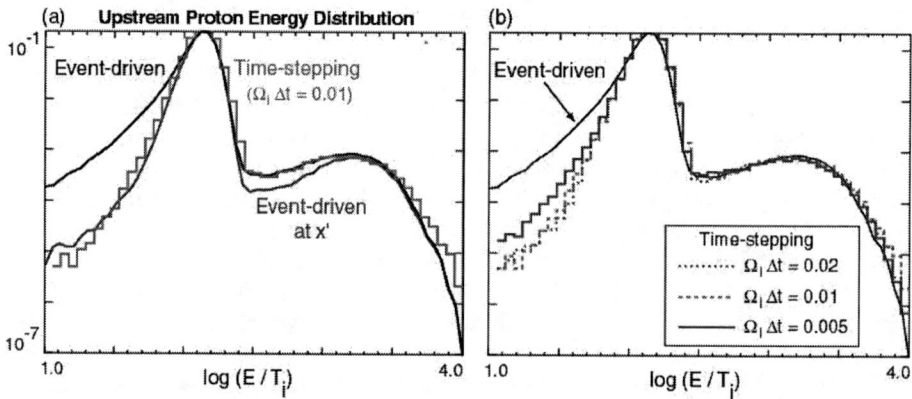

FIGURE 4. Comparison of upstream particle distributions in the DES and time-stepped simulations for the parameters defined in Figure 3. (a) In the case of DES, the distribution function is evaluated in two different intervals: the entire upstream, (black curve) and excluding the first 75 ion inertial lengths upstream, $x' = x_s - 75$ (red curve). Here x_s is the position of the shock (~500). (b) Comparison of distributions and particle acceleration in the DES and STS simulations for three time steps, as indicated. The time-stepped code does not properly account for deceleration in front of the shock (lowest energies) even at the smallest time step, and overestimates acceleration at the highest energies.

Solutions for three different STS time steps shown in Figure 4b demonstrate that the advanced predictor-corrector code[6,7] is accurate for most of the spectrum, but overestimates energization to the highest energies, and does not describe the near-upstream waves accurately enough to produce the ensuing pre-deceleration (low energies), even at the smallest time step. In figures 3-4, Ω is the ion gyrofrequency, T_i is the ion temperature, all distances are normalized to ion inertial length, and density and magnetic field are normalized to their upstream values.

SUMMARY

By combining techniques from two distinct fields of time-stepping and event-driven simulations, we have developed new and more efficient algorithms for modeling of systems described by partial differential equations. The DES algorithms are intelligent in that they self-adapt by adjusting local update rates in accordance with required accuracy, which results in more accurate and faster simulations. The technique is general and ideally suited for multi-scale problems characterized by disparate ranges in spatial and temporal scales. Given the ubiquitous nature of multi-scale problems in many areas of science and industry, we expect this technique to find broad application. We are currently extending this technique to several areas ranging from plasma physics to computational biology.

ACKNOWLEDGMENTS

This research was supported by NSF ITR grants 0325046, 0539106, and 0326431.

REFERENCES

1. Karimabadi, H., J. Driscoll, Y. A. Omelchenko, and N. Omidi, "A new asynchronous methodology for modeling of physical systems: breaking the curse of Courant condition", *J. Comp. Phys.* **205** (2), 755 (2005).
2. Karimabadi, H., J. Driscoll, J. Dave, Y. A. Omelchenko, K. Perumalla, R. Fujimoto, and N. Omidi, "Parallel discrete event simulation of grid-based models: asynchronous electromagnetic hybrid code", *Lecture Notes in Computer Science*, Springer-Verlag, in press, (2005).
3. Omelchenko, Y. A., and H. Karimabadi, "Self-adaptive time integration of flux-conserving equations with sources", *J. Comp. Phys.*, submitted, (2005).
4. Omelchenko, Y. A., and H. Karimabadi, "Event-driven hybrid particle-in-cell simulation: a new paradigm for multi-scale plasma modeling", *J. Comp. Phys.*, submitted, (2005).
5. Y. Tang, R. M. Fujimoto, K. Perumalla, H. Karimabadi, J. Driscoll, Y. Omelchenko, "Optimistic Parallel Discrete Event Simulations of Physical Systems using Reverse Computation," Principles of Advanced and Distributed Simulation, June (2005).
6. Karimabadi, H., D. Krauss-Varban, J. D. Huba, and H. X. Vu, On magnetic reconnection regimes and associated three-dimensional asymmetries: Hybrid, Hall-less hybrid, and Hall-MHD simulations, *J. Geophys. Res.*, Vol. 109, A09205, doi:10.1029/2004JA010478, (2004).
7. Krauss-Varban, D., From theoretical foundation to invaluable research tool: Modern hybrid simulations, Proceedings of ISSS-7, pp. 15-18, March, (2005).

Magnetohydrodynamics of Shocks with Reflected Particles: Rankine-Hugoniot Relations

B. Dasgupta, G. P. Zank, R. Bedros and G. M. Webb

Institute of Geophysics and Planetary Physics,
University of California at Riverside, Riverside, CA 92521

Abstract. Ion reflection at the transition layer by the cross-shock potential provides the fundamental dissipation mechanism for a quasi-perpendicular shock. Pickup ions (PUI) can be significantly energized by the multiple reflection (MRI) mechanism, which can resolve the injection problem for the anomalous cosmic rays [1][2]. In this work we investigate the role of reflected ions in modifying the Rankine-Hugoniot (RH) jump conditions in a simple one-fluid MHD formalism and deduce some important outcomes of these modified RH conditions.

INTRODUCTION

In a typical quasi-perpendicular shock, some of the ions of the incoming flow are reflected and gyrate back into the flow - depending on their incident energy - by the cross-shock potential developed at the shock front [3], [4]. The importance of ion reflection and their subsequent energization has long been recognized as a fundamental dissipation mechanism at a quasi-perpendicular shock. Models [5] and simulations [6], [7] have elucidated this basic understanding. Pickup Ions (PUI) created from interstellar atoms via charge-exchange or photoionization carry most of the thermal energy of the solar wind beyond ~ 20 AU. The shell distribution to which the PUIs relax [8] particularly contributes to ion reflection at the cross-shock potential. This led Zank et al [1] and Lee et al [2] to suggest that PUIs multiply reflected at the cross-shock potential could experience significant energization, and can resolve the so-called the injection problem [9], [10] for anomalous cosmic rays. The reflection of PUIs leads to the formation of an extended foot in addition to the small foot associated with cold reflected solar wind ions [11],[12].

The quasi-perpendicular shock is generally thought to be the best understood of the collisionless shocks yet we do not understand the fundamental scalings, particularly in the presence of a suprathermal population such as PUIs, or the stationarity of the shock, for example. Simulations have not yet resolved the basic ramp scale, because of the difficulties in accurately including the electron physics. Besides the fundamental questions of scaling, the role of PUIs at interplanetary shocks in the outer heliosphere or at the heliospheric termination shock (TS), all of which are likely to be highly perpendicular, the structure of the shock is still unclear. The efficiency of PUIs reflected by the cross-shock potential raises questions about the stationarity of quasi-perpendicular shocks. In this work we investigate the role of reflected particles in modifying the Rankine-

Hugoniot (RH) relations within a simple magnetohydrodynamical framework. It is seen that presence of reflected particles in the shock may alter some of the basic features of the shocks, like coplanarity, fast and slow mode propagation, switch-on and switch-off shocks, compression ratio, etc.

MHD EQUATIONS FOR SHOCKS WITH REFLECTED PARTICLES

We assume that at the location of the shock front particles are reflected by the cross-shock potential. Reflected particles are energized by the upstream motional electric field, and the momentum and energy gained by these particles are effectively injected continuously into the shocks and the flux of particles at the shock is unchanged. However, because of the particle reflection and subsequent energization at a quasi-perpendicular shock, the momentum and energy equations acquire source terms at the shock transition. The source terms can be understood as follows.

Reflection of ions from a thin shock-transition layer essentially introduces a non-ideal MHD feature in the usual MHD equations, due to the short timescale involved during the reflection process. This can be taken into account by retaining the displacement current experienced the particles undergoing reflection. Using the complete expression for the current density $\mathbf{j}$ as $\mathbf{j} = (c/4\pi)\nabla \times \mathbf{B} - \partial \mathbf{E}/\partial t$, one can write the momentum and energy conservation equations as,

$$\frac{\partial \rho \mathbf{u}}{\partial t} = \nabla \cdot \left[\rho \mathbf{v}\mathbf{v} + \left(p + \frac{B^2}{8\pi}\right)\mathbf{I} - \frac{\mathbf{B}\mathbf{B}}{4\pi} \right] = -\frac{\partial \mathbf{E}}{\partial t} \times \mathbf{B}, \quad (1)$$

$$\frac{\partial}{\partial t}\left(e + \frac{B^2}{8\pi}\right) + \nabla \cdot \left[(e+p)\mathbf{u} + \frac{uB^2}{4\pi} - \frac{\mathbf{u}\cdot\mathbf{B}}{4\pi}\mathbf{B}\right] = -\frac{1}{8\pi}\frac{\partial E^2}{\partial t}, \quad (2)$$

where $e = \rho u^2/2 + \rho\varepsilon$, with ρ is the fluid density, ε is the internal energy, $\mathbf{u}$ is the fluid velocity. The "non-MHD" change in the motional electric field immediately upstream of the shock as a consequence of ion reflection and energization is described by terms on the r.h.s. of (1) and (2). Rather than attempting to approximate these terms directly, we can instead estimate the corresponding momentum and energy gain experienced by an ensemble of reflected particles in the upstream motional electric field. If we assume a shock moving at speed s and located at $x - st$ after some time t, then we can write the modified MHD equations (in usual notation) as,

$$\frac{\partial \rho}{\partial t} + \frac{\partial}{\partial x}(\rho u_x) = 0, \quad (3)$$

$$\frac{\partial}{\partial t}(\rho u_x) + \frac{\partial}{\partial x}\left(\rho u_x^2 + p + \frac{B^2}{8\pi}\right) = S_x^m \delta(x - st), \quad (4)$$

$$\frac{\partial}{\partial t}(\rho \mathbf{u_t}) + \frac{\partial}{\partial x}\left(\rho u_x \mathbf{u_t} - \frac{B_x}{4\pi}\mathbf{B_t}\right) = \mathbf{S_t}^m \delta(x - st), \quad (5)$$

$$\frac{\partial}{\partial t}\left(e+\frac{B^2}{8\pi}\right) + \frac{\partial}{\partial x}\left(e+p)u_x + \frac{1}{4\pi}(u_x B^2 - B_x \mathbf{u_t}\cdot\mathbf{B_t})\right) = S^e \delta(x-st), \quad (6)$$

$$\frac{\partial \mathbf{B_t}}{\partial t} + \frac{\partial}{\partial x}(u_x \mathbf{B_t} - B_x \mathbf{u_t}) = 0, \quad (7)$$

where, S_x^m, $\mathbf{S_t}^m$, S^e are the source terms generated by the reflected particles for the x-component, transverse component of momentum, and the energy equations respectively, as shown in eqns. (1) and (2). We assume x-axis to be the direction of shock normal and the suffix t denotes the quantities along the transverse direction, which could be taken as the y-axis in specific situation.

We can estimate of the source terms from Zank et al. [1]. In the x-component of the momentum equation, we have,

$$S_x^m = \rho_r u_{rx}^2 + p_r \sim \rho_r u_{rx}^2,$$

where the suffix r refers to the reflected particles. We assume specular reflection, so that $u_{rx} \simeq v_{spec}$. With M_{A1} as the Alfvénic Mach number, the cross-shock potential ϕ is given as, (suffix 1 refers to the upstream quantities)

$$e\phi \simeq \frac{1}{M_{A1}^2}\frac{\delta B}{B_1} m u_1^2; \quad \text{so that} \quad v_{spec} \simeq \frac{4u_1^2}{M_{A1}^2}\frac{\delta B}{B_1} = \alpha^2 u_x^2,$$

which yields,

$$S_x^m = \epsilon_{ref}\rho\alpha^2 u_x^2 = \epsilon_{ref}\rho\Lambda_x u_x^2, \quad (8)$$

where $\epsilon_{ref} = \rho_r/\rho$, the fraction of the incident ions reflected. To obtain the source term for the transverse component we consider only the y-component; so $\mathbf{S_t^m} \equiv S_y^m$ and $S_y^m = \rho_r(u_r^y)^2$. We further assume

$$v_\parallel^r \equiv \mathbf{v_{tr}} \simeq e/m \mathbf{E} \tau_{foot},$$

where τ_{foot} is the characteristic time spent by the particle in the foot of a shock. The motional electric field $\mathbf{E} = -\mathbf{u}\times\mathbf{B} = u_x B_z \hat{y} \equiv E_y \hat{y}$. For cold upstream ions, $v_\parallel^r \simeq u_{ry}$, so that,

$$\rho_r(u_r^y)^2 = \left(\frac{e}{m}\right)^2 u_x^2 B_z^2 \tau_f^2 \epsilon_{ref}\rho = \epsilon_{ref}\beta^2 \rho u_x^2 = \epsilon_{ref}\Lambda_y \rho u_x^2. \quad (9)$$

The source term for energy equation, S^e, can be obtained as, $S^e = \frac{1}{2}\rho_r u_r^2 u_r^x$; $u_r^2 = (u_{rx})^2 + (u_{ry})^2$, so

$$S^e = \epsilon_{ref}\frac{\alpha(\alpha^2+\beta^2)}{2}\rho u_x^3 = \epsilon_{ref}\Lambda_e \rho u_x^3. \quad (10)$$

For a typical quasi-perpendicular shock at about 1AU, with $B \sim 5nT, B_z \sim 1.5nT$, [13] the ratio of the downstream magnetic field to the upstream magnetic field (jump) ~ 3 and the upstream $M_{A1} \sim 4$, we find; $\Lambda_x \sim 0.13$, $\Lambda_y \sim 6.3$ and $\Lambda_e \sim 1.14$.

The MHD Rankine-Hugoniot jump conditions for shocks with reflected particles are thus,

$$[\mathbf{B_n}] \equiv [B_x] = 0, \tag{11}$$

$$[\rho u_x] = 0, \tag{12}$$

$$[u_x \mathbf{B_t} - B_x \mathbf{u_t}] = 0 ; \tag{13}$$

$$\left[\rho u_x^2 + p + \frac{B^2}{8\pi}\right] = \epsilon_{ref} \Lambda_x \rho_0 u_{x0}^2 ; \tag{14}$$

$$\left[\rho u_x \mathbf{u_t} - \frac{B_x}{4\pi} \mathbf{B_t}\right] = \epsilon_{ref} \Lambda_y \rho_0 u_{x0}^2 \hat{\mathbf{e}}_\mathbf{t} ; \tag{15}$$

$$\left[(e+p)u_x + \frac{1}{4\pi}(u_x B^2 - B_x \mathbf{u_t} \cdot \mathbf{B_t})\right] = \epsilon_{ref} \Lambda_e \rho_0 u_{x0}^3 ; \tag{16}$$

where $\hat{\mathbf{e}}_\mathbf{t}$ is a unit vector along the transverse direction, which can be taken as the y-axis and we take $\mathbf{B_t} \equiv B_y$.

Non Coplanarity of the Upstream and Downstream Magnetic Field and Shock normal

Analysis of the R-H conditions for classical MHD shocks shows that the three vectors, the upstream magnetic field, shock normal and downstream magnetic field, are coplanar. However, one of the puzzling observational aspects of the structure of collisionless shock is the existence of a component that lies outside the coplanarity plane. This has been discussed by many authors [14], [15], [16], [17]. They ascribed non coplanarity arises due to (i) the two-fluid nature, and/or (ii) the tensorial nature of pressure. Newbury et al.[18] determined shock ramp width using the noncoplanar component of the magnetic field. From eqns.(11) and (13) we see that noncoplanarity arises due to the addition of transverse momentum by the reflected particles. Since,

$$\hat{n} \cdot (\mathbf{B_1} \times \mathbf{B_0}) \neq 0. \tag{17}$$

This shows that the upstream magnetic field $\mathbf{B_0}$ the downstream magnetic field $\mathbf{B_1}$ and the shock normal $\hat{n}$ do not lie in the same plane. It follows that addition of transverse component of momentum yields,

$$(M_{A1}^2 - 1)B_{y1} = (M_{A0}^2 - 1)B_{y0} - \epsilon_{ref} \Lambda_y B_n M_{A0}^2, \tag{18}$$

more generally, if we put $\mathbf{R_t} = \Lambda_y \hat{\mathbf{e}}_\mathbf{t}$ to denote the vector part of the transverse momentum added by the reflected particles, we can write,

$$(M_{A1}^2 - 1)\mathbf{B_{t1}} = (M_{A0}^2 - 1)\mathbf{B_{t0}} - \epsilon_{ref} \mathbf{R_t} B_n M_{A0}^2. \tag{19}$$

In the presence of reflected particles, the shock normal $\hat{n}$ is,

$$\hat{\mathbf{n}} = \frac{(\mathbf{B_1} - \mathbf{B_0}) \times (\mathbf{B_1} \times (\mathbf{B_0} - \epsilon_{ref}\mathbf{R_t}B_n M_{A0}^2))}{|(\mathbf{B_1} - \mathbf{B_0}) \times (\mathbf{B_1} \times (\mathbf{B_0} - \epsilon_{ref}\mathbf{R_t}B_n M_{A0}^2))|}; \neq \frac{(\mathbf{B_1} - \mathbf{B_0}) \times (\mathbf{B_1} \times \mathbf{B_0})}{|(\mathbf{B_1} - \mathbf{B_0}) \times (\mathbf{B_1} \times \mathbf{B_0})|}, \quad (20)$$

the last relation corresponding to the usual MHD shock normal condition.

SMALL AMPLITUDE PERTURBATION OF R-H CONDITIONS

To see the effects of the reflected particles on the flow, we can linearize the RH jump conditions, assuming ϵ_{ref} as a small parameter. For a normal incidence frame, and setting $B_z = 0$, with θ_0 the angle between $\mathbf{B_0}$ and shock normal, the linearized forms of eqns (10)-(14) are,

$$\rho_0 \delta u_x + u_{x0} \delta \rho = 0, \quad (21)$$

$$\delta p + \rho_0 u_{x0} \delta u_x + \frac{B_{y0}}{4\pi} \delta B_y = \epsilon_{ref} \Lambda_x \rho_0 u_{x0}^2, \quad (22)$$

$$\rho_0 u_{x0} \delta u_y - \frac{B_x}{4\pi} \delta B_y = \epsilon_{ref} \Lambda_y \rho_0 u_{x0}^2, \quad (23)$$

$$-B_x \delta u_y + u_{x0} \delta B_y + B_{y0} \delta u_x = 0, \quad (24)$$

$$\left(\rho_0 u_{x0}^2 + \frac{\gamma p}{\gamma - 1} + \frac{B_x^2 \tan^2 \theta_0}{4\pi} \right) \delta u_x + \frac{\gamma u_{x0}}{\gamma - 1} \delta p$$
$$+ \frac{2 u_{x0} B_{y0}}{4\pi} \delta B_y - \frac{B_x^2 \tan \theta_0}{4\pi} \delta u_y = \epsilon_{ref} \Lambda_e \rho_0 u_{x0}^3. \quad (25)$$

Expressing δu_y, δB_y, δp in terms of δu_x, and using these expressions in the energy equation we have,

$$-\frac{1}{(\gamma-1)(u_{x0}^2 - V_{A0}^2)} \{ u_{x0}^4 - u_{x0}^2 V_{A0}^2 - u_{x0}^2 C_{s0}^2 + C_{s0}^2 V_{A0}^2 - u_{x0}^2 V_{A0}^2 \tan^2 \theta_0 \}$$
$$= \epsilon_{ref} u_{x0}^3 \left\{ \Lambda_e - \frac{\gamma}{\gamma-1} \left(\Lambda_x - \Lambda_y \frac{\tan \theta}{M_{A0}^2 - 1} \right) - \Lambda_y \frac{\tan \theta_0}{M_{A0}^2 - 1} \right\}, \quad (26)$$

where we have used $e + p = \gamma/(\gamma-1)p/\rho$ for a classical gas. We now introduce the fast and slow magnetosonic speeds V_{f0} and V_{s0} for the upstream flow, so that $u_{x0}^4 - u_{x0}^2(C_{s0}^2 + V_{A0}^2 \sec^2 \theta_0) + C_{s0}^2 V_{A0}^2 = (u_{x0}^2 - V_{f0}^2)(u_{x0}^2 - V_{s0}^2)$, and eqn.(26) yields,

$$\frac{\delta u_x}{u_{x0}} = \epsilon_{ref} \cdot \frac{u_{x0}^2 (u_{x0}^2 - V_{A0}^2)}{(u_{x0}^2 - V_{f0}^2)(u_{x0}^2 - V_{s0}^2)} \left\{ \gamma \Lambda_x - (\gamma-1)\Lambda_e - \frac{\Lambda_y \tan \theta_0}{M_{A0}^2 - 1} \right\}. \quad (27)$$

From the above expression, the relation $V_{f0} > V_{A0} > V_{s0}$, shows that reflected particles can accelerate or decelerate the flow depending of the relative strength of the reflection parameters $(\Lambda_x, \Lambda_y, \Lambda_e)$ and the oblique angle θ_0 of the magnetic field. For large oblique angles with sub (slow) magnetosonic flow $u_{x0} < V_{s0}$, the flow accelerates for -ve value of the contributions from the curly bracket, whereas a super (fast) magnetosonic flow $u_{x0} > V_{f0}$, the flow decelerates for -ve value of the contributions from the curly bracket, and the flow accelerates for super Alfvénic $u_{x0} > V_{A0}$ but sub(fast)magnetosonic flow $u_{x0} < V_{f0}$.

HUGONIOT FOR EXACTLY PERPENDICULAR SHOCK

We can derive an expression for the Hugoniot function for an exactly perpendicular shock, (i.e., $B_n = 0$). Adopting a normal incident frame and putting $B_z = 0$, we first obtain from the momentum jump equation (12),

$$m^2 = \frac{1}{(1-\epsilon_{ref}\Lambda_x)\tau_0 - \tau_1}\left(p_1 - p_0 + \frac{1}{8\pi}(B_{y1}^2 - B_{y0}^2)\right). \tag{28}$$

For zero upstream transverse velocity, the downstream transverse velocity is, $u_{y1} = -\epsilon_{ref}\Lambda_y u_{x0}$. From the energy jump equation, (14), using the expression for m^2, one obtains, after some rearrangements of terms,

$$\varepsilon_0 - \varepsilon_1 + \frac{1}{2}(p_1+p_0)(\tau_0 - \tau_1) + \frac{1}{16\pi}(\tau_0 - \tau_1)(B_{y1} - B_{y0})^2$$
$$+\epsilon_{ref}m^2\tau_0\left(\frac{1}{2}\Lambda_x(\tau_0+\tau_1) - \frac{1}{2}\epsilon_{ref}\Lambda_y^2\tau_0 - \Lambda_e\tau_0\right) = 0. \tag{29}$$

Since for an exactly perpendicular shock, $B \equiv B_y$, we can now define the Hugoniot function for such shocks with reflected particles $H(\tau, p, B)$

$$H(\tau, p, B) = \varepsilon - \varepsilon_0 + \frac{1}{2}(p_1+p_0)(\tau - \tau_0) + \frac{1}{16\pi}(\tau - \tau_0)(B_1 - B_0)^2$$
$$+\epsilon_{ref}m^2\tau_0\left(+\frac{1}{2}\epsilon_{ref}\Lambda_y^2\tau_0 + \Lambda_e\tau_0 - \frac{1}{2}\Lambda_x(\tau_0+\tau_1)\right). \tag{30}$$

It is seen that the initial point (τ_0, p_0, B_0) does not lie on the Hugoniot curve/surface, as $H(\tau_0, p_0, B_0) \neq 0$. This necessitates a geometric definition of entropy for these types of shocks, and this is presented in subsequent work.

In terms of M_{A0}, M_{s0} the Alfvénic and sonic Mach numbers, the compression Ratio $\tilde{\rho} = \rho_1/\rho_0$, is given as

$$(\tilde{\rho}-1)\left\{\tilde{\rho}^2\frac{2-\gamma}{M_{A0}^2} + \tilde{\rho}\left(\frac{\gamma}{M_{A0}^2} + \frac{2}{M_{s0}^2} + \gamma - 1\right) - (\gamma+1)\right\}$$
$$-\epsilon_{ref}\{\tilde{\rho}^2(\gamma-1)(\epsilon_{ref}\Lambda_y^2 + 2\Lambda_e) - \tilde{\rho}\Lambda_x\} = 0. \tag{31}$$

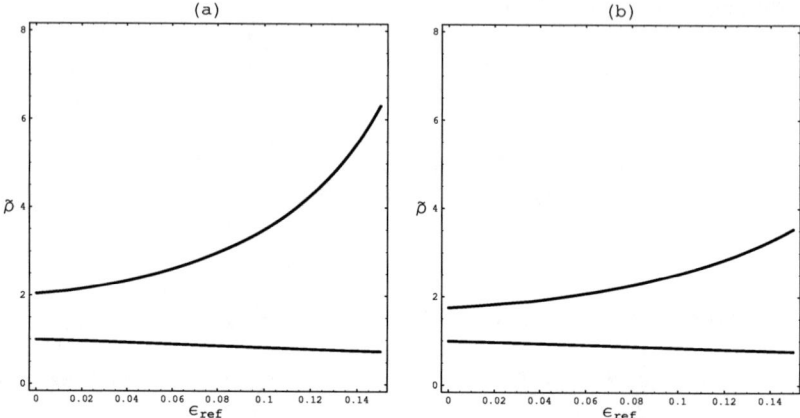

FIGURE 1. Two branches of the compression ratio $\tilde{\rho}$ obtained from numerical solution of eqn. (31) are plotted as a function of ϵ_{ref}. The upper curves in both (a) and (b) show the increase of $\tilde{\rho}$ with ϵ_{ref}, while lower curves show slight decrease from the classical value 1. For (a), $\Lambda_x = 0.13$, $\Lambda_y = 6.3$, $M_{A0} = 4$, $M_{s0} = 2$ and for (b), $\Lambda_x = 0.08$, $\Lambda_y = 5.5$, $M_{A0} = 3$, $M_{s0} = 1.8$, and these correspond to typical shock parameters at 1 AU [13].

This is a cubic equation, but $\tilde{\rho} = 1$ is not a root. To show the dependence of $\tilde{\rho}$ on ϵ_{ref}, we have plotted both the values of $\tilde{\rho}$, obtained from a numerical solution of eqn. (31), against ϵ_{ref} for two sets of values for Λ_x and Λ_y, with $\gamma = 5/3$ in Fig.1. It is seen that $\tilde{\rho}$ always increases with ϵ_{ref}, starting from its classical value.

For the high Mach number limit, $M_{A0} \to \infty, M_{s0} \to \infty$, $\tilde{\rho}$ satisfies a quadratic equation,

$$\tilde{\rho}^2(\gamma - 1)\{1 - \epsilon_{ref}(2\Lambda_e + \epsilon_{ref}\Lambda_y^2)\} - 2\{\gamma + \epsilon_{ref}\Lambda_x\}\tilde{\rho} + (\gamma + 1) = 0, \quad (32)$$

so that,

$$\tilde{\rho} = \frac{1}{2(\gamma-1)\{1 - \epsilon_{ref}R_1\}} \left[2\gamma\{1 + \epsilon_{ref}R_2\} \right. \\ \left. \pm \sqrt{4\gamma^2\{1 + \epsilon_{ref}R_2\}^2 - 4(\gamma^2 - 1)\{1 - \epsilon_{ref}R_1\}} \right], \quad (33)$$

with with, $R_1 = 2\Lambda_e + \epsilon_{ref}\Lambda_y^2$, $R_2 = \Lambda_x/2\gamma$. It is easy to see from eqn.(33) that for $\epsilon_{ref} = 0$, $\tilde{\rho}$ go to the classical limits, $(\gamma+1)/(\gamma-1) = 4$ (for $\gamma = 5/3$) and 1.

CONCLUSION

We have presented a preliminary investigation of quasi-perpendicular MHD shocks in the presence of ions reflected at the cross-shock potential. We see that reflected particles can profoundly alter shock characteristics and generate a noncoplanar component of the downstream magnetic field. This can effect fast and slow mode shock propagation and modify conditions for the existence of switch-on and switch-off shocks and rotational discontinuities. A detail investigations on these different aspects of these shocks are under progress.

ACKNOWLEDGMENTS

This work is supported in part by a NASA grant NNG4GF83G.

REFERENCES

1. Zank, G. P., H. L. Pauls, I. H. Cairns and G. M. Webb, Interstellar pickup ions and quasi-perpendicular shocks: Implication for the termination shock and intermediary shock, J. Geophys. Res. **101**, 457, (1996)
2. Lee, M.A., V.D. Shapiro and R.Z. Sagdeev, Pickup ion energization by shock surfing, J. Geophys. Res. **101**, 4777, (1996)
3. Woods, L.C., On the structure of collisionless magneto-plasma shock waves at super-critical Alfvén-Mach numbers, J. Plasma Phys., **3**, 437, (1969)
4. Gosling, J.T, M.F. Thomsen, S.J. Bame and W.C. Feldman, Evidence for specularly reflected ions upstream from the quasi-parallel bow shock, Geophys. Res. Lett. **9**, 1333, (1982)
5. Moiseev, S.S. and R.Z. Sagdeev, Collisionless shock waves in a plasma in a weak magnetic field, J. Nuclear Energy, Pt. C, **5**, 43, (1963)
6. Leroy, M, Structure of Perpendicular shocks in collisionless plasma, Phys. Fluids, **25**, 2742, (1983)
7. Schöler,M., I Shinohara and S. Matsukiyo, Quasi-perpendicular shock: Length scale of cross-shock potential, shock formation, and implication for shock surfing, J. Geophys. Res. **108**, doi: 10.1029/2002/JA009515, SSH 4-1 (2003)
8. Williams, L.L, G.P. Zank, Dissipation of pickup-induced waves: A solar wind temperature increase in the outer heliosphere, J. Geophys. Res. **100**, 17,059, (1995)
9. Jokipii, J.R., Rate of energy gain and mazimum energy in diffusive shock acceleration, Astrophys. J., **313**, 842, (1987)
10. Zank, G.P., W.K.M. Rice, J.A. le Roux and W.T. Mattaehaus, The injection problem for anomalous cosmic rays, Astrophys. J., **556**, 494, (2001)
11. Liewer, P.C., B. Goldstein, and N. Omidi, Hybrid simulations of the effects of interstellar pickup hydrogen on the solar wind termination shock, J. Geophys. Res. **98**, 15,211, (1993)
12. Lipatov, A. and G.P. Zank, Pickup ion acceleration at low-β_p perpendicular shocks,Phys. Rev. Lett, **82**, 3609, (1999)
13. Berdichevsky, D., A. Szabo, R. P. Lepping and A. F. Viñas, Interplanetary fast shocks and associated drivers observed through the 23rd solar minimum by Wind over its first 2.5 years, J. Geophys. Res. **105**, 27,289, (2000)
14. Jones, F.C. and D.C. Ellison, Noncoplanar magnetic fields, shock potentials and ion deflection, J. Geophys. Res. **92**, 11,205, (1987)
15. Gosling, J.T., D. Winske, and M.F. Thomsen, Noncoplanar magnetic fields at collisionless shocks: a test of newapproach, J. Geophys. Res. **93**, 2735, (1988)
16. Friedman, M.A., C.T. Russell, J.T. Gosling and M.F. Thomsen, Noncoplanar component of the magnetic field at low Mach number shocks, J. Geophys. Res. **95**, 2441, (1990)
17. Gedalin, M., Noncoplanar magnetic field in the collisionless shock front, J. Geophys. Res. **101**, 11,153, (1996)
18. Newbury, J.A., C.T. Russel and M. Gedalin, The determination of shock ramp width using the noncoplanar magnetic field component, Geophys. Res. Lett. **24**, 1975, (1997)

Electrons at Shocks

K.W. Ogilvie

Goddard Space Flight Center, Greenbelt MD.

Abstract. Moderate strength quasi-perpendicular shocks have been observed by the Wind Spacecraft in the interplanetary medium. Both electron heating and acceleration have been inferred from these observations. We discuss electron heating and accleration in this paper.

INTRODUCTION

Natural collisionless shocks can be very well studied using spacecraft, because there are no edge effects, and the small size of the average spacecraft in comparison with the Debye radius in space plasmas, ensures a point measurement. Spacecraft instruments, especially those measuring electrons, have time resolutions of seconds or fractions of seconds, so that both the low and high entropy sides of a shock can be resolved well enough to bring out details, but sometimes the ramps can not be distinguished. Planetary bow shocks, usually quite strong, are curved around the corresponding magnetosheath. Their Mach number vary and their obliquity vary from parallel to perpendicular along the shock fronts. These shocks are standing reverse shocks. Besides these shocks, propagating forward shocks are also observed in the interplanetary medium. These are often weak to moderate in strength and might be curved (usually with large radii). In any case, a wide variety of collisionless shocks is available for study using spacecraft instrumentation (magnetometers, plasma and energetic particle instruments, etc.).

The distribution function of electrons can be observed in detail on both sides of the shock. Using such observations, successful attempts have been made to predict the downstream plasma properties from the upstream measurements. For example, using measurements made by the SWE instrument on the Wind spacecraft (figure 1), where 576 point 3-D distributions are produced every 9 seconds, mapping using a Vlasov – Liouville procedure including mirroring has been carried out across a number of shocks.

The measured downstream distributions can then be compared with these predictions. (Hull, 1998; 2000; Fitzenreiter,et al.2003) The shocks shown in Figure 1 were of moderate strength, and the work led to three interesting results:

1. Upstream distributions show that shock acceleration are accompanied by Langmuir waves.
2. In the Downstream region, broader distributions with flat tops accompanied by beams appear (as shown in Figures 2, 3) and there were inaccessible trapping regions downstream of the strongest shocks.
3. The appearance of beams indicates both heating and acceleration.

Thus the distributions are modified in passing through the shocks, and the reflected beams produce the Langmuir waves.

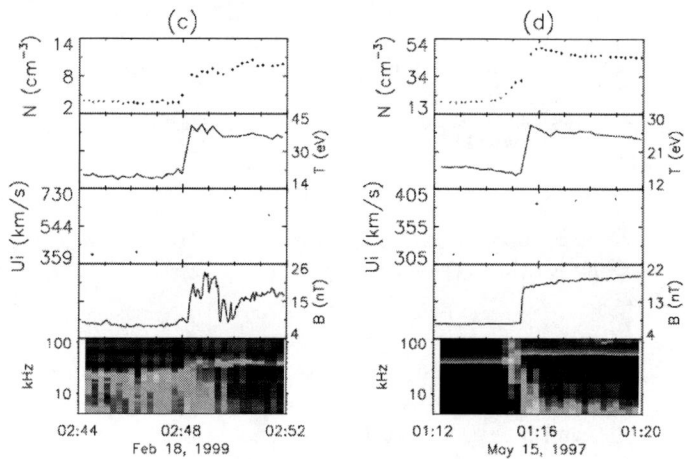

Figure 1: Interplanetary propagating shocks. Top, electron density, electron Temperature, ion speed and magnetic field; bottom, plasma wave observations.

The heating, broadening of the distribution, and the acceleration, displacement of the maximum, can be seen in Figure3. The heating, (T2-T1) was approximately proportional to the shock strength (B2/B1). An example with B2/B1=3 gave T2-T1=30eV. How far this proportionality holds is not known yet. The downstream-directed beam can be understood in terms of the de-Hoffman-Teller potential. In the de-Hoffman-Teller frame there is no electric field associated with the shock except along the B direction, in the ramp. Since this is the direction about which electrons are gyrating, and highly magnetized, the axis of their pitch angle distribution, can be accelerated across the shock into the high entropy region by falling through the de-Hoffman-Teller potential which has the correct sign. If the electrons cross the shock more than once they will gain more energy.

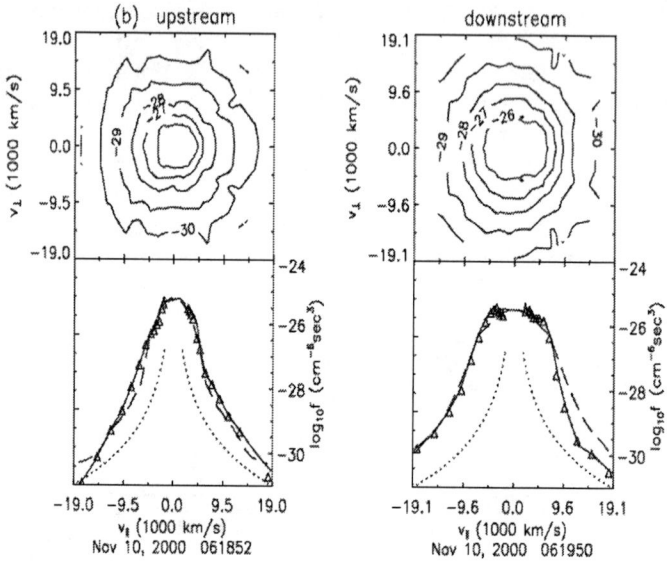

Figure 2: 3-D distribution functions derived from shock observations. Left is upstream, right is downstream.

Thus the pitch angle distribution of the electrons has a large effect on the acceleration process. The de-Hoffman-Teller potential has a magnitude of, at most, a few hundred volts, and it depends on the magnetic field strength (Hull et al. 1988). The extent to which external parameters could increase the efficiency of the acceleration of electrons at shocks is not well understood at present. The small energy gain by an electron which falls through the cross-shock potential, although the equivalent of a high temperature, is not spectacular. Many gains in series are required for this. Ions, with much larger gyro-radii can acquire much more energy at shocks. Another important difference between ions and electrons at shocks results from the pickup process, in which ions acquire much larger gyro-radii in the solar wind. Acceleration of pickup ions is an important process at the termination shock.

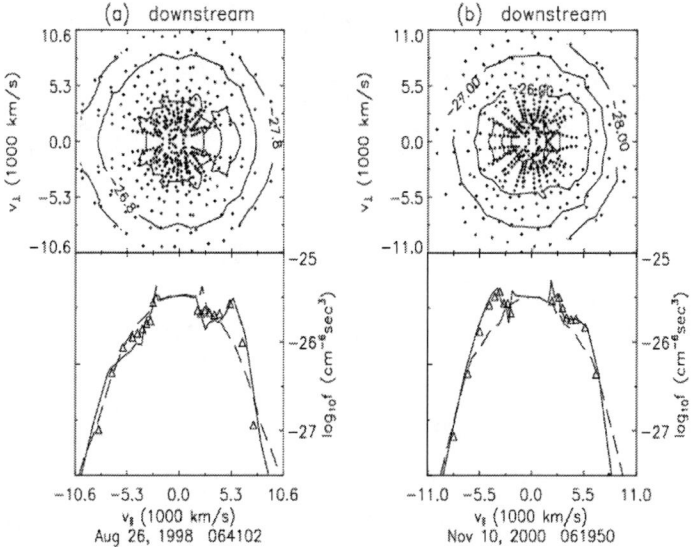

Figure 3: 3-D distributions downstream of shocks shown in Fig.1, as a function of observing positions. The lower curves are reduced distributions along the magnetic field.

SUMMARY AND CONCLUSIONS

The existence of the de-Hoffman-Teller frame allows us to understand the heating and acceleration of electrons at weak and moderate shocks, and the existence of beams which occur in the upstream and downstream regions, for the moderate strength quasi-perpendicular shocks we are discussing. Mapping distribution functions across shocks to deduce the downstream distribution is a successful technique. "Blast Waves" resulting from point singularities are special shocks, which run ahead of a high-pressure driver. Here the down stream region is the driver; processes taking place there, (for example the emission of photons of all kinds) can pass through the shock to the upstream side and have a large effect. Such shocks are hard to study in detail.

ACKNOWLEDGMENTS

The author acknowledges conference organizers for their hospitality.

REFERENCES

1. Hull,A.J. ,J.D.Scudder, L.A.Frank, W.R. Patterson,and M.G.Kivelson, *J.Geophys. Res.* 103,2041,1998.
2. Gurnett, HullA.J.,J.D.Scudder,R.J.Fitzenreiter,*J.Geophys. Res.*105,20957, 2000.
3. Fitzenreiter,R.J.,K.W.Ogilvie,S.D.Bale and A.F.Vinas *J. Geophys. Res.*,108,1415, 2003.

SESSION 3
PLANETARY BOW SHOCKS, STRUCTURE AND WAVES

The electric potential at the Earth's quasi-parallel bow shock: Initial Cluster results

R. Behlke*, H. Kucharek[†], S. D. Bale**, M. André* and E. A. Lucek[‡]

*Swedish Institute of Space Physics, Uppsala division, Box 537, 75121 Uppsala, Sweden
[†]Space Science Institute, University of New Hampshire, Durham, USA
**Space Science Laboratory, University of California, Berkeley, USA
[‡]Blackett Laboratory, Imperial College, London, UK

Abstract. We present multi-spacecraft measurements of a quasi-parallel shock crossing obtained by the Cluster satellites. Observations commonly show the presence of Short Large-Amplitude Magnetic Structures (SLAMS) at the Earth's quasi-parallel bow shock. These structures are thought to play an important role for decelerating, heating and deflecting the incident solar wind. We investigate the electric potential over SLAMS in the Normal Incidence Frame Φ_{NIF} as a possible means for achieving thermalisation in the foreshock region of the Earth's quasi-parallel bow shock. We show that SLAMS exhibit a substantial electric potential on the order of a few hundred Volt (V). Thus at the Earth's quasi-parallel bow shock, these structures may be important for dissipation that transforms the ram energy of the solar wind bulk flow into thermal energy. SLAMS might be responsible for returning particles, that have been reflected at the shock and whose guiding centre motion is pointing upstream, back to the shock.

Keywords: Quasi-parallel collisionless shock, electric cross-shock potential
PACS: 94.30.Va, 96.50.Fm, 96.50.Bh

INTRODUCTION

A major controlling factor for collisionless shocks is the angle θ_{Bn} between the upstream magnetic field and the shock normal. At quasi-perpendicular shocks ($\theta_{Bn} > 45°$), a significant part of the incoming solar wind ions is immediatly convected downstream after the first gyration where they are subsequently thermalised.

Under quasi-parallel conditions ($\theta_{Bn} < 45°$), the guiding centre motion of specularly reflected particles is pointed upstream and these particles may easily escape from the shock and propagate far upstream along magnetic field lines [1]. These specularly reflected particles cannot directly contribute to downstream thermalization in this shock geometry. Dissipation has to be achieved by other means, e.g., by waves that are generated by the backstreaming, reflected ions and shock-generated waves. These waves in their turn may point the guiding centre of reflected ions downstream again due to a slowly changing θ_{Bn}.

Incoming solar wind ions may interact with backstreaming, reflected ions. These interactions lead to the generation of low-frequency (ULF) waves [2] via an ion-ion beam instability. Some of the ULF waves steepen and form so-called Short Large-Amplitude Magnetic Structures (SLAMS) during their approach towards the shock [3]. A model of the quasi-parallel bow shock emerged in which the shock is considered a patchwork of SLAMS [4]. The shock is cyclically reforming as these structures are

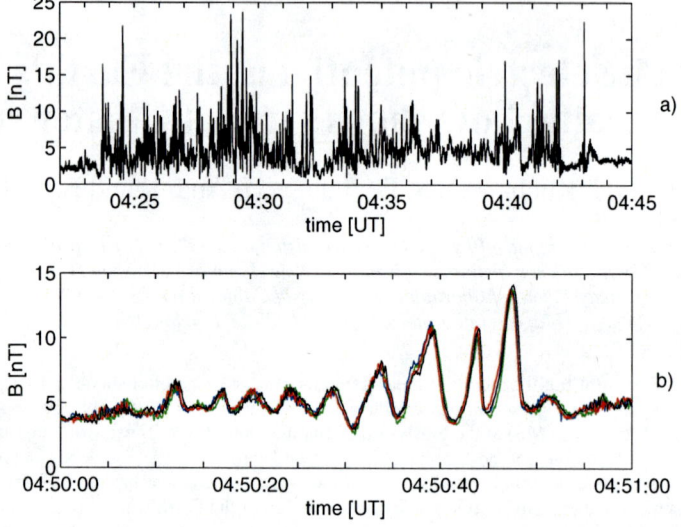

FIGURE 1. Cluster observations of a quasi-parallel shock crossing on February 3, 2002. Panel (a): The magnetic field amplitude B for a 23 minutes period (UT 04:22-04:45) obtained by Cluster 1 is shown. Panel (b) displays B for a 1 minute period (UT 04:50-04:51) for all four spacecraft. Note the Cluster colour notation: black, red, green, blue for spacecraft 1, 2, 3, 4, respectively.

convected into the shock by the solar wind. Note that SLAMS are produced far upstream in this scenario. However, some questions concerning the importance of SLAMS for pre-dissipation in the foreshock region of the quasi-parallel shock have not yet been investigated. The electric potential can, in combination with other mechanisms, be responsible for the dissipation of solar wind energy at SLAMS. It was thus an obvious question to investigate the electric potential over SLAMS.

INSTRUMENTATION

The data for this study are obtained by the Electric Field and Wave (EFW) instrument [5] onboard the four Cluster spacecraft [6]. It consists of two pairs of spherical probes on wire booms in the spin plane of each satellite. The separation between each probe pair is 88 m. For the data presented here, the electric field data are sampled at 25 Hz using a lowpass filter at 10 Hz. We also present data from the fluxgate magnetometer (FGM) instrument [7] in the analysis.

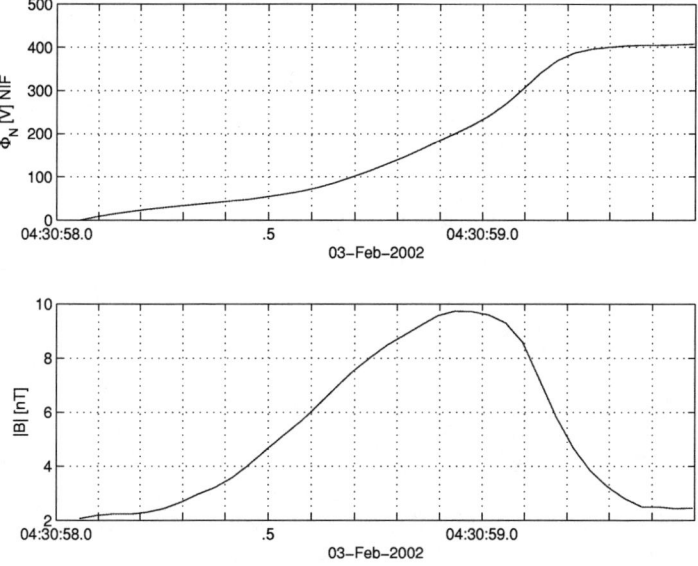

FIGURE 2. Cluster observations on February 3, 2002, UT 04:30:58.0 - 04:30:59.5. Panel (a) shows an example of the cross-SLAMS potential in the Normal Incidence Frame Φ_{NIF} for spacecraft 2. Panel (b) displays the corresponding magnetic field magnitude B.

OBSERVATIONS

Figure 1 shows magnetic field data during a supercritical, quasi-parallel shock crossing on February 3, 2002, as observed by the Cluster spacecraft. Note that the spacecraft separation during these observations was on the order of 100 km. The satellites were positioned at $(11, 2, -8) R_E$ (R_E=Earth radii) in the GSE (Geocentric Solar Ecliptic) system. Panel (a) displays a 23 minutes interval within the shock crossing. It can clearly be seen that quasi-parallel shock crossings are characterised by an extented transition region consisting of magnetic pulsations. Panel (b) reveals that these pulsations are built up by ULF semicycles with periods of $T \sim 4$ sec and $B_{max}/B_0 \sim 1$ and SLAMS with $T \sim 2$ sec and $B_{max}/B_0 > 1$, where B_{max} is the peak magnetic field amplitude and B_0 the background magnetic field. The magnetic field enhancement is accompanied by a density enhancement, i.e., SLAMS are fast-mode structures [8]. Within SLAMS a change from the quasi-parallel to the quasi-perpendicular shock configuration is generally observed [9], i.e., whereas the inter-SLAMS regions are characterized by quasi-parallel conditions, a change to the quasi-perpendicular regime is observed in association with the leading edge. In association with the trailing edge, the magnetic field returns to quasi-parallel conditions. SLAMS generally have a direction of propagation that is quasi-parallel to the ambient magnetic field [10] and exhibit transverse scale lengths of the order of

1000 km [9]. Simulations revealed that SLAMS might change significantly on the order of seconds, i.e., the time it takes the SLAMS to move over the spacecraft [11]. Recent observations indicated indeed a growth rate of SLAMS of the order of 10% per second.

For a complete description of the methodology of the calculation of the electric potential in the Normal Incidence Frame, we refer to [13]. Figure 2 (a) displays the calculated cross-SLAMS potential Φ_{NIF} for a single SLAMS during this shock crossing. A net potential $\Phi_{NIF} \sim 400$ V is calculated. Note that the potential is evenly divided between the trailing edge (left hand side) and leading, but more steepend, edge (right hand side) of the SLAMS. Panel (b) shows the corresponding magnetic field profile.

DISCUSSIONS AND CONCLUSIONS

We have shown that Short Large-Amplitude Magnetic Structures (SLAMS) are associated with a substantial electric potential of the order of a few hundred V. These structures are thought to play an essential role in quasi-parallel shock physics. Simulations have indeed shown that the potential over SLAMS is associated with a deceleration of the solar wind flow [14]. For supercritical shocks in general it is well-accepted that the velocity dispersion which results from ion reflection and subsequent transmission through the cross-shock potential accounts for a substantial part of effective dissipation [15]. The observed potential over SLAMS might, in combination with other processes, be responsible for deceleration, deflection and heating of the incident solar wind.

Future studies need to investigate the dependence of the cross-SLAMS potential on different parameters and its impact on particle distributions. Work in progress includes a statistical approach on the cross-SLAMS potential and associated ion distributions [13].

ACKNOWLEDGMENTS

The authors thank the Cluster software and engineering teams. We are thankful to the PI of the CIS-instrument, H. Remé and the PI of the FGM-instrument, A. Balogh, for making the data of their instruments available. RB thanks D. Burgess, B. Lembège, A. Vaivads, A.I. Eriksson and A. Tjulin for enlightening discussions, S.C. Buchert for help with FGM data and Y. Hobara for help with CIS data. RB is grateful for the hospitality at the SSL, UC Berkeley, at which this study was initiated during a visit.

REFERENCES

1. J. T. Gosling, M. F. Thomsen, S. J. Bame, and C. T. Russell, *J. Geophys. Res., 94*, 10027–10037 (1989).
2. M. T. Hoppe, and C. T. Russell, *J. Geophys. Res., 88*, 2021–2028 (1983).
3. S. J. Schwartz, D. Burgess, W. P. Wilkinson, R. L. Kessel, M. W. Dunlop, and H. Lühr, *J. Geophys. Res., 97*, 4209–4227 (1992).
4. S. J. Schwartz, and D. Burgess, *Geophys. Res. Lett., 18*, 373–376 (1991).
5. G. Gustafsson, R. Boström, B. Holback, G. Holmgren, A. Lundgren, K. Stasiewicz, L. Åhlen, F. S. Mozer, D. Pankow, P. Harvey, P. Berg, R. Ulrich, A. Pedersen, R. Schmidt, A. Butler, A. W. C. Fransen,

D. Klinge, M. Thomsen, C.-G. Fälthammar, P.-A. Lindqvist, S. Christenson, J. Holtet, B. Lybekk, T. A. Sten, P. Tanskanen, K. Lappalainen, and J. Wygant, *Space Sci. Rev., 79*, 137–156 (1997).
6. C. P. Escoubet, R. Schmidt, and M. L. Goldstein, *Space Sci. Rev., 79*, 11–32 (1997).
7. A. Balogh, C. M. Carr, M. H. Acūna, M. W. Dunlop, T. J. Beek, P. Brown, K.-H. Fornacon, E. Georgescu, K.-H. Glassmeier, J. Harris, G. Musmann, T. Oddy, and K. Schwingenschuh, *Ann. Geophys., 19*, 1207–1217 (2001).
8. R. Behlke, M. André, S. C. Buchert, A. Vaivads, A. I. Eriksson, E. A. Lucek, and A. Balogh, *Geophys. Res. Lett., 30*(4), 10.1029/2002GL015871 (2003).
9. E. A. Lucek, T. S. Horbury, M. W. Dunlop, P. J. Cargill, S. J. Schwartz, A. Balogh, P. Brown, P. Carr, K.-H. Fornacon, and E. Georgescu, *Ann. Geophys., 20*, 1699–1710 (2002).
10. G. Mann, H. Lühr, and W. Baumjohann, *J. Geophys. Res., 99*, 13315-13323 (1994).
11. K. Tsubouchi, and B. Lembége, *J. Geophys. Res., 109*, doi:10.1029/2003JA010014 (2004).
12. E. Lucek, et al., these proceedings, (2005).
13. R. Behlke, H. Kucharek, S. D. Bale, M. André, and E. A. Lucek, *Ann. Geophys.*, submitted (2005).
14. M. Scholer, H. Kucharek, and I. Shinohara, *J. Geophys. Res., 108*, doi:10.1029/2002JA009820 (2003).
15. D. Sherwell, and R. A. Cairns, *J. Plasma Phys., 17*, 265–279 (1977).

On increasing accuracy of bow shock shape and position predictions

Jan Merka

L-3 Communications GSI, Vienna, VA

Abstract. In more than three decades, spacecraft have crossed the terrestrial bow shock tens of thousand times providing enough data to construct many bow shock models. However, several studies have demonstrated severe limitations in the predictive capabilities of the most used bow shock models due to inadequate parameterization and/or limited amount of data employed. The accuracy of bow shock models is also hindered by the fact that bow shock models predict an equilibrium position for given conditions while spacecraft generally observe a moving shock. Furthermore, *Merka and Szabo* [1] recently pointed out that the dipole tilt angle and the solar wind/IMF angle significantly influence bow shock position but are not accounted for in any of the existing bow shock models. The present paper discusses the limitations of the current approaches in bow shock modeling and outlines the ingredients required for improving predictive capabilities of bow shock models.

Keywords: bow shock, magnetopause, models
PACS: 94.30.Di, 96.50.Fm

INTRODUCTION

The Earth's bow shock has been studied for more than four decades and many models predicting its shape and position have been created [2, 3, 4, 5, 6, 7, 8, 9, 10]. Although the current bow shock models provide satisfactory predictions of average shock position and shape, their accuracy typically decreases for certain solar wind and interplanetary magnetic field (IMF) conditions [11, 12, 13]. This paper presents a list of the most significant factors limiting predictive capabilities of the current bow shock models and, thus, offers a general recipe for a more accurate next-generation model.

FACTORS LIMITING ACCURACY OF SHOCK MODELS

A global 3-dimensional (3D) Earth bow shock model could be, in principle, derived from equations describing interaction between the solar wind/IMF and the magnetosphere. Unfortunately, this means that both magnetopause and bow shock shapes/positions would have to be solved simultaneously which is not currently possible without significant simplifications of the problem [2, 7]. On the other hand, a (3D) functional form describing bow shock shape can be easily fitted to observed positions of the shock wave. Due to non-simultaneity of the observed shock crossings, such empirical models provide relatively good statistical predictions for average conditions but tend to fail for unusual conditions [3, 4, 6, 14]. Development of a global bow shock model can be also aided by numerical simulations of the solar wind/IMF interaction with the magnetosphere, which

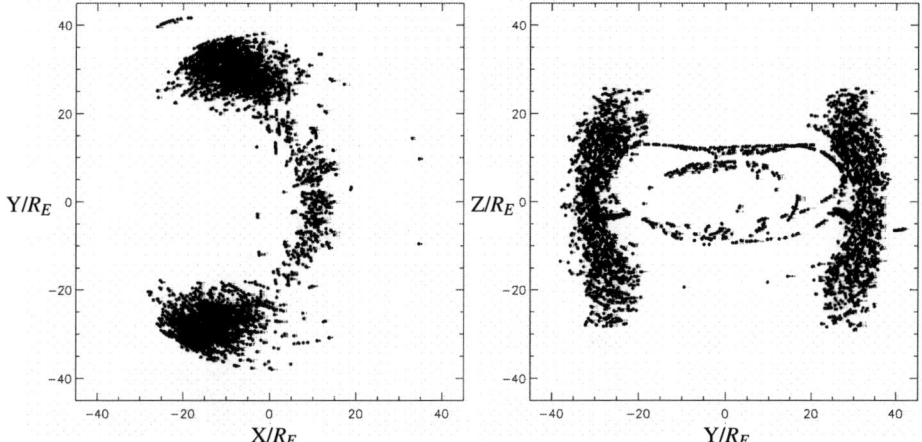

FIGURE 1. Positions in GSE coordinates of 9320 bow shock crossings observed by Cluster, Geotail, IMP 8, and Interball-1 spacecraft.

resolve the entire shock shape and position at once. However, accuracy of resulting bow shock models is hindered due to assumptions and approximations involved in the simulations and due to generalization and/or simplification of the simulation results for the purpose of creating a bow shock model [8, 10]. Bow shock models usually combine several of the three approaches by using their strengths and eliminating their weaknesses.

Spacecraft observations of bow shock crossings are crucial for an evaluation of models' performance and for the design of empirical models. However, observational coverage of the bow shock is rather irregular. Figure 1 showing positions of 9320 bow shock crossings observed by Cluster, Geotail, IMP 8, and MAGION-4 spacecraft clearly demonstrates lack of observations around the magnetospheric tail, at high latitudes and even at the subsolar point. More data from those regions would certainly improve accuracy of shock models.

A good bow shock model should be accurate in a wide range of conditions and area, from the subsolar point down the tail, while still be physics-based in order to allow extrapolations outside of conditions for which it was designed. However, most of the current models predict bow shock position with varying degree of accuracy for different conditions even though they were often designed specifically for such conditions [12, 13]. As a particular example, Figure 2 compares performance of two bow shock models for low and high upstream Mach numbers [9, 14] where the model performance is expressed by the ratio of modeled and observed radial distances (R_M and R_O) to the bow shock. The ratio R_M/R_O changes less between the low and high Mach number bins in the right panel of Figure 2 and, therefore, the right-side model is considered better because it performs with similar accuracy in both Mach number bins, i.e. the model is applicable in a wider range of conditions. A detailed description of the Figure 2 format can be found in [12].

Merka and Szabo [1] recently investigated bow shock properties in the near-tail region $-15R_E < X < -10R_E$ and found that bow shock cross-section moves in the North-

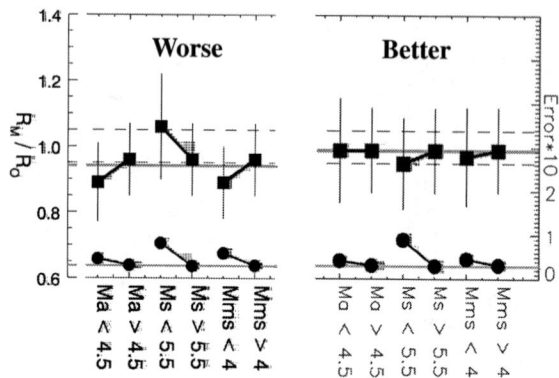

FIGURE 2. Predictive accuracy of bow shock models often changes for unusual conditions as, for example, low upstream Mach numbers (left) although the accuracy should not change (right) if the model is applicable for wide range of conditions. Displayed models are [9, 14].

South direction by 3.8 R_E when the dipole tilt angle changes from negative ($\lambda < -15°$) to positive ($\lambda > 15°$). *Merka and Szabo* [1] confirmed the observational result by performing global 3D MHD simulations of the magnetosphere (the BATSRUS model) which showed a shift of ~ 4 R_E between dipole tilts $\lambda = -35°$ and $\lambda = 35°$. The MHD simulations also suggested 1–2 R_E shifts at the dayside magnetopause and bow shock. Therefore, both magnetopause and bow shock models should be parameterized by the dipole tilt angle λ. However, there is currently no global bow shock nor magnetopause model taking the dipole axis orientation into account in spite of theoretical calculations of the magnetopause shape considering this effect as early as four decades ago [15].

Merka and Szabo [1] also demonstrated, using IMP 8 bow shock observations, that the statistical bow shock cross-section shrinks, i.e. the semi-axes get shorter by 6–8%, when the IMF orientation with respect to the solar wind flow changes from quasi-perpendicular ($\theta_{Bv} > 45°$) to quasi-parallel ($\theta_{Bv} < 20°$). The magnitude of this effect is comparable or even slightly larger than the increase in cross-section size when the upstream Alfvénic Mach number decreases from $M_A > 10$ to $5 < M_A < 8$ [1]. Thus, this IMF orientation effect should be also considered when predicting bow shock position.

Bow shock models predict an equilibrium position of the shock which could be correct if the magnetosphere/bow shock system had enough time to reach this state. However, permanent variations of the solar wind/IMF conditions cause that the bow shock wave is not static relative to Earth. On the other hand, models move the predicted shock instantaneously, they don not restrain the speed of bow shock motion relative to Earth (e.g. [16] and Figure 3). Earth's bow shock is a fast magnetosonic wave so we can surmise that it cannot propagate upstream faster than fast magnetosonic speed. Indeed, this insight is supported by MHD simulations and observations [17]. In the opposite direction (antisunward), the bow shock wave maximum speed relative to Earth can be expected to equal the solar wind bulk speed. Even with this crude understanding of bow shock movement, we could improve upon existing bow shock models by limiting the maximum distance the shock can reach in a given time step. We

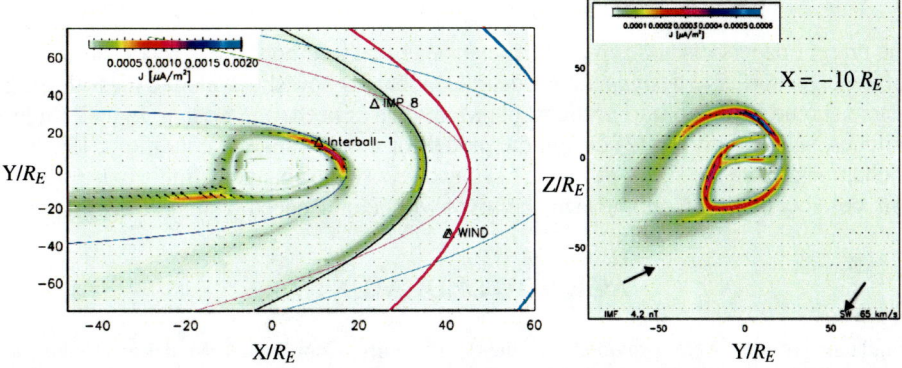

FIGURE 3. MHD simulation of an extremely low-density solar wind interacting with the magnetosphere observed on May 11, 1999 by the Interball-1, IMP 8, and WIND spacecraft [20]. For comparison, empirical models of the magnetopause (blue) and bow shock (purple and cyan) are displayed [7, 8, 18].

can expect a significant decrease in the spread of bow shock crossing around a model shock surface if the shock movement is constrained. For a detailed understanding of bow shock movement, a systematic study of bow shock velocity under representative set of conditions and locations is required, for example using multi-point observations from the Cluster spacecraft.

The shape of an obstacle influences both the position and shape of a bow shock and, thus, we need to understand how the magnetopause changes in response to the solar wind/IMF and internal magnetospheric (dipole tilt, ring current, etc.) parameters. It has been already discussed above that current magnetopause models ignore the influence of varying dipole tilt axis orientation. Furthermore, MHD simulation (the BATSRUS model) show that magnetopause can become strongly asymmetric under certain conditions and, thus, strongly deviate from model predictions as shown in Figure 3. Figure 3 presents two snapshots (1425 UT and 1830 UT) of 10 hour-long magnetospheric simulation using the solar wind/IMF conditions observed on May 11, 1999 when the solar wind density dropped down to ~ 0.2 cm^{-3} and several spacecraft observed the magnetopause (Interball-1) and the bow shock (IMP 8 and WIND) move far upstream [20]. The left panel compares the simulation results, the total current density J, with a magnetopause model [18] (blue) and two bow shock models [7, 8] using two different shock shape parameterizations [19]. It is immediately apparent that a model prediction can agree locally with an observation although the global shape is dubious. A comparison with previous and subsequent snapshots (not displayed) would show the bow shock models *jumping* by many Earth radii from one location to another while the numerical simulation move the shock boundary continuously. The dawn-side magnetopause strongly deviates from the model prediction [18] and, in the right panel at a later time, a gross deformation of the magnetospheric cross-section is obvious. We should note that it is not clear whether the simulations capture magnetospheric configuration correctly in this case due to the lack of observation on the dawn-side and behind the terminator.

SUMMARY

The present papers discusses major deficiencies of current bow shock models and concludes that, in order to design a more accurate global 3D bow shock model, the following factors should be taken into account: (1) More data is needed in certain regions (high-latitudes, subsolar point); (2) design the model for a wide range of conditions; (3) IMF orientation affects the shock position; (4) obstacle's shape changes (dipole axis orientation, possibly other factors); (5) shock moves with a finite speed relative to Earth.

ACKNOWLEDGMENTS

Simulation results were provided by the Community Coordinated Modeling Center at NASA/GSFC. The CCMC is a multi-agency partnership between NASA, AFMC, AFOSR, AFRL, AFWA, NOAA, NSF, and ONR. The BATSRUS Model was developed by the Center for Space Environment Modeling at the University of Michigan with funding support form NASA ESS, NASA ESTO-CT, NSF KDI, and DoD MURI.

REFERENCES

1. J. Merka, and A. Szabo, *J. Geophys. Res.*, **109**, doi: 10.1029/2004JA010567 (2004).
2. J. R. Spreiter, A. L. Summers, and A. Y. Alksne, *Planet. Space Sci.*, **14**, 223–253 (1966).
3. D. H. Fairfield, *J. Geophys. Res.*, **76**, 6700–6715 (1971).
4. V. Formisano, *Planet. Space Sci.*, **27**, 1151–1161 (1979).
5. J. A. Slavin, and R. E. Holzer, *J. Geophys. Res.*, **86**, 11401–11418 (1981).
6. Z. Němeček, and J. Šafránková, *J. Atmos. Terr. Phys.*, **53**, 1049–1054 (1991).
7. M. H. Farris, and C. T. Russell, *J. Geophys. Res.*, **99**, 17681–17689 (1994).
8. I. H. Cairns, and J. G. Lyon, *J. Geophys. Res.*, **100**, 17173–17180 (1995).
9. M. Verigin, G. Kotova, A. Szabo, J. Slavin, T. Gombosi, K. Kabin, F. Shugaev, and A. Kalinchenko, *Earth Planets Space*, **53**, 1001–1009 (2001).
10. J. F. Chapman, and I. H. Cairns, *J. Geophys. Res.*, **108**, doi: 10.1029/2001JA000219 (2003).
11. J. Šafránková, Z. Němeček, and M. Borák, "Bow Shock Position: Observations and Models," in *Interball in the ISTP Program, Studies of the Solar Wind-Magnetopause-Ionosphere Interaction*, edited by D. G. Sibeck, and K. Kudela, Kluwer Academic Publishers, Dordrecht, 1999, vol. 537 of *NATO Science Series*, pp. 187–201.
12. J. Merka, A. Szabo, T. W. Narock, J. H. King, K. I. Paularena, and J. D. Richardson, *J. Geophys. Res.*, **108**, doi: 10.1029/2002JA009384 (2003).
13. J. Merka, A. Szabo, T. W. Narock, J. D. Richardson, and J. H. King, *Planet. Space Sci.*, **53**, 79–84 (2005).
14. J. Merka, A. Szabo, J. A. Slavin, and M. Peredo, *J. Geophys. Res.*, **110**, doi: 10.1029/2004JA010944 (2005).
15. J. R. Spreiter, and B. R. Briggs, *J. Geophys. Res.*, **67**, 37–51 (1962).
16. J. F. Chapman, and I. H. Cairns, *J. Geophys. Res.*, **109**, doi: 10.1029/2004JA010540 (2004).
17. M. El-Alaoui, R. L. Richard, M. Ashour-Abdalla, and M. W. Chen, *Geophys. Res. Lett.*, **31**, doi: 10.1029/2003GL018788 (2004).
18. J.-H. Shue, P. Song, C. T. Russell, J. T. Steinberg, J. K. Chao, G. Zastenker, O. L. Vaisberg, S. Kokubun, H. J. Singer, T. R. Detman, and H. Kawano, *J. Geophys. Res.*, **103**, 17691–17700 (1998).
19. J. Merka, A. Szabo, J. Šafránková, and Z. Němeček, *J. Geophys. Res.*, **108**, doi: 10.1029/2002JA009697 (2003).
20. D. H. Fairfield, I. H. Cairns, M. D. Desch, A. Szabo, A. J. Lazarus, and M. R. Aellig, *J. Geophys. Res.*, **106**, 25361–25376 (2001).

Field-aligned and Gyrating Ion Beams in a Planetary Foreshock

C. Mazelle[1], K. Meziane[2], M. Wilber[3] and D. Le Quéau[4]

[1]*Centre d'Etudes Spatiales des Rayonnements, CNRS, 9 Avenue du Colonel Roche, Toulouse, 31400, France (christian.mazelle@cesr.fr)*
[2]*Physics Department, University of New Brunswick, Fredericton, NB, Canada*
[3]*Space Sciences Laboratory, University of California, Berkeley, USA* [4]*Observatoire Midi-Pyrénées, Toulouse, France*

Abstract. The foreshock region is the first signature of the interaction of the solar wind with a planet's plasma environment when approaching its collisionless bow shock. Part of its structure and dynamic is determined by instabilities, which are created by the interaction of the solar wind with backstreaming ion populations. The interaction of the reflected ions with the solar wind drives ion/ion beam instabilities, which generate waves that are then convected towards the shock by the solar wind. Subsequently they may mediate the shock structure and its reflection properties. The most well-know examples are the field aligned ion beams (FABs), produced by reflection processes in the quasi-perpendicular and oblique regions of the shock. Other prominent examples are the gyrating ions with well-defined pitch-angle and gyrophase organization around the local magnetic field observed downstream of the FABs region. These gyrophase-bunched ions are always associated with large amplitude quasi-monochromatic right-hand mode low-frequency waves. Different mechanisms have been put forward to explain these ion features. This paper will discuss recent advances on this topic from multi-spacecraft observations (Cluster) as well as theoretical considerations.

Keywords: Planetary bow shock, ion foreshock, wave-particle interaction.
PACS: 96.59.Ek, 96.50.Pw

1. INTRODUCTION

Several types of ion populations have been observed upstream of the Earth's bow shock and these population have been extensively studied and hypotheses have been put forward to explain their origin [1, and references therein]. Ion beams of several keV collimated along interplanetary field lines have been observed upstream from the quasi-perpendicular shock. Downstream of the field-aligned beam region, distributions characterized by a gyromotion around the magnetic field, *i.e.* a non-vanishing perpendicular bulk velocity with respect to the background magnetic field, have been reported. These gyrating ion distributions are nongyrotropic or nearly-gyrotropic. Numerous studies concerning gyrating ions have been reported in earlier investigations mainly from ISEE 1 and 2 [2,3,4,5,6,7], AMPTE [8] and WIND [9,10,11]. Gyrating ions are often observed in association with ULF waves having substantial amplitude [7]. The waves are right-handed and propagate nearly along the ambient magnetic field [4,9,10,11]. It is believed that the ULF waves are excited through a beam plasma instability resulting from the propagation of field-aligned ions which precede them closer to the foreshock boundary [12]. The produced waves can in turn trap the ions and cause the phase bunching of the distribution in what is called a beam disruption mechanism [13].

2. PREVIOUS OBSERVATIONS

A quantitative analysis of particle and monochromatic waves was made from ISEE data [6] which strongly suggested that there was a coherent wave-particle interaction. They obtained a phase relationship between the gyrovelocity $\mathbf{v}_\perp$ and the transverse wave field $\delta\mathbf{B}_\perp$ so that energy transfer occurred between the particles and the waves and gyrophase trapping by the wave was possible. Since the field-aligned beams propagate into regions deep within the foreshock, the local production of gyrating ions through this process should be observed very far from the shock contrary to directly shock-produced gyrating ions which are subject to rapid gyrophase-mixing [14]. First observations of several gyrating ion distributions and their association with low frequency waves at distances larger than 20 R_E from the shock were reported from WIND data [9]. There was again a clear indication of coherent wave-particle interaction. A more detailed study of the three-dimensional ion distributions with a large data set and the highest available time resolution (3s) has shown that these observational features can be found up to more than 80 R_E from the shock [11]. An investigation of the non-linear wave trapping mechanism has shown that it can explain the properties of such gyrating ion distributions registered at large distances from the shock [10]. It has been shown that the particles are not only bunched in gyrophase but also trapped in pitch-angle in velocity space around a value which is directly related to the amplitude of the wave self-consistently generated by the original field-aligned ion beam.

3. CLUSTER OBSERVATIONS

This local production mechanism has been recently investigated to explain the existence of well-defined gyrating ion distributions reported from the Cluster CIS measurements in the Earth's foreshock [15]. One example is displayed on Figure 1. For this event, Cluster s/c 1 was connected to the bow shock, during a much larger interval. At 2334:30 UT, energetic ions are revealed in the second energy spectrogram corresponding to measurements by the High side of the HIA instrument (the difference with the first panel showing the solar wind population is quite obvious). High fluxes are then continuously observed until 2344 UT, followed by two small patches. These ions are mostly propagating sunward, as revealed from the analysis of their guiding center velocities, i.e. they are backstreaming ions. Before 2334:30 UT, the IMF was nearly quasi-steady.

Conversely, prominent large amplitude low frequency waves are observed after 2335:45 UT both on the magnetic field and on the solar wind velocity. Figure 2 displays three-dimensional 4-s representation of the ion distribution functions registered by CIS-CODIF. Nine consecutive distributions are shown for one energy channel (~8 keV) for which the observed backstreaming fluxes are maximum for a time interval inside the event displayed on Figure 1. Each frame in Figure 2 is a projection in gyrophase and pitch-angle with the **B**-direction located at the center.

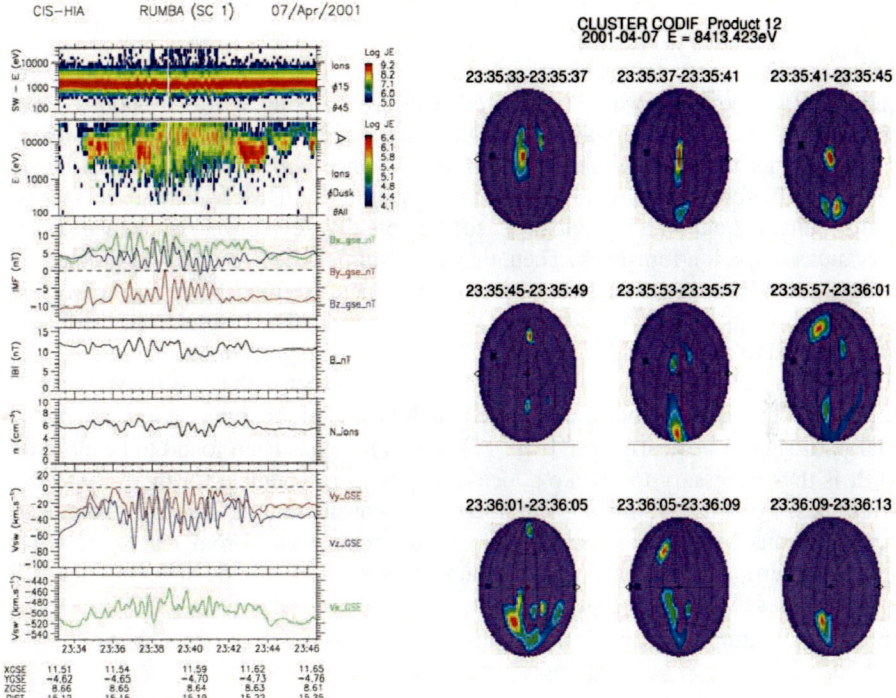

FIGURE 1. Observations from CLUSTER CIS and FGM on satellite 1 between 23:33-23:46 UT on April 7, 2001 : energy-time spectrograms of all ions from CIS/HIA for "solar wind sectors" (sunward looking direction - upper panel) and "dusk" solid angle (duskward looking direction - second panel), respectively; dc magnetic field components in GSE coordinates and its magnitude; ion density and bulk velocity in GSE coordinates derived from HIA measurements.

FIGURE 2. Fig. 2. Sequence of consecutive three-dimensional 4-s display of the proton angular distributions registered by CIS-CODIF for an energy of ~8 keV (flux maximum). Each frame represents the normalized distribution function on a surface of constant energy in the solar wind frame of reference projected to display 4π-coverage. The $\mathbf{B_0}$ vector is located at the center of each plot (background field shown by a '+'identical for all frames) and the '*' sign indicates the solar wind direction. For each frame, the maximum value of the normalized phase space density is shown in red.

The three first snapshots indicate an ion beam propagating along the +**B** direction with a parallel velocity of 1,100 km/s but the third one also shows a second peak for a large pitch-angle of about 60°. Then after 2335:45, the spacecraft has entered a gyrating ion region. Gyrating ions are identified by their gyrophase-restricted distribution peaked off the magnetic field direction. The interplanetary magnetic field used to plot the distributions is averaged over the spin interval (4 seconds) while the local proton cyclotron period is 7 seconds (i.e., about two ion sampling intervals). The gyrating distributions show a clear rotation of their maximum phase density in the left-handed sense around the magnetic field with alternating values separated by about 180°. Such gyrating ion distributions are observed up to ~2344 UT.

4. WAVE-PARTICLE INTERACTION

Using multi-spacecraft analysis techniques [e.g., 16], the properties of large amplitude low frequency waves associated with the gyrating ion distributions have been analyzed by [15]. The waves are right-hand mode waves ('30-s waves'). They have shown that these wave are in cyclotron resonance with the field-aligned beam observed just before the spacecraft entered the gyrating ion/ULF wave region. This is the first direct quantitative evidence so far of this cyclotron resonance from observations in the ion foreshock. Then, they have studied the possibility of resonantly driving these waves unstable from the electromagnetic ion/ion beam instability by field-aligned beam ions also observed in the same region. The results from the linear theory has lead to a very good agreement with the observed wave mode.

The gyrating feature of the ion distribution is inconsistent with a specular reflection at the Earth's bow shock since the observed pitch-angles of the gyrating ions are much too large (it should be nearly θ_{Bn} [e.g., 17], which here has been found to be close to 30°). It is thus necessary to invoke a local production mechanism for these upstream distributions. For this, we make some theoretical considerations about the nonlinear trapping of ions by a monochromatic electromagnetic wave. From the equation of motion of a particle of velocity **v** in the frame moving along the dc magnetic field $\mathbf{B_0}$ (//z) at the phase velocity V_ϕ (<<c) of a monochromatic wave, propagating along $\mathbf{B_0}$ with a constant amplitude B_1, it is easy to deduce two constants of the motion [e.g., 18,19,20,21]:

$$T = w_{//}^2 + w_\perp^2 = C_1 \qquad (1)$$

$$S = (w_{//}-1)^2 - 2\frac{\Omega_1}{\Omega_0} w_\perp \sin\psi = C_2 \qquad (2)$$

where $\quad \mathbf{w} = \frac{k_{//}}{\Omega_0}\mathbf{v}, \quad \Omega_{0,1} = \frac{qB_{0,1}}{m}, \quad \psi = \varphi + k_{//}z$

and φ is the gyrophase angle. The invariance of T is simply the conservation of total particle energy in the wave frame (no electric field in this frame). The invariant S relates the parallel and perpendicular motion of the particles. Using the particle equations of motion with (4) and (5), it is possible to show that S can be used as a Hamiltonian of the particle motion and that the system is solvable by a quadrature

$$\frac{dw_{//}}{dt} = \pm\frac{\Omega_0}{2}\sqrt{4\left(\frac{\Omega_1}{\Omega_0}\right)^2 \sin^2\psi(C_1 - w_{//}^2) - \left[(w_{//}-1)^2 - C_2\right]^2} \qquad (3)$$

which can be solved in terms of elliptic integrals as for a pendulum equation. The particle is thus trapped in a potential well. Using the pitch-angle α such as $\tan\alpha = w_\perp/w_{//}$, the equations of motion

$$\frac{d\alpha}{dt} = -\delta\cos\psi \quad \text{and} \quad \frac{d\psi}{dt} = \delta\cot\alpha\sin\psi + \sqrt{T}\cos\alpha - 1, \text{ where } \delta = \frac{\Omega_1}{\Omega_0} = \frac{\delta B_\perp}{B_0} \qquad (4)$$

can be derived from the Hamiltonian $S(\alpha,\psi)$.

This Hamiltonian has a singularity for $\psi_0 = \pi/2$ which is the only to be considered since the pitch angle α is defined in the interval [0 π]. As a first step, we consider a mono-energetic parallel ion beam, i.e. this means from equations (2), (4) and (6) that we have T=1. Then, by linearizing the trapping potential, around $\psi_0 = \pi/2$, it is straightforward to show that this singularity corresponds to a value α_0, defining the center of the trapping cells [for small values of α_0, it is possible to use the approximation $\alpha_0 \approx (2\delta)^{1/3}$]. If $V_{//0}$ is the initial velocity of the cyclotron resonant beam (i.e., T=1), the nonlinear interaction will tend to create a peak in the distribution around the center of the trapping cell in phase space associated with the pitch angle α_0.

We have derived the experimental parallel and perpendicular velocities for some observed gyrating ion distributions corresponding to the time intervals where we have analyzed the low frequency waves. Then we have computed the associated pitch-angles in the wave frame (using the experimental wave phase speed). To illustrate this, for the event described above, the experimental pitch-angle is $\alpha_{0exp}=60\pm5°$ while the theoretical value is $\alpha_{0theory}=59.8°$ using the mean value $\delta=0.85$ from the observations. The very good agreement obtained strongly suggests the possible scenario that the quasi-monochromatic wave generated from the ion/ion beam instability could then have non-linearly trapped the ions to produce the resulting gyrating distributions.

5. CONCLUSIONS

The possibility of producing the observed gyrophase-bunched ion distributions from the disruption of the beam by the excited wave has led to a good quantitative agreement from a nonlinear trapping theory which predicts that the pitch-angle of the final gyrating ion distribution is related to the wave amplitude [10]. This result is very similar to those obtained from previous studies in the distant foreshock (up to 80 R_E) from WIND data with lower backstreaming ion densities and wave amplitude [9,10], which could mean that the present case study corresponds to the same mechanism observed by Cluster closer to the bow shock. Other gyrating ion events have been identified in the Cluster data [22]. Most appear consistent with this trapping mechanism while only one event with gyro-phased bunched ions produced by specular reflection at the bow shock surface has also been identified [17].

The analytical test-particle calculations of the nonlinear interaction between a field-aligned beam and the single self-generated cyclotron-resonant wave briefly described here cannot tell anything about the physical description during the wave growth. Numerical simulations are necessary for that. A previous study [13] led to pitch-angle diffusion though the production mechanism of gyrophase-bunched ion distributions is a coherent process and not a diffusive one. It would be therefore strongly necessary to conduct new numerical kinetic simulations to better understand this trapping process. It could help to quantify the life-time of the gyrophase-bunched distribution and compare it with observations.

ACKNOWLEDGMENTS

The authors thank colleagues of the ISSI International Team on the study of the Ion Foreshock for useful discussions and the International Space Science Institute, Bern for its support to this team.

REFERENCES

1. Fuselier, S. A., *Geophysical Monograph 81*, (1994), pp.107-119.
2. Eastman, T.E., R.R. Anderson, L.A. Frank, G.K. Parks, *J. Geophys. Res.*, **86**, 4379-4395 (1981).
3. Gosling, J. T., M. F. Thomsen, S. J. Bame, W. C. Feldman, G. Paschmann, and N. Sckopke, *Geophys. Res. Lett.*, **87**, 1333 (1982).
4. Thomsen, M. F., J. T. Gosling, S. J. Bame, and C. T. Russell, *J. Geophys. Res.*, **90**, 267 (1985).
5. Fuselier, S. A., M. F. Thomsen, J. T. Gosling, S. J. Bame, and C. T. Russell, *J. Geophys. Res.*, **91**, 91–, (1986).
6. Fuselier, S. A., M. F. Thomsen, S. P. Gary, S. J. Bame, C. T. Russell, and G. K. Parks, *Geophys. Res. Lett.*, **13**, 60–63 (1986).
7. Fuselier, S. A., J. T. Gosling, and M. F. Thomsen, *J. Geophys. Res.*, **91**, 4163–4170 (1986).
8. Fazakerley, A.N., A.J. Coates, and M.W. Dunlop, *Adv Space Res*, **15**, (8/9)103-106 (1995).
9. Meziane, K., *et al.*, *Adv. Space Res.*, **20**, n°4, pp. 703-706 (1997).
10. Mazelle, C., D. Le Quéau, and K. Meziane, *Nonlinear Processes in Geophysics,* **7**, 185-190 (2000).
11. Meziane, K., C. Mazelle, R. P. Lin, D. LeQuéau, D. E. Larson, G. K. Parks, and R. P. Lepping, *J. Geophys. Res.*, **106**, 5731–5742 (2001).
12. Gary, S. P., J. T. Gosling, and D. W. Forslund, *J. Geophys. Res.*, **86**, 6691 (1981).
13. Hoshino, M., and T. Terasawa, *J. Geophys. Res.*, **90**, 573 (1985).
14. Gurgiolo, C., G. K. Parks, and B. H. Mauk, *J. Geophys. Res.*, **88**, 9093 (1983).
15. Mazelle, C., K. Meziane, D. Le Quéau, M. Wilber, J.P. Eastwood, *et al.*, *Planet. Sp. Science*, doi:10.11016/j.pss.2003.05.002, **51**, 785-795 (2003).
16. Schwartz, S. J., in *Analysis Methods for Multi-spacecraft Data*, G. Paschmann and W. Daly (Eds.), *ISSI Scientific Report* (1998), pp. 249-270.
17. Meziane, K., C. Mazelle, M. Wilber, D. Le Quéau, J.P. Eastwood, *et al.*, *Annales Geophys.* **22**: 1–11, SRef-ID: 1432-0576/ag/2004-22-1 (2004a).
18. Roux, A., and J. Solomon, *Ann. Geophys.*, **2**, 279-297 (1970).
19. Gendrin, R., *Astrophysics and Space Science*, **28**, 245-266 (1974).
20. Matsumoto, H., *Space Science Reviews* **42**, 429-448 (1985).
21. Le Quéau, D., and A. Roux, *Solar Physics* **111**, 59-80 (1987).
22. Meziane, K., M. Wilber, C. Mazelle, D. Le Quéau, H. Kucharek, *et al.*, *J. Geophys. Res.*, **109**, A05107, doi:10.1029/2003JA010374 (2004b).

The Locations and Shapes of Jupiter's Bow Shock and Magnetopause

Raymond J. Walker[1,2], Steven P. Joy[1,2], Margaret G. Kivelson[1,2], Krishan Khurana[1], Tatsuki Ogino[3], Keiichiro Fukazawa[3]

[1] *Institute of Geophysics and Planetary Physics, University of California, Los Angeles, CA. 90095-1567*

[2] *Department of Earth and Space Science, University of California, Los Angeles, CA 90095-1567*

[3] *Solar Terrestrial Environment Laboratory, Nagoya University, Toyokawa, Aichi, Japan*

Abstract. The shape and location of the Jovian bow shock and magnetopause have been studied by using magnetic field observations and global magnetohydrodynamic (MHD) simulations. MHD simulations in which the interplanetary magnetic field (IMF) was set to zero were used to define the boundary shapes and positions and how they depend on solar wind dynamic pressure. Polynomial fits to the simulated boundaries along with spacecraft observations were used to determine the probability of a given position being outside of the bow shock or inside of the magnetopause. The magnetopause and possibly the bow shock have two preferred locations, one representing a compressed magnetosphere and the other an expanded magnetosphere. Variations in the solar wind parameters near Jupiter also show a bimodal distribution but the changes in the solar wind dynamic pressure are not sufficient to account for the observed bimodal distribution of observed magnetopause positions. Internal pressure changes at Jupiter are required. The interplanetary magnetic field also influences the location and shape of the boundaries. In particular, when the IMF is in the B_y direction or northward magnetopause reconnection acts to reduce polar flattening. Higher internal pressure at dusk leads to a dawn-dusk asymmetry in the magnetopause position with the boundary being farther from Jupiter on the dusk side. For all the simulations the ratio of the bow shock stand-off distance to that of the magnetopause was less than that at the Earth.

INTRODUCTION

Solar wind plasma is heated and diverted around planetary obstacles by bow shocks that form upstream of the planets. Studies of the Earth's bow shock indicate that the solar wind magnetosonic Mach number, the interplanetary magnetic field (IMF) and plasma beta influence both the strength and location of the shock [1]. Many years ago Spreiter and colleagues [2, 3] used gas dynamic calculations to demonstrate that the shape of the Earth's bow shock depends on the shape of the obstacle. At the Earth the magnetospheric obstacle is determined by pressure balance at the magnetopause between the solar wind and the Earth's magnetic field. At Jupiter on the other hand the magnetospheric obstacle is dominated by an azimuthal equatorial current sheet containing a hot plasma sheet with flows that are atmospherically driven toward corotation. As a result at Jupiter the thermal and dynamic plasma pressures also are important at the magnetopause [4]. The presence of the equatorial current sheet stretches dayside magnetic field lines [5] leading to a more sharply pointed magnetopause at Jupiter than at the Earth. This is thought to result in Jupiter's bow shock being relatively much closer to the magnetopause than is the case at the Earth [6, 3]. The Jovian magnetopause and bow shock locations are highly variable [7, 6].

Jupiter provides us with a relatively handy example of a rapidly rotating magnetospheric obstacle that is very different than that at Earth. Seven spacecraft (Pioneer 10 and 11, Voyager 1 and 2, Ulysses, Galileo and Cassini) have provided observations of Jupiter's magnetopause and bow shock. Together they provide us with a substantial database with which to study the Jovian boundaries. Joy et al. [8] used all of the observations available up to that time to develop probabilistic models of the bow shock and magnetopause. The models were based on a combination of this large data base with results from magnetohydrodynamic (MHD) simulations of the effects of different solar wind dynamic pressures on the bow shock and magnetopause shapes and locations. Surprisingly both the bow shock and the magnetopause had bimodal distributions of location with two most probable positions. In this paper we will review both the simulation and data studies used to determine the probabilistic models of the bow shock and magnetopause shapes and locations. We also will use results from additional simulations to evaluate the possible effects of IMF parameters not included in the previous simulations. In section 2 we briefly review the MHD simulation model and in section 3 we examine simulated boundaries as a function of solar wind dynamic pressure. We review the probabilistic model in section 4. Section 5 contains the simulations of the effects of the IMF on the boundaries. Finally in section 6 we summarize our understanding of the overall configuration of the Jovian magnetopause and bow shock.

SIMULATION MODEL

Our simulation model of Jupiter's magnetosphere has been described in [9]. In this section we briefly review the simulation model and discuss the runs used to support this study. Starting from a model of the plasma and field configuration near Jupiter, at time t = 0 we placed an image dipole upstream of Jupiter to hasten the formation of the magnetopause and help assure $\nabla \bullet \mathbf{B} = 0$ throughout the simulation box [10]. We launched an unmagnetized solar wind with a dynamic pressure of $\rho v_{sw}^2 = 0.75$ nPa (v_{sw}=300 km s^{-1}) and a temperature of 2×10^5 K from the upstream boundary of the simulation box and solved the resistive MHD equations as an initial value problem by using the Modified-Leap Frog Method described by [11].

The Jovian magnetosphere was modeled on either a 602 × 402 × 202 point, 602×402×402 point or 452×302×152 point Cartesian grid with grid spacing of 1.5R$_J$ (1R$_J$ = 71,492km). In the simulation the magnetic field (**B**), velocity (**v**), mass density (ρ) and thermal pressure (p) are maintained at solar wind values at the upstream boundary (x

= 300R$_J$ or 225R$_J$) while free boundary conditions through which waves and plasmas can freely leave the system are used at the downstream, side, and top boundaries. Symmetry boundary conditions are used at the equator (z = 0) for the simulations with zero, northward and southward IMF. For the case with an IMF y-component a full three dimensional box was used and free boundary conditions were used at the bottom boundary. The dipole tilt is set to zero in all of the calculations. At the inner magnetosphere boundary all of the simulation parameters (B, v, ρ, p) are fixed for r < 15R$_J$. The simulation quantities are connected with the inner boundary through a smooth transition region (15<r<21R$_J$). The numerical stability criterion is $v_g^{max} \Delta t / \Delta x < 1$ where v_g^{max} is the maximum group velocity in the calculation domain and Δt is the time step. Since the Alfvén velocity becomes very large near Jupiter we placed the inner boundary of the simulation at 15R$_J$ in order to keep the time step from getting too small. The simulation parameters are fixed at the inner boundary. In particular the azimuthal velocity is set to corotate and the pressure and density are set to values determined from the Voyager 1 flyby of Jupiter [12]. This reservoir of plasma provides a source for the Jovian magnetosphere. Typically about 1×10^{30} AMU s^{-1} pass through a surface at 22.5 R$_J$ and enter the Jovian magnetosphere. The source values for each of the simulations are given in Table 2 of [13]. We have not included mass loading terms in the MHD equations for these simulations. The magnetic field is fixed to values from Jupiter's internal dipole.

THE EFFECTS OF SOLAR WIND DYNAMIC PRESSURE ON THE LOCATION AND SHAPE OF THE BOW SHOCK AND MAGNETOPAUSE

In Figure 1 we have plotted the thermal pressure in the noon-midnight meridian plane (left) and dawn-dusk meridian plane (right) for four different solar wind dynamic pressures. The IMF was set to zero for these simulations. The pressures range from 0.045nPa to 0.36nPa and were selected by scaling the mean dynamic pressure at Jupiter's orbit, 0.092nPa [6, 4, 8], by factors of two. Both the bow shock and magnetopause move toward Jupiter with increasing pressure. We have used the sharp pressure gradients at the boundaries to tabulate their positions in Table 1. In both the observations and the simulations the distance to the magnetopause and bow shock vary with pressure as a power law between $P_{Dyn}^{-1/4}$ and $P_{Dyn}^{-1/5}$ [6, 4, 9, 8]. At noon the ratio of the bow shock distance to the magnetopause distance is between 1.31 and 1.24. It decreases with increasing pressure. The magnetopause has a marked dawn-dusk asymmetry in which the boundary is closer to Jupiter at dawn than at dusk. The asymmetry increases (Y$_{dawn}$/Y$_{dusk}$ decreases) with increasing pressure. Within the magnetosphere the equatorial plasma sheet is much thinner at dawn than dusk. This is reflected in an irregularly shaped magnetopause in the X=0 plane (Figure 1, right). However the dawn-dusk asymmetry and irregular shape of the magnetopause are much less evident in the bow shock. The dawn-dusk asymmetry may actually reverse for small dynamic pressure (Table1).

Table 1. Distances to the bow shock and magnetopause for the case $B_z = 0$.

P_{Dyn} (nPa)	Magnetopause					Bow Shock					Standoff Ratio
	X	Y_{avg}	Z	Z/Y_{avg}	Y_{dawn}/Y_{dusk}	X	Y_{avg}	Z	Z/Y_{avg}	Y_{dawn}/Y_{dusk}	X_b/X_m
0.045	90	136	125	0.919	0.985	118	207	204	0.986	1.029	1.311
0.090	76	112	107	0.951	0.940	100	174	173	0.994	1.012	1.316
0.180	67	92	92	1.000	0.906	84	146	145	0.990	1.007	1.254
0.360	58	78	80	1.026	0.867	72	124	123	0.988	0.992	1.241

At Jupiter in addition to the magnetic field, hot internal current sheet plasma and magnetospheric flows contribute to the pressure balance at the equatorial magnetopause. Thus we would expect the magnetopause to be relatively further from Jupiter at the equator than at the poles. This is frequently called polar flattening. We have listed the ratio of the boundary locations along the z-axis to the average value along the y-axis. For low dynamic pressure the simulated magnetopause exhibits polar flattening ($Z/Y_{avg} < 1$). However this effect goes away for higher pressure ($Z/Y_{avg} \sim 1$). Again the bow shock shape is not sensitive to the structure found in the magnetopause.

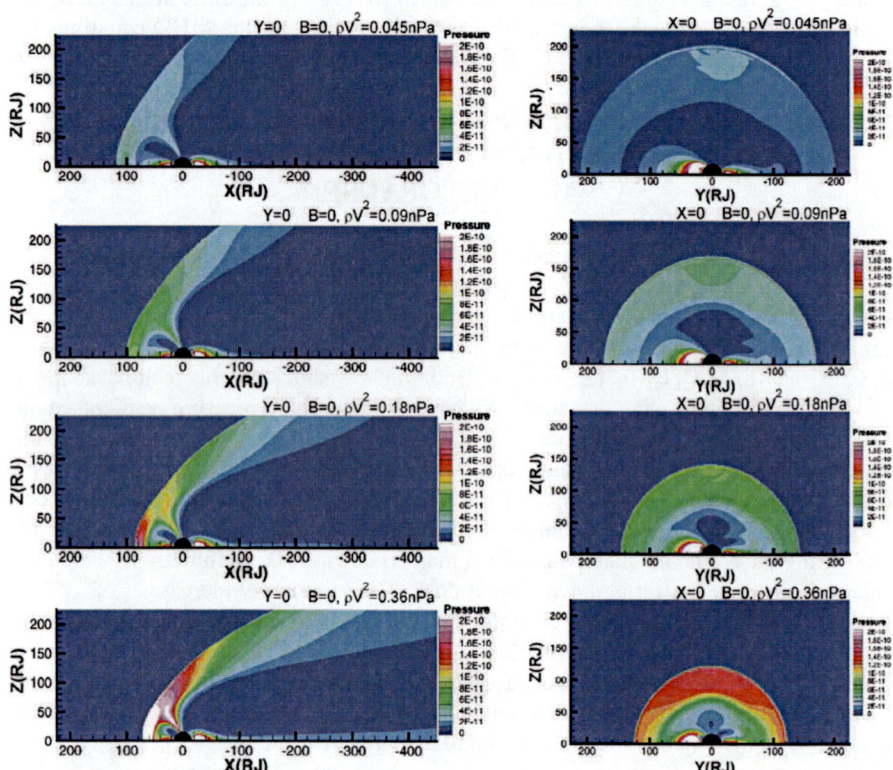

Figure 1. Pressure contours in the noon-midnight meridian plane (left) and dawn-dusk meridian plane (right) for solar wind dynamic pressures of 0.045nPa, 0.09nPa, 0.18nPa, and 0.36nPa. The IMF was set to zero for these simulations.

EMPIRICAL MODELS OF THE BOUNDARY SHAPES AND POSITIONS

There have been several attempts to model Jupiter's bow shock and magnetopause empirically. One approach is to fit actual boundary crossings by assuming that the boundaries are conic sections of revolution [14, 6, 4]. The conic section models are symmetric about the x-axis. In [8] we used the simulation results to argue that observable dawn-dusk asymmetry and polar flattening were probable. We argued even with Galileo orbiter data there are too few boundary crossings to make reliable fits to the data including asymmetries. However, spacecraft collect more information about the boundary locations than just the locations of the actual boundary crossings. For instance if a spacecraft is in the magnetosheath, the bow shock must lie further from the planet and the magnetopause must lie closer to the planet. Therefore we developed probabilistic models of the boundaries that include all of the observations.

In [8] we used the global MHD simulations to organize the boundary model. First we identified the boundary locations in the simulations for different solar wind dynamic pressures. We then fit the boundary shapes to a functional form ($z^2 = A + Bx + Cx^2 + Dy + Ey^2 + Fxy$) that does not assume dawn-dusk or x-axis symmetry. We assumed that the parameters A through F were functions of dynamic pressure (P_{dyn}). A linear function of $P_{dyn}^{-1/4}$ gives an excellent fit to A, B and C while D, E and F are well fit with a linear function of P_{dyn}. We fit this functional form to boundary positions determined from the simulation results plotted in Figure 1. A number of parameters (the magnetic field, the current density, velocity, pressure etc.) can be used to determine the bow shock and magnetopause positions from the simulations. All of these worked well for the bow shock but the gradient of the speed worked slightly better for the magnetopause along the flanks of the magnetosphere. The boundary fits were optimized for the dayside and within $50R_J$ of the equator. The fits should not be used for x<-250R_J [8]. Figure 2 shows the results of the fits to the simulation. The plot contains views in the XY, XZ and YZ planes of both boundaries for three dynamic pressures (0.02nPa, 0.098nPa, 0.227nPa). These represent the 10th, 50th and 90th percentiles of the observed solar wind dynamic pressure at Jupiter's orbit. The plots show the extreme variability in the boundary locations at Jupiter. For instance the magnetopause the standoff distance at the subsolar point varies from over 100R_J to ~50R_J.

Figure 3 contains the trajectories of 35 Galileo orbits and the Pioneer 10 and 11, Voyager 1 and 2, Ulysses and Cassini flybys. We used magnetometer and plasma wave observations to determine the times at which the Galileo satellite crossed the boundaries while for the other spacecraft we used published crossing times [15, 16, 17, 18]. Then we shaded the trajectories according to whether the spacecraft was in the solar wind (blue), the magnetosheath (green) or the outer magnetosphere (red). Trajectories from which data are not available are black. Even with orbiter data there are relatively few boundary crossings. Despite the relatively few boundary crossings, the Galileo orbiter observations significantly improve the probabilistic determination of the boundary locations. However, spacecraft collect more information about the boundaries than just the location of the actual crossings. For instance if a spacecraft is in the magnetosheath, the shock must lie further from the planet and the magnetopause must lie closer. All of the data can be used

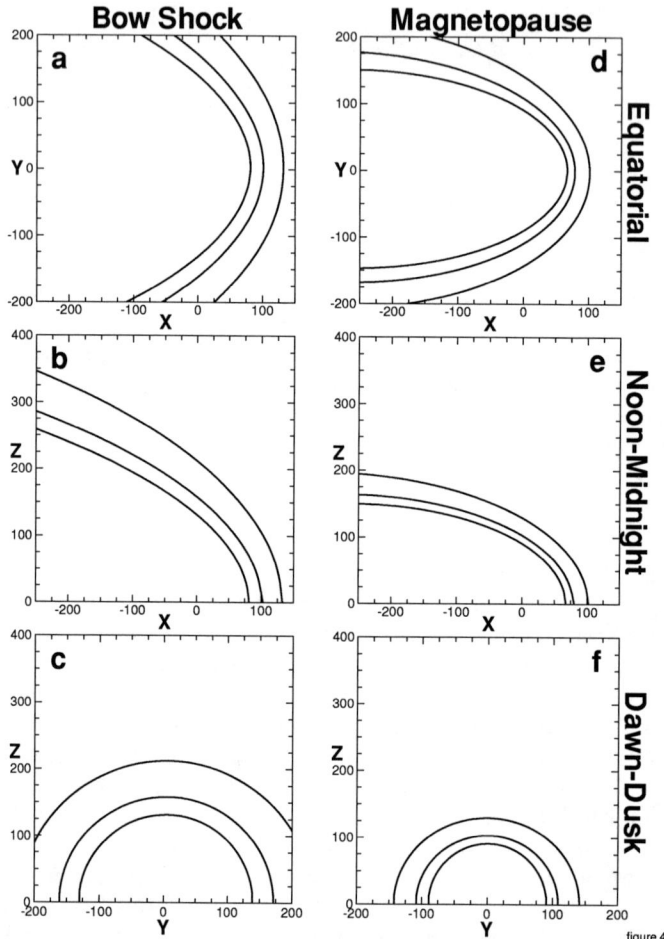

Figure 2. Fits to the bow shock and magnetopause shapes in Figure 1 evaluated at the 10th (outer), 50th(middle) and 90th percentiles (inner) of the observed solar wind dynamic pressure in the three axis planes (rows). The distances are in Jovian radii. [8].

to establish the probability of finding the bow shock or magnetopause at different locations.

Ten minute samples of the data from all of the spacecraft were collected into bins whose shapes were determined by the fits to the simulation results in Figure 2 and whose subsolar standoff distances varied by $4R_J$. The fraction of data inside or outside of a boundary was determined for each bin. In Figure 4 the fraction of observations outside the bow shock is plotted in the left column while the fraction inside the magnetopause is plotted on the right. The error bars mark the actual data points. The error bars that are closest together give the probable error of the mean. The outer error bars were determined by randomly selecting 10 subsets of the data each with 10% of the data and repeating the analysis. The error bars give the spread in the results from these calculations. Surprisingly both the bow shock and magnetopause positions have bimodal distributions with two preferred boundary positions. Fits to a bimodal distribution (sum of two Gaussian distributions) give peaks in the bow shock position at $73R_J$ and $108R_J$ with a standard deviation of $10R_J$ in both cases. The magnetopause positions are $63R_J$ and $92R_J$ with standard deviations of $4R_J$ and $6R_J$ respectively. Single Gaussian fits were used to create the solid lines. An F-test was used to compare the variances of the deviations from the Gaussian distribution to those of the bimodal distribution. For the magnetopause the improvement in the fit of the bimodal distribution over the Gaussian distribution was at the 99.9% confidence level while for the bow shock it was at the 89.8% confidence level. We have plotted the equatorial intercepts of the model boundaries in Figure 3. The dark shading shows the region between the 25 and 75 percentile contours of being in the solar wind. The lightly shaded regions denote one sigma (standard deviation) bands about the two peaks in the magnetopause distribution.

Finally we analyzed all of the solar wind observations (interplanetary magnetic field, dynamic pressure and Alfvén Mach number) near Jupiter [8]. We found a bimodal distribution of solar wind parameters however the magnitude of the bimodal solar wind pressure distribution was too small to account for the bimodal distributions in the boundary positions. Internal pressure changes also are required.

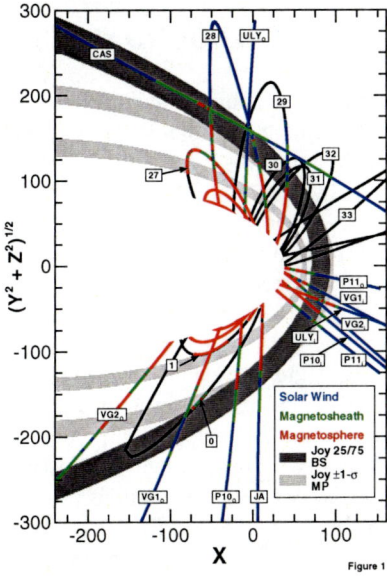

Figure 3. Trajectories of spacecraft near Jupiter shaded to show the region through which the spacecraft were traveling. Trajectory segments in the solar wind are blue, those in the magnetosheath are green and those in the magnetosphere are red. Trajectories for which data were not available at the time of the *Joy et al.* [2002] study are black. The dark gray shading shows the region between the 25 and 75 percentile probabilistic models for the bow shock. The light gray regions on the right show plus or minus one standard deviation about the two preferred locations of the magnetopause. (After [8])

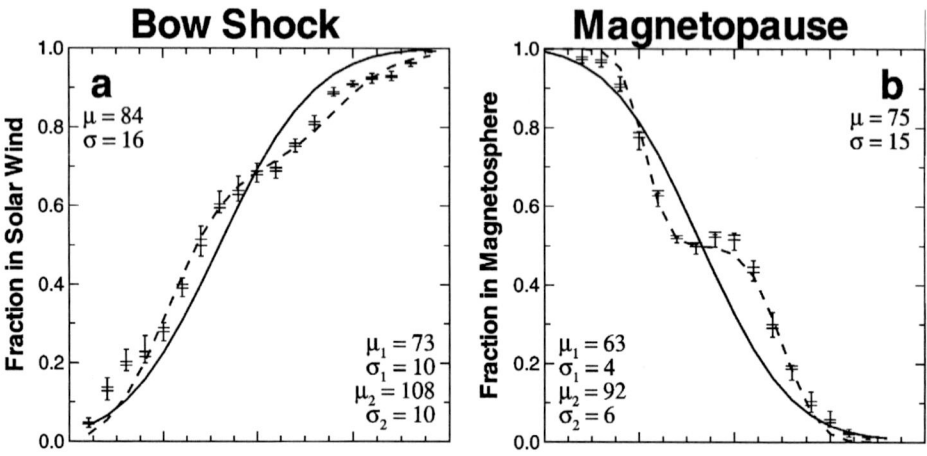

Figure 4. The fraction of observations outside of the bow shock boundary surfaces (left column) and the fraction of observations inside of the magnetopause boundary surfaces (right column). Solid lines are single distribution fits and the dashed lines are bimodal distribution fits. The error bars are discussed in the text. (After[8])

THE EFFECT OF THE IMF ON THE SHAPES AND POSITIONS OF THE BOUNDARIES

By the time the solar wind reaches Jupiter the spiral of the IMF has wound up so tightly that the magnetic field is mainly in the Jovicentric Solar Equatorial (JSEq) y-direction. However, to simplify our investigation of the influence of the IMF on the Jovian boundaries we will start by assuming that the IMF is oriented in the north-south direction. In Figure 5 the thermal pressures for weak southward (B_z=-0.105nT) and northward (B_z= 0.105nT) IMF have been plotted in the Y=0 (top two panels) and X=0 (bottom two panels) planes. The locations of the magnetopause and bow shock are tabulated in Table 2. Recall in the following that Jupiter's intrinsic magnetic field is opposite to that of the Earth. For northward IMF reconnection at the dayside magnetopause moves both it and the bow shock closer to Jupiter than when B_{IMF}=0. Conversely for southward IMF the boundaries move away from Jupiter. However the bow shock is relatively closer to the magnetopause for both southward and northward IMF than for the zero IMF case. For northward IMF the dawn magnetopause is closer to Jupiter than the dusk magnetopause while the opposite is true for southward IMF. In both cases Y_{avg}>Z but it is less so for the northward IMF case. In Figure 5 the high latitude magnetosheath for B_Z>0 has a "hat" like region of increased thermal pressure. A close examination of the magnetic field lines in the "hat" shows that this region is on field lines that have been opened by dayside reconnection.

Since some of the most dramatic changes occurred when the IMF was northward we have carried out a pair of numerical experiments in order to quantify better the effects of the IMF. In the first experiment we set the dynamic pressure to the mean at Jupiter (0.09nPa) and modeled the magnetosphere by assuming that the mean IMF (0.8nT), [8] was entirely northward. In Table 3 we have listed the distances to the boundaries from this experiment and from one with the same dynamic pressure but half the magnetic field.

For the larger IMF the subsolar magnetopause is further eroded while the distance to the northern magnetopause increases dramatically. For the simulation with B_Z=0.84nT the bow shock exits the top of the simulation box just sunward of X=0 so the Z value is a lower limit. The most dramatic change with larger IMF is that Z/Y_{avg}>1 for both the magnetopause and the bow shock. This is a direct result of increased dayside reconnection and the addition of open flux to the magnetosheath.

Table 2. Boundary positions for southward and northward IMF

B_Z(nT)	Magnetopause					Bow Shock					Standoff Ratio
	X	Y_{avg}	Z	Z/Y_{avg}	Y_{dawn}/Y_{dusk}	X	Y_{avg}	Z	Z/Y_{avg}	Y_{dawn}/Y_{dusk}	X_b/X_m
0.105	117	159	149	0.94	0.95	144	231	250	1.08	0.96	1.23
-0.105	130	178	152	0.85	1.0	165	280	245	0.88	1.0	1.27
0	119	170	137	0.81	0.98	155	261	250	0.96	1.0	1.30

Table 3. Boundary locations for constant dynamic pressure and decreasing northward IMF.

B_Z(nT)	Magnetopause					Bow Shock					Standoff Ratio
	X	Y_{avg}	Z	Z/Y_{avg}	Y_{dawn}/Y_{dusk}	X	Y_{avg}	Z	Z/Y_{avg}	Y_{dawn}/Y_{dusk}	X_b/X_m
0.42	90	137	119	0.87	0.91	114	181	191	1.05	0.98	1.31
0.84	87	110	117	1.06	0.92	106	199	>225	>1.13	0.91	1.23

Next we held the northward IMF constant at 0.42nT and decreased the dynamic pressure by a factor of ~4 from 0.09nPa to 0.02nPa. The boundary locations can be found in Table 4. The largest effect of lowering the pressure for constant magnetic field is to increase Z/Y_{avg} at both the magnetopause and bow shock.

Table 4. Boundary locations for constant IMF and decreasing dynamic pressure.

P_{Dyn} (nPa)	Magnetopause					Bow Shock					Standoff Ratio
	X	Y_{avg}	Z	Z/Y_{avg}	Y_{dawn}/Y_{dusk}	X	Y_{avg}	Z	Z/Y_{avg}	Y_{dawn}/Y_{dusk}	X_b/X_m
0.02	84	124	150	1.24	0.90	110	200	245	1.23	0.93	1.31
0.09	90	137	119	0.87	0.91	114	181	191	1.05	0.98	1.31

In Figure 6 we have plotted the pressure in the YZ plane from a simulation for which the IMF was in the Y-direction (B_y=0.42nT) pointing toward dusk. The solar wind dynamic pressure was 0.09nPa. The corresponding fits to the B_{IMF}=0 bow shock and magnetopause positions are shown with solid and dashed lines respectively [8]. The entire magnetosphere rotates about the sun-Jupiter line for $B_y \neq 0$. At high latitudes the boundaries are farther from Jupiter than for B_{IMF}=0 while nearer the equator they are closer to Jupiter. Reconnection can occur near the equator on the flanks of the magnetopause when the IMF points in the y-direction. This can change the shape of the obstacle. The addition of IMF B_y does not change the standoff distance at the bow shock or the magnetopause. The standoff ratio remains 1.31.

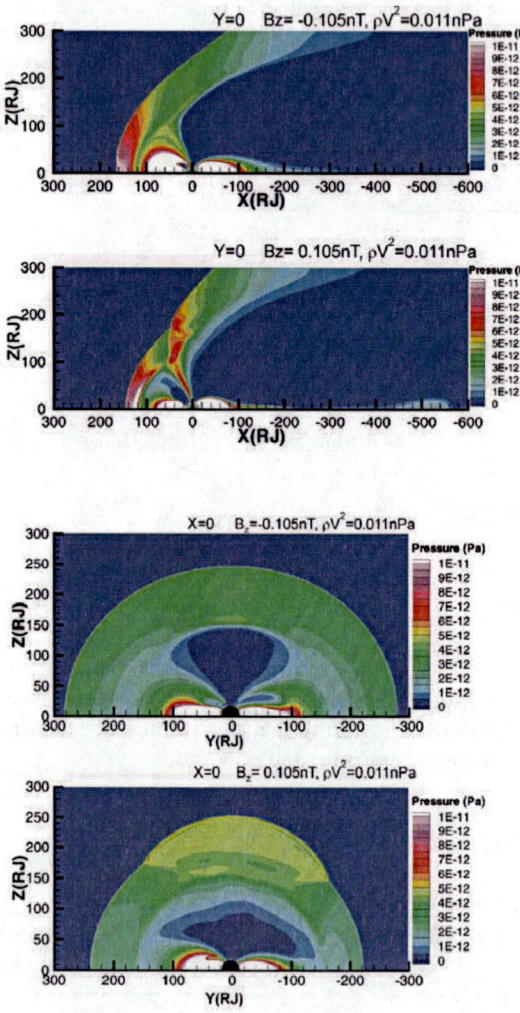

Figure 5. Pressure contours in the noon-midnight meridian plane (top two plots) and the dawn-dusk meridian plane (bottom two plots) from simulations with the northward and southward IMFs of 0.105nT and dynamic pressure of 0.011nPa.

SUMMARY AND DISCUSSION

We have used a combination of global magnetohydrodynamic simulations and observations to form models of Jupiter's magnetopause and bow shock. Rather than fitting observed boundary crossings, we used all of the spacecraft observations at Jupiter to determine the boundary positions in terms of the probability of being outside of the bow shock or inside of the magnetopause. We used the global MHD models to define boundary models, to define the boundary shapes and locations and to determine how they vary with solar wind dynamic pressure.

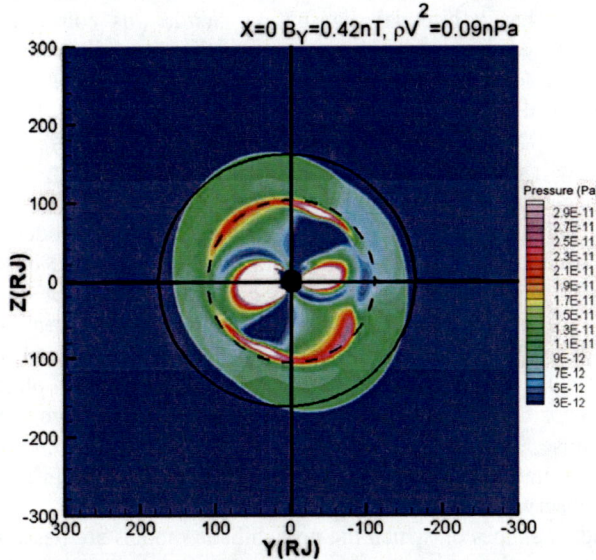

Figure 6. Pressure contours in the dawn-dusk meridian plane for a simulation with a 0.42nT IMF in the y-direction and solar wind dynamic pressure of 0.09nPa. The solid line gives the bow shock position and the dashed line gives the magnetopause position from the fit to the MHD simulations for 0.09nPa and zero IMF.

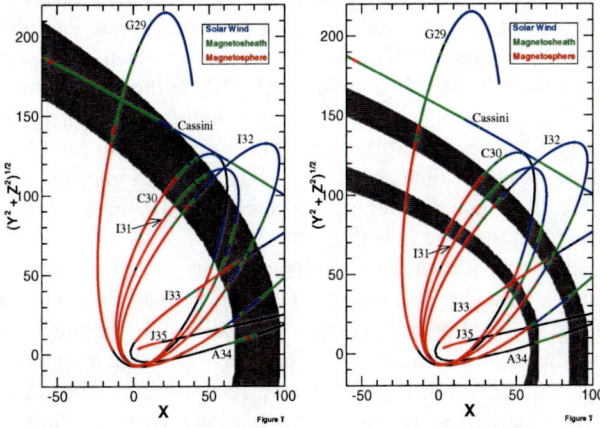

Figure 7. Trajectories of spacecraft near Jupiter not included in the Joy et al. [8] study. The trajectories have been labeled to indicate the region in which the spacecraft was flying. Trajectory segments in the solar wind are blue, those in the magnetosheath are green and those in the magnetosphere are red. There were no data along the black trajectory segments. The gray shading on the left shows the region between the 25 and 75 percentile probabilistic models for the bow shock. The gray regions on the right show plus or minus one standard deviation about the two preferred locations of the magnetopause. Distances are in Jovian radii.

The magnetopause at Jupiter and possibly the bow shock have two preferred locations, one representing a compressed magnetosphere and the other an expanded magnetosphere. The solar wind dynamic pressure in the neighborhood of Jupiter during the time interval under study also has a bimodal distribution. While this contributes to the bimodal distribution observed in the boundary positions the dynamic pressure changes are too small to account for the large variation in the magnetopause position [8]. Internal pressure changes also are required. The bimodal distribution is less clear for the bow shock. The speed with which the bow shock can adjust to changes in either the solar wind or the obstacle will smear out the observed distribution.

In the left column of Figure 7 we have shaded the region between the 25 and 75 percentiles of being in the solar wind from the probabilistic model along with Galileo observations from orbits not included in the original study and observations from the Cassini flyby of Jupiter. Most of the magnetosheath observations are from the region between the 25% and 75% curves. However, some of the observations especially those from the Galileo G29 orbit suggest that the bow shock may extend farther from Jupiter than in the MHD simulations. In the right panel we have plotted two bands of magnetopause positions. The shaded areas are centered on the two preferred locations of the magnetopause. The shading extends plus or minus one sigma (standard deviation) about the two preferred locations. Here the models seem to be in reasonable agreement with the new observations.

The boundary shapes on which the probabilistic models are based did not include the IMF. We have examined simulations with both a purely north-south IMF and with an IMF in the y-direction. Although the dynamic pressure has the largest effect the inclusion of a non-zero IMF can make smaller changes in the location of the boundaries. For instance for northward IMF dayside reconnection erodes the position of the magnetopause and as a result the bow shock moves toward Jupiter as well. For southward IMF the reconnection site moves to high latitudes and the boundaries move away from Jupiter. However for all of the simulations the ratio of the standoff distances remains significantly less than the typical values at the Earth. Both the solar wind Mach number and the shape of the obstacle can influence the standoff distance. We compared simulations of the Earth and Jupiter at the same Mach number and found that the ratio at Jupiter (1.23-1.31) was smaller than at Earth (1.4-1.5). This indicates that the obstacle shape is responsible for the differences in the standoff ratio and that Jupiter's magnetopause is less blunt than the Earth's.

We would expect Jupiter's boundaries to have strong polar flattening because of the equatorial current sheet. However, the simulations suggest that this is not always the case. Clear polar flattening is evident in zero IMF simulations for below average dynamic pressure and when the IMF is southward. For above average dynamic pressure the flattening decreases. For northward IMF dayside reconnection adds flux to the lobes thereby moving the boundaries in the z-direction away from Jupiter reducing or eliminating the polar flattening. At Jupiter's orbit the IMF is primarily in the y-direction. For an IMF in the y-direction we also find that the polar flattening is reduced.

In the simulations the magnetopause is generally found closer to Jupiter at dawn than at dusk. This effect becomes smaller for smaller dynamic pressure when the magnetopause is farthest from Jupiter. The dawn-dusk asymmetry seems to be related to higher thermal pressure on the dusk side of the magnetosphere (Figures 1, 5 and 6).

The bow shock shape and position also are influenced by the IMF. The largest effect seems to be related to changes in the obstacle shape caused by reconnection. This can be seen most dramatically in Figures 5 and 6 and Table 3. The addition of open magnetic flux to the tail lobes and magnetosheath caused the "hat" in Figure 5 and the large increase in the z position of the shock in Table 3. Similarly the inclusion of an IMF B_y caused the magnetopause to twist (Figure 6) and that resulted in a twisted bow shock.

For given solar wind dynamic pressure and IMF conditions we have analyzed the bow shock and magnetopause locations and shapes by assuming steady-state conditions. In particular we ran the simulations until quasi-steady configurations resulted. However, the Jovian magnetosphere in the simulations can be very dynamic [19]. Changes in the dynamic pressure and IMF lead to large amplitude waves which distort the boundary shapes as the system responds to the changes.

Acknowledgements: We would like to thank Mr. Joseph Mafi for help with the data processing and display. Helpful comments by Dr. Lee Bargatze are gratefully acknowledged. The work at UCLA was supported by grant NAG5-12769. The work at Nagoya University was supported by grants in aid from the Ministry of Education, Science and Culture. Computing support was provided by the Computer Center of Nagoya University.

REFERENCES

1. Farris, M. H., and C. T. Russell, Determining the standoff distance of the bow shock: Mach number dependence and use of models, J. Geophys. Res., 99, 17,681, 1994.
2. Spreiter, J. R., A. L. Summers, and A. Y. Alksne, Hydromagnetic flow around the magnetosphere, Planet. Space Sci. 14, 223, 1966.
3. Stahara, S. S., R. R. Rachiele, J. R. Spreiter, and J. A. Slavin, A three dimensional gas dynamic model for the solar wind past non-axisymmetric magnetospheres: Application to Jupiter and Saturn, J. Geophys. Res., 94(A10), 13,353, 1989.
4. Huddleston, D. E., C. T. Russell, M. G. Kivelson, K. K. Khurana, and L. Bennett, Location and shape of the Jovian magnetopause and bow shock, J. Geophys. Res., 103(E9), 20,075, 1998.
5. Engle, I. M., and D. B. Beard, Idealized Jovian magnetopause shape and field, J. Geophys. Res, 85(A2), 579, 1980.
6. Slavin, J. A., E. J. Smith, J. R. Spreiter, and S. S. Stahara, Solar wind flow about the outer planets: Gas dynamic modeling of the Jupiter and Saturn bow shocks, J. Geophys. Res., 90(A7), 6275, 1985.
7. Smith, E. J., R. W. Fillius, and J. H. Wolfe, Compression of Jupiter's magnetosphere by the solar wind, J. Geophys. Res., 83, 4733, 1978.
8. Joy, S. P., M. G. Kivelson, R. J. Walker, K. K. Khurana, C. T. Russell and T. Ogino, Probabilistic models of the Jovian magnetopause and bow shock locations, J. Geophys. Res., 107(A10), 1309, doi 10.10290/2001JA009146, 2002.
9. Ogino, T., R. J. Walker, and M. G. Kivelson, A global magnetohydrodynamic simulation of the Jovian magnetosphere, J. Geophys. Res., 103, 225, 1998.

10. Watanabe, K., and T. Sato, Global simulation of the solar wind-magnetosphere interaction: The importance of its numerical validity, J. Geophys. Res., 95, 75, 1990.
11. Ogino, T., R. J. Walker, and M. Ashour-Abdalla, A global magnetohydrodynamic simulation of the magnetosphere when the interplanetary magnetic field is northward, IEEE Trans. Plasma Sci., 20, 6817, 1992.
12. Belcher, J. W., The low-energy plasma in the Jovian magnetosphere, Physics of the Jovian magnetosphere, A. J. Dessler, editor, Cambridge University Press, New York, p. 68, 1983.
13. Walker, R. J., and T. Ogino, A simulation study of currents in the Jovian magnetosphere, Planet. Space Sci., 51, 295, 2003.
14. Lepping, R. P., L. F. Burlaga, and L. W. Klein, Jupiter's magnetopause, bow shock, and 10-hour modulated magnetosheath: Voyagers 1 and 2, Geophys. Res., Lett., 8(1), 99, 1981a.
15. Intriligator, D. S., and J. H. Wolfe, Results of the plasma analyzer experiment on Pioneers 10 and 11, in Jupiter, edited by T. Gehrels, pp. 848-869, Univ. of Ariz. Press, Tucson, 1976.
16. Bame, S. J., B. L. Barraclough, W. C. Feldman, G. R. Gisler, J. T. Gosling, D. J. McComas, J. L. Phillips, M. F. Thomsen. B. E. Goldstein, and M. Neugebauer, Jupiter's magnetosphere: Plasma description from the Ulysses flyby, Science, 257, 1539, 1992.
17. Lepping, R. P., M. J. Silverstein, and N. F. Ness, Magnetic field measurements at Jupiter by Voyagers 1 and 2: Daily plots of 48 second averages, NASA Tech., 83, 864, 1981b.
18. Achilleos, N., M. K. Dougherty, D. T. Young, and F. Crary, Magnetic signatures of Jupiter's bow shock during the Cassini flyby, J. Geophys. Res., 109, A09s04, doi:10.1029/2003JA010258, 2004.
19. Fukazawa, K., T. Ogino, and R. J. Walker, Dynamics of the Jovian magnetosphere for northward interplanetary magnetic field (IMF), Geophys. Res. Lett., 32, 3, L03202, doi:10.1029/2004GL021392, 2002.

Bow Shock and Upstream Waves at Jupiter and Saturn: Cassini Magnetometer Observations

C. Bertucci[1], N. Achilleos[1], C.T. Russell[2], M.K. Dougherty[1], E.J. Smith[3], M. Burton[3], B.T. Tsurutani[3] and C. Mazelle[4].

[1] *Space and Atmospheric Physics Group, Imperial College London, UK.* [2] *IGPP / UCLA, Los Angeles, CA.* [3] *Jet Propulsion Laboratory, Pasadena, CA.* [4] *CESR CNRS/UPS, Toulouse, France.*

Abstract. The measurements obtained by the Cassini magnetometer investigation (MAG) during the first orbits of its mission around Saturn yielded a rich set of observations of the bow shock and the upstream region at that planet. In this work we comment on the properties of the Kronian bow shock as implied by MAG data. We also study the properties of non-linear, low frequency waves observed in the foreshock of Saturn and we discuss their origin and their role in the formation of the quasi parallel shock. Finally, we compare these recent results with similar observations at Jupiter obtained by Cassini MAG during the 2000-2001 Jupiter flyby.

Keywords: Planetary Bow Shocks, Foreshock, Nonlinear Plasma Waves, Jupiter, Saturn, Cassini.

PACS: Planets, interaction with solar wind, 96.50.Ek, Plasma waves, 52.35.–g, Shock waves, interplanetary, 96.50.Fm, Jupiter, 96.30.Kf, Saturn, 96.30.Mh.

1. INTRODUCTION

It is well known that the presence of non-absorbing obstacles in the supersonic and superalfvenic solar wind generates a collisionless bow shock. This shock wave will slow, heat, and deflect the incoming flow around the planetary obstacle. In the case of Jupiter and Saturn, the properties of the upstream plasma give supercritical bow shocks as the magnetosonic Mach numbers are expected to be very high. In the vicinity of these planets occurrence of upstream waves is mainly concentrated in the foreshock, i.e., the sector where the interplanetary magnetic field intersects the shock surface (e.g. Greenstadt et al., 1995). In this region, the presence of charged backstreaming particles resulting from either the reflection at supercritical bow shocks (Gosling and Robson, 1985) or from leakage from the magnetosheath makes the upstream plasma unstable to the growth of a wide variety of electromagnetic oscillations. In particular, the implantation of ions produces ion-resonant low frequency electromagnetic waves. The study of the properties of these upstream waves is essential to understand very important processes such as the formation of quasi parallel shocks (e.g., Thomsen et al., 1998), solar-wind particle reflection, particle leakage, their contribution to the turbulence in the magnetosheath, etc. Due to the atypical plasma conditions upstream from outer planets such as Jupiter and Saturn processes like these deserve more study.

The observations made by the Cassini spacecraft in the last 5 years around Jupiter and more recently at Saturn have greatly contributed to the study of the bow shock and upstream phenomena in environments with the highest mach numbers accessible to us in the solar system. In this work, we present observations by the Cassini magnetometer on the physical properties of the bow shocks and upstream low frequency waves in the plasma environments of Jupiter and Saturn. Then, we discuss these observations in order to gain new insights into the similarities and differences with the Earth.

2. OBSERVATIONS

The dual technique magnetometer experiment (MAG) onboard Cassini is extensively described in Dougherty et al. (2004). It consists of a vector helium magnetometer (VHM) and a fluxgate magnetometer (FGM) mounted respectively at the end and halfway along the 11-meter spacecraft boom. The FGM provides fast vector measurements (up to 32 vectors/s), whereas the VHM provides measurements with a resolution of 2 vectors per second.

2.1. The Jovian Bow Shock

The Cassini spacecraft visited the Jovian system from November 2000 to February 2001 on its way to Saturn. The geometry of the flyby was such that Cassini grazed the Jovian bow shock on the dusk-side, leading to more than 40 crossings for solar zenith angles (SZA, the angle between the spacecraft's position vector and the planet-sun line) greater than 47° and distances over 141 Jovian radii (Jovian radius = 71,492 km). This permitted us to constrain the shape of the Jovian shock deduced from Voyager and Pioneer observations (Slavin et al., 1985; Achilleos et al., 2004).

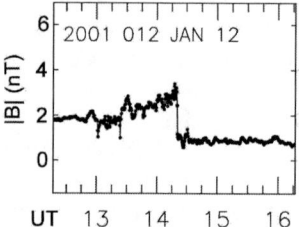 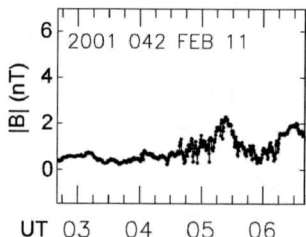

FIGURE 1. Examples of quasi-perpendicular and quasi-parallel bow shock crossings at Jupiter (Achilleos et al., 2004).

During the encounter, MAG and other Cassini plasma instruments provided information on the structure of the Jovian bow shock. In agreement with the expected direction of the Parker spiral at Jupiter's orbit, Achilleos et al., (2004) showed that the angle between the shock normal vector from coplanarity theorem and the interplanetary magnetic field decreases with increasing SZA. Figure 1 illustrates two examples of bow shock crossings as recorded by Cassini MAG. The left panel shows a highly structured quasi-perpendicular shock close to 18h LT (left) with a noticeable overshoot and a well-developed foot indicative of a supercritical shock. This is compatible with the high magnetosonic mach numbers expected for Jupiter (Russell et al., 1982). The left panel shows an example of a quasi-parallel shock in the far downstream region. It can be noted that the ramp is not clearly resolved because of the presence of a conspicuous wave activity.

2.2. Upstream Waves at Jupiter

The Jupiter flyby geometry also permitted the exploration of the upstream space close to the Jovian bow shock. In this region, low frequency nonlinear waves were observed. Figure 2 shows the magnetic field measured by Cassini MAG upstream from the Jovian bow shock on December 28, 2000, between 00:29:40 and 00:40:20 UT. The left panel shows one period of these waves in the magnetic field magnitude. The shape of the wave resembles a saw tooth: an elongated sinusoidal profile with a sharp edge on the left (around 00:35). The period of these oscillations in the spacecraft frame is of the order of several minutes (~ 6 minutes in a 0.3 nT background field) and varies with the background magnetic field intensity. Examples of these oscillations can be found during most of the flyby phase, either as isolated events or in wave trains. In all these examples, the direction of the background magnetic field intersects the nominal position of the shock, suggesting that these waves are located in the foreshock.

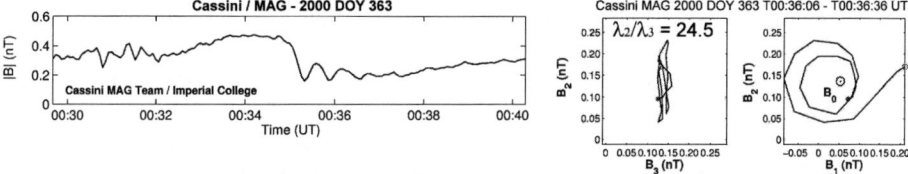

FIGURE 2. Left: Nonlinear steepened waves at Jupiter with dispersive wave packet. Right: Hodogram with polarization of a dispersive wave packet (Bertucci et al., 2005a).

We applied minimum variance analysis (MVA) on MAG data in numerous intervals where the waves were observed. This technique provides an estimate of the direction of propagation **k** for a plane wave by calculating the eigenvectors of the covariance matrix of the magnetic field within a given interval. The direction of propagation is associated with the eigenvector that corresponds to the minimum eigenvalue λ_3 (the maximum and intermediate eigenvalues are respectively λ_1 and λ_2). For these waves, the ratio λ_2/λ_3 is systematically smaller than λ_1/λ_2, indicating that the polarization of these steepened waves is either highly elliptical or linear, and that they propagate rather perpendicularly to the background magnetic field $\mathbf{B}_0$. As a result, these oscillations occur typically along the direction of the mean magnetic field, confirming their compressive character. Moreover, the oscillations are almost restricted to the $\mathbf{k}\times(\mathbf{k}\times\mathbf{B}_0)$ direction in coincidence with magnetosonic waves, whereas the so-called helical component (i.e., along $\mathbf{k}\times\mathbf{B}_0$) is systematically small.

MAG measurements also reveal the presence of quasi monochromatic wave packets at a higher-frequency and attached to the steepening front. In the left panel of Figure 2, such a packet extends from 00:35 to approximately 00:37 UT. The packets are evident in the magnitude as well as in each component of the field. The period of the oscillations within the wave packet is of the order of few seconds and their amplitude decreases with increasing distance from the steepening front. In coincidence with the decrease in the amplitude, a comparison of the period of consecutive oscillations shows a decrease in the periods of the oscillations (in this case, from 37.6 to 28.3 s) with increasing distance from the steepened front. This reveals the presence of dispersive effects. The magnetic structure of the packets can be also studied by applying MVA. The right panel in Figure 2 contains two hodograms revealing very plane waves (λ_3/λ_2 = 24.5) with a left-hand polarization with respect to the mean magnetic field in the spacecraft frame (o indicates the beginning of the hodogram and * indicates the end).

2.3 The Saturnian Bow Shock

Cassini entered into orbit around Saturn on July 1, 2004. During the approach and the Saturn orbit insertion phase (SOI), Cassini explored the Kronian bow shock on the dawn side and its surroundings. Figure 3 shows an example of the multiple bow shock crossings by Cassini during SOI as seen by MAG. The magnetic field magnitude measurements reveal a highly structured supercritical quasi-perpendicular shock with a well defined foot, ram and a large overshoot (8 to 10 times the upstream field). Preliminary analyses of the magnetic field and plasma data yield M_{MS} ~ 13 and $\beta \geq 2$ in agreement with the highly fluctuating magnetic field profile observed by MAG. Recently, Achilleos et al., (2005) reported strong bow shock compressions/expansions during SOI (up to ~40% change in the stand-off distance) and characterized the fine structure of the shock for first crossings in order to estimate its velocity. As a result, they reported velocities of the same order of those at the Earth bow shock and ram sizes of the order of a few tenths of upstream ion inertial lengths.

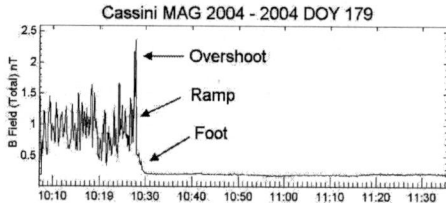

FIGURE 3. Example of a bow shock crossing.

2.4. Upstream Waves at Saturn

The first observations of upstream waves at Saturn take place during SOI (Bertucci et al., 2005b). Figure 4 shows MAG measurements from 09:25 to 09:38 UT on July 13 2004 at 81.6 R_S distance (1 R_S = Saturn radius = 60,330 km), 5h 41m AM local time. On the figure, two compressive oscillations are evident between 09:33:09 and 09:34:17 UT. These oscillations have also a saw-tooth shape with the steepening front to the right. In the spacecraft frame, the period of these oscillations for this example is 270 s for a background magnetic field magnitude of 0.57 nT. As in the case of Jupiter these steepened waves present higher frequency wave packets attached to the steepening front. Similar waves are also observed in the successive orbits of Cassini, sometimes within long (~ 5 hours) wave trains. Also as in the previous case, these steepened waves are followed by a packet of higher frequency waves attached to the right.

As in the case of the upstream waves observed at Jupiter, The steepened waves show a polarization which ranges from highly elliptical (marginally right-hand and left-hand polarized) to linear. Similarly, the higher frequency wave packets are characterized by two features: first, the amplitude of the oscillations decreases with increasing distance from the steepening front; second, a decrease with increasing distance from the steepening front. The right panel on Figure 4 shows the results of the MVA for the interval 14:20:46 – 14:21:44 UT on November 15, 2004. The hodograms reveal the plane character of these waves (λ_2/λ_3 = 79.9) and show a left-hand polarization with respect to the mean magnetic field. The angle between the direction of propagation and the mean magnetic field is θ_{kB} = 12.2° ± 0.8°.

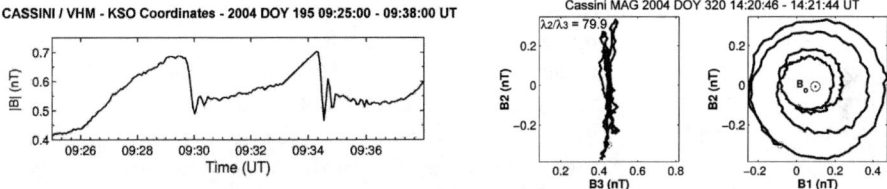

FIGURE 4. Left: Nonlinear steepened waves at Saturn with dispersive wave packet. Right: Hodogram with polarization of a dispersive wave packet (Bertucci et al., 2005b).

Interestingly, the magnetic field orientation at the beginning of the low frequency wave is recovered after the high frequency wave train, as if the high frequency packet helps to reorient the magnetic field distorted by the low frequency wave. Once again, these waves are observed when Cassini is in the nominal foreshock of Saturn.

3. DISCUSSION AND CONCLUSIONS

The observations obtained by Cassini MAG at the bow shock of Jupiter and Saturn have contributed to the comparative study of these boundaries at the outer planets and the Earth's bow shock. First, Cassini could be able to study the properties of the Jovian bow shock for large SZA. The magnetic field profiles revealed that the quasi perpendicular crossings become less frequent as SZA increases, in agreement with the direction of the interplanetary magnetic field deduced from the Parker spiral model. For quasi-perpendicular crossings, the supercritical shock features (foot ramp and overshoot) tend to less evident as SZA increases, since the normal component of the solar wind flow becomes smaller. Unfortunately, the availability of plasma measurements for most of the bow shock crossings has precluded further analyses on its physical size and structure. On the other hand, the Cassini observations at Saturn revealed the presence of a highly dynamic bow shock with quasi-perpendicular crossings showing large overshoots and a well-defined shock foot. These structures confirm the supercritical character deduced from plasma data. In addition, the crossings display a higher degree of turbulence with respect to the Jovian shock. This is perhaps due to the higher β values for Saturn. In the upcoming years, Cassini will continue to explore the Saturnian shock at different local times providing a clearer global description.

In addition, low-frequency (T ~ several minutes) nonlinear plasma waves were observed in the foreshocks of these two planets. In spite of the differences in scale, these waves share most of their characteristics: a) phase steepened shape; b) highly elliptical / linear polarization; c) perpendicular propagation. These waves are accompanied by higher frequency (T ~ tens of seconds) dispersive wave packets with the following characteristics: a) decrease of amplitude and period with increasing distance from steepening front; b) left-hand circular polarization in the spacecraft frame. Comparisons with the measurements by the Cassini RPWS instrument show that in the case of the low frequency steepened oscillations and the higher frequency wave packets, the magnetic field magnitude is systematically correlated to the variations in the electron density (Bertucci et al., 2005b).

Several theoretical works have investigated the origin and growth of nonlinear steepened waves. As an example, Omidi and Winske (1990) simulated the interaction between the solar wind (protons) and a field-aligned beam of protons representing a population of specularly reflected protons at the bow shock as the one we can find in planetary foreshocks. As a result, the ion/ion right-hand instability gives rise to fast magnetosonic waves propagating in the direction of the beam (sunward) at velocities smaller than the solar wind. As the amplitude increases beyond the limit of linear theory, the waves steepen and develop a precursor whistler attached to the steepening front. This suggests that these waves could be originated by field aligned ion beams in the foreshocks of Jupiter and Saturn, however, the origin of these ions remains to be determined.

In the linear regime, the wave's phase velocity V_{ph} only depends on the frequency and the wave number ($V_{ph} = \omega/k$) and is independent from the phase. However, when the amplitude is important V_{ph} becomes dependent on the amplitude. In that case, V_{ph} will be maximum at the crest of the wave, whereas it will be minimum at the troughs. As a result, the maxima will progressively catch up each of their preceding minima, producing the steepening of the wave in the direction of the propagation. This alteration in the shape of the wave will cause a change in the initial circular polarization since the steepening "stretches" the shape of the wave. As a result, the polarization of the steepened wave will be quasi-linear with a remnant rotation of the wave magnetic field near the front. According to MHD theory (Tsurutani et al., 1987), the steepening rate is controlled by a variation in density, and therefore there will be a rapid steepening for high amplitude compressive waves as the ones described in this work. The steepening will continue until the dispersion and/or dissipation effects can be strong enough to stop this process (Lembege, 1990).

The fact that the higher frequency packets appear most of the time as attached to the steepened front suggests that their group velocity is similar. Therefore, this velocity will be also smaller than the solar wind velocity and the polarization as perceived in the spacecraft frame will be the opposite to that in the solar wind frame. As the polarization in the spacecraft frame is left-handed, the polarization in the solar wind frame will be right handed in agreement with the whistler mode for frequencies above the proton cyclotron frequency. As the dispersion relation for this mode is $\omega \propto k^2$, V_{ph} will be proportional to $\omega^{1/2}$ and therefore higher frequencies will travel faster from their source than the lower frequencies. The nature of the decrease in the amplitude is explained in Tsurutani et al., (1989). In this case, the appearance of a dispersive whistler shows that the system tries to neutralize the steepening using dispersion rather than dissipation.

Waves with similar properties have been found in different planetary foreshocks (e.g., Russell et al., 1971; Tsurutani et al, 1993), These steepened waves with precursor whistlers were called "shocklets" because of the similarity with subcritical fast-shock magnetic field profiles.

These waves are usually associated to diffuse ion distributions. This should be also the case at least for Jupiter, since these waves coincide with bursts in the fluxes of ions between 36 and 56 KeV measured by the Cassini MIMI/LEMMS instrument (Bertucci et al., 2005a).

ACKNOWLEDGMENTS

The authors thank the members of the Cassini CAPS, MIMI and RPWS Teams who contributed to the discussion of the results presented in this paper.

REFERENCES

1. Greenstadt, E.W. "Oblique, parallel, and quasi-parallel morphology of collisionless shocks", in *Collisionless shocks in the heliosphere: Reviews of current research*, edited by B.T. Tsurutani and R.G. Stone, Washington, DC, American Geophysical Union, 1985, p. 169-184.
2. Gosling, J.T. and Robson, A.E., "Ion reflection, gyration, and dissipation at supercritical shocks", in *Collisionless shocks in the heliosphere: Reviews of current research*, edited by B.T. Tsurutani and R.G. Stone, Washington, DC, American Geophysical Union, 1985, p. 141-152.
3. Thomsen M., et al., *Adv. Space Res.*, **8**, Issues 9-10 , 1988, Pages 175-178.1998
4. Dougherty, M. K., et al., *Space Sci. Rev.*, **114**, Issues 1-4, 2004, 331-383.
5. Slavin J.A., et al., *J. Geophys. Res.*, **90**, 6275, 1985.
6. Achilleos, N., et al., *J. Geophys. Res.*, 109, A9, 2004, CiteID A09S04.
7. Russell, C.T., et al., *Nature*, **296**, Mar. 4, 1982, pp. 45-48.
8. Bertucci C., et al., 2005a, in preparation.
9. Achilleos N.A., et al., 2005, submitted to *J. Geophys. Res.*
10. Bertucci C., et al., 2005b, submitted to *Geophys. Res. Lett.*
11. Omidi N., and Winske, D., *J. Geophys. Res.*, **95**, A3, 1990, p. 2281.
12. Tsurutani, B.T., et al., J. Geophys. Res., **92**, 1987, p. 11074-11082.
13. Lembège, B., "Numerical Simulations of Collisionless Shocks", in *Physical Processes in Hot Cosmic Plasmas*, edited by W. Brinkmann et al., Netherlands: Kluwer, 1990, 81-139.
14. Tsurutani, B.T., et al., *Planet. Space Sci*, **37**, 1989, p. 167-182.
15. Russell, C.T., et al., *J. Geophys. Res.*, **76**(4), 1971, p. 845.
16. Tsurutani, B.T., et al, J. Geophys. Res., **98**, A12, 1993, p. 21,203-21,216.

A review of field-aligned beams observed upstream of the bow shock

Karim Meziane*, M. Wilber†, C. Mazelle**, G.K. Parks† and A.M. Hamza*

*University of New Brunswick, Fredericton
†Space Sciences Laboratory, U. California, Berkeley
**Centre d'Etude Spatiale des Rayonnements, Toulouse

Abstract. For more than two decades the Earth's bow shock and traveling interplanetary shocks have attracted much attention as researchers have attempted to understand the collisionless mechanisms that thermalize transmitted particles and accelerate those that are observed propagating away from the shock into the upstream. We are concerned here with the class of particles emerging from the shock that are field-aligned and have energies of a few to several keV, and base our results on observations primarily from the Earth's foreshock. While the basic empirical picture has been known for some time, fundamental questions about the underlying mechanisms producing them have resisted a comprehensive explanation. This review talk will begin with an overview of the observational framework, along with selected new results. The latter include recent refinements in the characterizations of upstream field-aligned beams as a function of the shock geometry parameter θ_{Bn}. Other observations from the Cluster spacecraft have shown the occurence of a very sharp boundary separating FABs and gyrating ion populations in the foreshock. The Wind spacecraft has seen FABs at distances in excess of $\sim 100\ R_E$ from the Earth, indicating lifetimes greater than expected from linear theory of the ion-ion streaming instability. These observations prompt new questions. Some analytic calculations will be reviewed briefly. Models based upon the guiding center approximation and those which introduce diffusion as a means of enhancing the fluxes of upstream beams fail to produce the properties observed.

Keywords: foreshock, ion, acceleration, quasiperpendicular, shock
PACS: 96.50.Ek, 96.50.Pw

INTRODUCTION

It is now well-known that broad classes of upstream distributions associated with sunward propagating ions are commonly observed upstream of the Earth's bow shock [1]. Our interest in the present paper is with field-aligned ion beams (FABs), which consist of ions collimated along the interplanetary magnetic field (IMF). Although significant observations and investigations, including theoretical studies [2, 3] and simulations [4] have been carried out in the previous years, their production mechanisms remain an open issue [5]. Understanding the production mechanism responsible for coherent particle distributions is of fundamental importance in shock acceleration since it is intimately associated with the particle injection problem at quasi-perpendicular shocks. High quality data are now available from recent spacecraft such as Cluster that allow for detailed quantitative studies and the deconvolution of spatial and temporal variations. In this paper we present new observations briefly, and indicate how production mechanisms currently proposed cannot explain these observations.

EARLY OBSERVATIONS

Observations describing the basic properties of FABs, based on 2-D particle instrumentation, have been extensively reported during the ISEE era [6]. Typically, field-aligned ion beams are observed upstream of quasi-perpendicular shocks characterized by $40° \lesssim \theta_{Bn} \lesssim 70°$, where θ_{Bn} is the angle between the interplanetary magnetic field direction and the local bow shock normal. Their bulk energy is typically few keV and the energy spectrum rarely extends beyond ~ 15 keV. Their bulk speeds, ranging from one to several times the solar wind speed, have been satisfactorily measured and are well correlated with the angle θ_{Bn} [7]. Downstream of the FAB region intermediate ion distributions are observed to have a kidney-shape that is symmetric across the magnetic field direction. These distributions are always observed in association with large amplitude ($\Delta B/B \sim 1$), weakly compressive, nearly monochromatic, ULF ($\omega/\Omega_i \ll 1$) waves [8]. These waves propagate along the ambient magnetic field direction [9, 10]. The large amplitude ULF waves have not been observed with FABs. However, whistler-like waves (commonly called foreshock 1-Hz waves) having small-amplitude $\Delta B/B \sim 0.1$ are occasionally observed in association with FABs [9, 11]. Carefull examination of the ion energy-spectrogram showed that the foreshock 1-Hz waves are more often observed with "spread" FABs rather than with the "narrow" ones [11]. From case examples, it seems that the energy spread occurs at a lower energy. Early field aligned beam temperature determinations provided an average value of 345 km s^{-1}, which may extend up to ~ 800 km s^{-1} [6]. The temperature anisotropy $T_\perp/T_\parallel$ is in the range of 4–9 [1]. Based upon particle features alone, the distinction between the FABs and intermediate distributions is quite arbitrary; the main criterion to distinguish the two populations is the lack or presence of large-aplitude ULF waves. However, it is not unusual to see very small amplitude ULF waves in association with the FABs.

It is worthy of mention that the FABs are considered the most important source of free energy in the foreshock region. The resulting ion distribution (FAB superposed on the solar wind) is susceptible to numerous instabilities. The electromagnetic ion-ion cyclotron instability, discussed in detail by Gary et al. [12], has the highest linear growth rate. Both the parallel and the oblique modes are unstable for a large range of FAB speeds, and both are in cyclotron resonance with FABs; the parallel case has the maximum growth rate. The resonant parallel mode has been successfully indentified in detailed case studies, which showed that the ULF waves observed in association with gyrating ion distributions are actually in cyclotron resonance with adjacent FABs [10, 13].

RECENT OBSERVATIONS

Theory and simulation models [14] predict that, due to electromagnetic instabilities briefly mentioned above, a typical FAB is heated and becomes an intermediate-like distribution after $\sim 25/\Omega_i$. We may then expect a cut-off distance on the order of $\sim 30\, R_E$ from the shock. Figure 1 from WIND/3DP shows examples of particle distributions having 15 keV FABs observed at $\sim 93\, R_E$ from Earth. (Presented in the plasma frame, the main beam in the center is the solar wind.) The FABs are travelling opposite

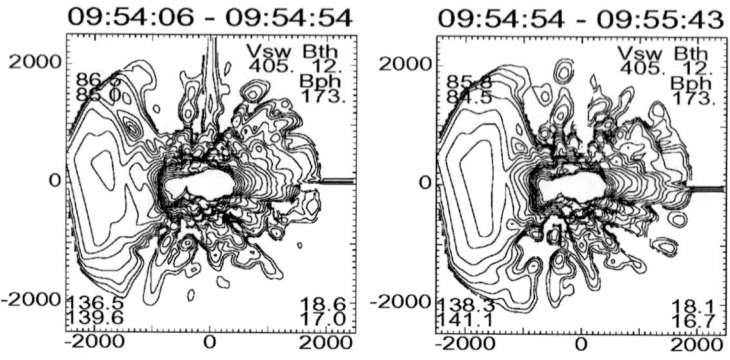

FIGURE 1. Wind/3DP ion measurements of FABs on 28 August 1995 at 93 R_E.

to **B** (left of center, with pitch angles extending to 45°. Several similar distributions have been observed at $\sim$ 100 R_E and beyond. Clearly, these beams are unexpected at such large distances from the shock. Work is now in progress to establish their source characteristics and to understand how they reach far upstream regions.

We have examined in detail the properties of field aligned beams as a function of shock geometry, taking special care to isolate the influences of θ_{Bn} from other parameters [15]. Simple kinematic arguments lead to expectations that densities and beam speeds will vary with θ_{Bn}, although no temperature relation immediately follows. Figure 2 shows moments computed for successive FABs observed by the Cluster/CIS experiment on 23 April 2001, 0647–0651 UT. During the interval of interest no ULF waves were observed and the IMF direction was slowly rotating toward a less radial configuration. As selected, the only parameter that changed significantly for this event was the angle θ_{Bn}. Successive panels show θ_{Bn} variation of the beam density normalized to the solar wind density (Figure 2a), the beam speed normalized to the solar wind speed (Figure 2b), and beam parallel and perpendicular temperatures (Figure 2c). Clearly the beam properties are remarkably well correlated with the shock geometry. The strong decrease in beam densities with θ_{Bn} shown in Figure 2a is expected, since only those particles moving upstream in the de Hoffman-Teller frame will avoid re-encounter with the shock. The increase in the frame transformation velocity as θ_{Bn} increases requires that the beams originate from further in the tail of the source distribution. The plateau in n seen toward the left of Figure 2a suggests that the beam production mechanism breaks down for a critical value of θ_{Bn}. Detailed study of other events are now underway to verify the occurence of this density-plateau relation.

Figure 2b shows that the acceleration increases linearly with $\cos\theta_{vn}/\cos\theta_{Bn}$, where θ_{vn} is the angle the shock normal makes with the direction of the solar wind flow. Similar results have been obtained in an early study [2]. We mention here that the beam speed is given in the plasma frame of reference. The $\cos\theta_{vn}/\cos\theta_{Bn}$ factor appears in the expression for the de Hoffman-Teller frame speed V_S expressed in the plasma rest frame: $V_S = -v_{sw}\cos\theta_{Bn}/\cos\theta_{vn}$. During the time interval presented above, θ_{vn} remains nearly constant. According to the kinematic description of ion reflection by Sonnerup [16], the post encounter speeds are given by $V_B = -(1+\delta)V_S$, where δ is a coefficient indicating

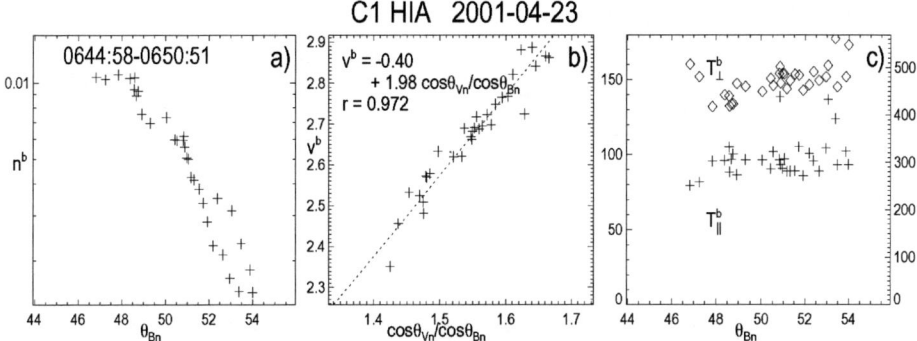

FIGURE 2. Moments determined for FABs on 23 April 2001. a) $n = n_b/n_{sw}$ vs. θ_{Bn}.; b) $v = v_b/v_{sw}$ vs. $\cos\theta_{vn}/\cos\theta_{Bn}$; c) $T_\parallel$ vs. θ_{Bn} ('+'s, left axis with units of eV) and $T_\perp$ vs. θ_{Bn} (diamonds, right axis).

the degree to which the particle magnetic moment is conserved ($\delta = 1$ for the adiabatic case).

In Figure 2c (which has the scale for $T_\parallel$ on the left axis, and that for $T_\perp$ on the right) the temperature anisotropy $T_\perp/T_\parallel$ is a little lower than 5, and within the range of 4–9 quoted above. Most notable is that there is no clear dependence of either temperature upon θ_{Bn}. This and other cases suggest, if anything, a weak increase in $T_\perp$ with θ_{Bn}. A straightforward view might be that the perpendicular temperature should reflect the free-escape conditions for particles in a source distribution just exterior to the shock. In that case, however, we should expect [15] $T_\perp \sim 1/\sin^2\theta_{Bn}$, which, rather than increasing, should decrease by $\sim 20\%$ over this range of θ_{Bn}. This is a topic of ongoing investigation.

The different types of populations observed upstream of the bow shock are located in distinct foreshock regions. FABs are seen in a layer of $\sim 0.4\ R_E$ thickness, followed further downstream by a $\sim 3.5 R_E$-wide layer of intermediate ions [7]. Gyrophase-bunched ions, characterized by a bulk motion at non-zero pitch angles, can be found along the upstream edge of the intermediate region. This implies the existence of a spatial boundary separating the different types of populations. Previously Greenstadt and Baum [17] showed a spatial boundary separating regions where ULF waves are present from those where thay are absent. Subseqently, Le and Russell [18] found a similar boundary, but its statistical average slope differed significantly. However, it is difficult to resolve transitions between different foreshock regions due to the rapid rate of IMF rotations.

More recently, we reported the presence of a sharp spatial boundary separating FABs from gyrating ions using observations from the Cluster spacecraft. Figure shows successive spectra from CIS/CODIF on spacecraft 1, which indicate an apparent merging of an energetic population with FABs. Detailed examination of the 3D distributions showed that the energetic component was a remotely-sensed gyrating population[19]. Figure is a schematic illustrating the remote sensing model accounting for the two-peak spectra seen near 2111 UT. The spacecraft is located at S and observes FABs travelling along the field line threading it. Simultaneously, it detects gyrating ions at high energy having

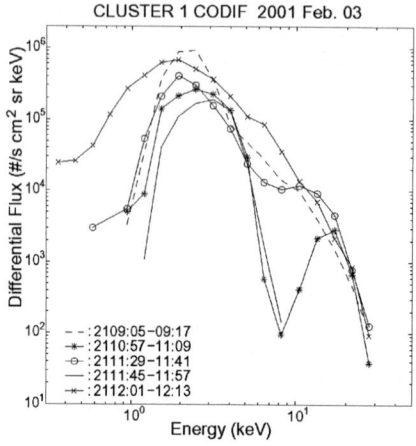 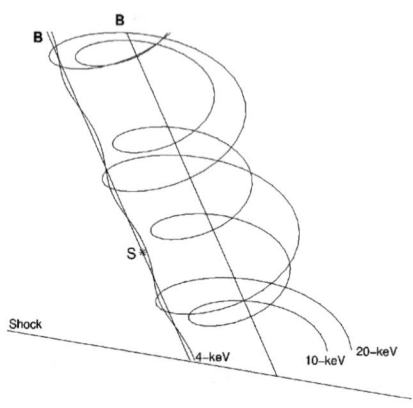

FIGURE 3. Spectra from CIS/CODIF for different times on 3 February 2001.

FIGURE 4. Remote-sensing schematic for 3 February 2001.

guiding centers along adjacent field lines separated by less than one ion gyroradius ρ. Figure 3 of Meziane et al. [19] shows that the energetic component initially appears in narrow range of gyrophase angles consistent with guiding centers downstream of the spacecraft. A direct inference from these observations was that the transition between the two types of populations occured within ~ 1 gyrating ion gyroradius. This boundary agreed with the ULF wave boundary observed by Le and Russell [18], and predicted theoretically [20]. We note, however, that the FABs observed along the boundary do not match quantitatively the emission mechanisms posited to occur at the shock.

PRODUCTION MECHANISMS FOR FABS

In this section we briefly discuss guiding center production mechanisms. A straightforward application of kinematic reflection theory, where energy and the first adiabatic invariant are conserved in the de Hoffman-Teller frame, leads to an expression relating a required initial particle energy to the observed final (beam) particle parameters, all in the plasma frame: $E_i = E_f[1 + 4(\frac{V_s^2}{v_f^2} - \frac{V_s}{v_f}\cos\alpha)]$, where α is the pitch angle. Decker [21] showed that the energy gain for adiabatic reflection depends exclusively upon the shock strength B_2/B_1. Typical upstream values indicate an initial energy of 0.2–0.5 keV, far greater than the typical energies of the presumed solar wind source. Similar reasoning can be applied to adiabatic leakage of particles from the magnetosheath, but this produces fluxes lower than those observed by an order of magnitude [22]. Generalizing the problem to allow for non-conservation of first adiabatic invariant or loss of energy does not improve the situation.

Researchers have attempted to reconcile the inadequacies of guiding center models by considering diffusive processes. One scenario that is receiving a lot of attention includes strong cross-field diffusion of particles within magnetic field turbulence [23].

Scattering occuring on time scales smaller than a gyroperiod repeatedly return particles to the upstream, permitting additional opportunities to sample the strong electric field and consequent acceleration in the shock ramp. Hybrid and 1-D full-particle simulations [23] using thermal seed populations show that this approach produces upstream particle fluxes lower than observed by several orders of magnitude. Additional difficulties are a lack of physical motivation for the diffusion coefficients, which up until now have been applied in an *ad hoc* manner, and the unlikelihood of diffusion producing beams that are field aligned. Tanaka [24] modeled upstream ion production by noting that the temperature anisotropy in the magnetosheath drives strong electromagnetic ion cyclotron waves. Non-thermal gyrating particles, initially specularly reflected in the shock ramp, and subsequently crossing into the sheath were scattered by these waves, and those redirected back upstream were adiabatically folded in pitch angle as they cross the shock. While these models were able to produce fluxes comparable to those of the observed beams, the dependence upon θ_{Bn} found was the reverse of what was observed [15]. This dependence in turn was determined by the assumed gyrating particle profile. A recent idea [25] is that strong scattering of the gyrating particles instead occurs within the shock ramp, diffusing these particles into free-escape regions of velocity space, which are sufficient in number to provide the FABs. See Kucharek (this volume) for additional details.

ACKNOWLEDGMENTS

We thank the International Space Science Institute in Bern for their support of this work, and members of the ISSI Upstream Ions Collaboration for useful discussions. Work at UNB is supported by the Canadian Natural Science and Engineering Council. MW and GKP acknowledge support from NASA Grant NAG5-10131.

REFERENCES

1. G. Paschmann, N. Sckopke, I. Papamastorakis, J. R. Asbridge, S. J. Bame, and J. T. Gosling, *Journal of Geophysical Research*, **86**, 4355–4364 (1981).
2. G. Paschmann, N. Sckopke, J. R. Asbridge, S. J. Bame, and J. T. Gosling, *Journal of Geophysical Research*, **85**, 4689–4693 (1980).
3. M. F. Thomsen, S. J. Schwartz, and J. T. Gosling, *Journal of Geophysical Research*, **88**, 7843–7852 (1983).
4. D. Burgess, *Journal of Geophysical Research*, **92**, 1119–1130 (1987).
5. S. J. Schwartz, M. F. Thomsen, and J. T. Gosling, *Journal of Geophysical Research*, **88**, 2039–2047 (1983).
6. C. Bonifazi, and G. Moreno, *Journal of Geophysical Research*, **86**, 4381–4396 (1981a).
7. C. Bonifazi, and G. Moreno, *Journal of Geophysical Research*, **86**, 4397–4404 (1981b).
8. S. A. Fuselier, M. F. Thomsen, J. T. Gosling, S. J. Bame, and C. T. Russell, *Journal of Geophysical Research*, **91**, 91– (1986).
9. M. M. Hoppe, C. T. Russell, L. A. Frank, and E. W. Eastman, T. E.; Greenstadt, *Journal of Geophysical Research*, **86**, 4471–4492 (1981).
10. K. Meziane, C. Mazelle, R. P. Lin, D. LeQuéau, D. E. Larson, G. K. Parks, and R. P. Lepping, *Journal of Geophysical Research*, **106**, 5731–5742 (2001).
11. M. M. Hoppe, C. T. Russell, T. E. Eastman, and L. A. Frank, *Journal of Geophysical Research*, **87**, 643–650 (1982).
12. S. P. Gary, J. T. Gosling, and D. W. Forslund, *Journal of Geophysical Research*, **86**, 6691– (1981).

13. C. Mazelle, K. Meziane, D. Le Quéue, M. Wilber, J. P. Eastwood, H. Rème, J.-A. Sauvaud, J.-M. Bosqued, I. Dandouras, M. McCarthy, L. M. Kistler, B. Klecker, A. Korth, M. B. Bavassano-Cattaneo, R. Lundin, and A. Balogh, *Planetary and Space Science*, pp. 785–795 (2003).
14. D. Winske, and M. M. Leroy, *Journal of Geophysical Research*, **89**, 2673–2688 (1984).
15. M. Wilber, K. Meziane, G. K. Parks, L. M. Kistler, I. Dandouras, H. Rème, J.-A. Sauvaud, J.-M. Bosqued, M. McCarthy, B. Klecker, A. Korth, M.-B. Bavassano-Cattaneo, R. Lundin, and E. Lucek, *Annales Geophysicae* (2005), submitted, May 2005.
16. B. U. O. Sonnerup, *Journal of Geophysical Research*, **74**, 1301–1304 (1969).
17. E. W. Greenstadt, and L. W. Baum, *Journal of Geophysical Research*, **91**, 901– (1986).
18. G. Le, and C. T. Russell, *Planetary and Space Science*, **40**, 1203–1213 (1992).
19. K. Meziane, M. Wilber, C. Mazelle, D. LeQuéau, H. Kucharek, E. Lucek, J. A. Sauvaud, A. M. Hamza, H. Rème, J. M. Bosqued, I. Dandouras, G. K. Parks, M. McCarthy, B. Klecker, A. Korth, M.-B. Bavassano-Cattaneo, and R. Lundin, *Journal of Geophysical Research*, **109**, A05107, doi:10.1029/2003JA010374 (2004).
20. G. Skadron, R. T. Holdaway, and M. A. Lee, *Journal of Geophysical Research*, **93**, 11,354–11,362 (1988).
21. R. B. Decker, *Journal of Geophysical Research*, **88**, 9959–9973 (1983).
22. J. P. Edmiston, C. F. Kennel, and D. Eichler, *Geophysical Research Letters*, **9**, 531 (1982).
23. J. Giacolone, J. R. Jokipii, and J. Kóta, *Journal of Geophysical Research*, **99**, 19,351–19,358 (1994).
24. M. Tanaka, C. C. Goodrich, D. Winske, and K. Papadopoulos, *Journal of Geophysical Research*, **88**, 3046–3054 (1983).
25. H. a. Kucharek, *Annales Geophysica* (2004).

Methods of Plasma Turbulence Analysis: Application to Shock Studies

M. A. Balikhin* and S. N. Walker*

*ACSE, University of Sheffield, Sheffield, U.K.

Abstract. The availability of multisatellite observations (e.g. ISEE, AMPTE, and Cluster) has triggered the development of new methods of analysis to shed light on the complex dynamics inherent in the solar wind and magnetosphere. This paper looks at the results of two such methods. Firstly, the phase differencing method is used to determine the properties of waves observed upstream of a quasiperpendicular bow shock. The resulting dispersion relation is then interpreted as evidence that the waves are generated as a result of the dynamics of the shock front. The second, NARMAX, is used to investigate the linear and nonlinear processes if the plasma observed at a antiparallel shock. The results show that for a small amplitude whistler wavetrain, third order nonlinear interactions are only important at the interface between the shocklet and the wavetrain. For higher amplitude wavetrains, the phase of the linear term describing the plasma is shifted.

Keywords: Bow shock, plasma turbulence, wave analysis techniques
PACS: 94.30.Gm,94.30.Kq,94.30.Va,94.30.Tz,95.75.Wx,96.50.Fm

INTRODUCTION

Plasma waves are key to many processes that take place at a collisionless shock front. In an ideal situation, experimental data must allow the identification of all observed modes of plasma waves and quantitatively assess energy exchange processes between the plasma waves and particle populations and energy redistribution processes within the plasma turbulence. Modern field instruments are capable of returning high time resolution spectra and waveform data that contain key information that can be used to characterise the processes occurring within the plasma. The main task in the analysis of such data sets is to unlock this information and produce a complete description of the plasma dynamics.

With the advent of multipoint mission data sets such as ISEE, AMPTE, and Cluster new methods of analysis have been either developed or adopted from other branches of science and engineering to shed light on the complex dynamics inherent in the plasma. This paper discusses and presents the results of the application of two different classes of methodology to data from the quasiperpendicular bow shock. The first, based on measuring the relative phases of waves at two points is used to determine the wave propagation vector hence its dispersion relation. The second, originally developed for systems engineering, treats the plasma as a black box whose dynamics are determined from two point (i.e. input/output) measurements. This technique is able to characterise the dynamics of the processes within the plasma. Linear processes within the black box correspond wave propagation and energy exchange with the plasma. Nonlinear processes, for example the decay and modulational instabilities, correspond to the redistribution of energy within the plasma turbulence between different scales.

K-VECTOR AND DISPERSION DETERMINATION

The simplest task is the identification of the composition of plasma turbulence i.e. the determination of the observed plasma modes. The method we will describe is based on the straight forward assumption that any wave field can be described by a supposition of plane waves (1) then the phase differences measured at two closely spaced points is related to the projection of the **k**-vector of the wave along the measurement separation direction (2) [1]. In some cases, it is then possible to estimate the complete **k**-vector by applying other analysis techniques such as a minimum variance analysis (MVA) [2].

$$B(r,t) = \Sigma_{mode}\Sigma_\omega \exp(ikr - i\omega t) + CC \qquad (1)$$

$$\Delta\Phi(\omega) = \Psi_1 - \Psi_2 = k(\omega)r_1 - k(\omega)r_2 = k(\omega)\Delta r \qquad (2)$$

This method has been further extended to determine the **k**-vector using the multisatellite measurements available from the Cluster mission. In this case, **k** may be determined from its projections along 3 independent directions, and hence is not restricted by the use of MVA [3].

This method has been applied to the identification of turbulence in the foot region of a quasiperpendicular bow shock [4]. The left hand panel of Fig 1 shows a crossing from of a quasi-perpendicular shock measured by AMPTE-IRM on December 29th, 1984. Initially upstream of the shock, the satellite encounters a wave packet at 0429:25-0429:35 before eventually crossing the shock at around 0429:47UT. The data from these wavetrains were used to generate a histogram of the distributions of phase difference between the IRM and UKS measurements at each frequency (the joint ω-k spectrogram) and hence compute the magnitude of the **k**-vector projected along the satellite separation vector. By using minimum variance analysis the direction of the **k**-vector may be determined. Combining these results enables the full **k**-vector and hence the plasma frame frequency and wave dispersion relation to be determined. The right panel in Fig. 1 shows the plasma frame dispersion (crosses) compared with the whistler mode dispersion relation. This comparison demonstrates that the waves propagate in the whistler mode and also enables information regarding their generation to be deduced.

This result was used to differentiate between a number of competing theories that were proposed to explain the generation of such waves. The discussion in Balikhin et al. [4] can be summarised as follows. Wong and Goldstein [6] proposed that the whistler waves were the result of a proton distribution instability. These authors predict that the wave frequency was of the order $20\Omega_i \approx 20 rads^{-1}$. However, the plasma frame frequency determined from the analysis was much higher as shown in Fig 1. Orlowski and Russell [7] suggested that the whistler waves were generated as the result of instabilities within ramp. In this case, the wavelength of the waves should be less than the ramp width. The experimentally determined wavelength was found to be around $125 km$ which is greater than the experimentally determined ramp width ($< 60 km$). The models of Tidman and Northrop [8] and Krasnosel'skikh [9] based on shock dynamics were determined to be the most compatible with the observations.

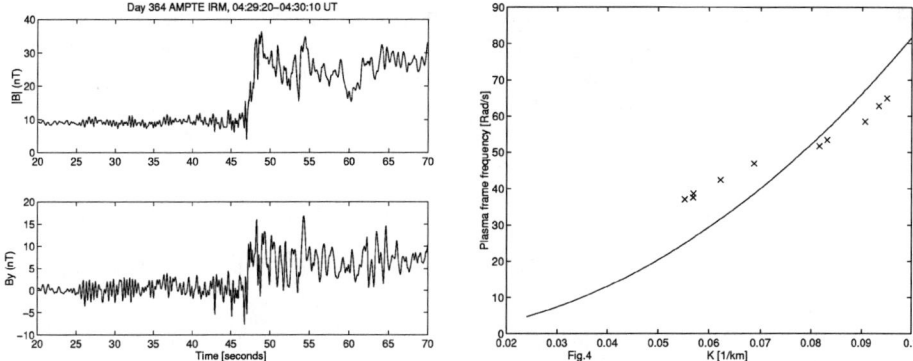

FIGURE 1. The magnitude (top) and By component (bottom) of the magnetic field measured by AMPTE-IRM (left panel) and the plasma frame dispersion relation (right) determined using the phase differencing method. Reprinted from Balikhin et al. [5] ©(1997), with permission from Elsevier.

IDENTIFICATION OF LINEAR AND NONLINEAR PROCESSES IN PLASMA TURBULENCE BASED ON THE BLACK BOX APPROACH

From a systems identification point of view, multipoint measurements of plasma phenomena that propagate between the measurement points represent the input/output measurements of a system (the plasma) whose state is unknown and is thus treated as a black box. In the overview that follows, the vector $x(t)$ represents the input to the system and $y(t)$ its output.

In the simple case when the black box represents a linear system the input $x(t)$ and output $y(t)$ are related by the impulse response function (3) in the time domain or by the linear frequency response function (4) in the frequency domain.

$$y(t) = \int_{-\infty}^{\infty} h_1(\tau)x(t-\tau)d\tau \qquad (3)$$

$$Y_f = H_1(f)X_f \qquad (4)$$

System identification in the frequency domain is performed by simply calculating the ratio of the spectra of the input and output data series i.e. $H(f)=Y(f)/X(f)$. In the time domain, the task is to fit a linear function such as that shown in (5) to the input and output series using a method such as LSM.

$$y_k = \Sigma_{i=0}^{N} a_i x_{k-i} + \Sigma_{i=0}^{N} b_i y_{k-i} \qquad (5)$$

The generalisation of the convolution integral for nonlinear systems may be formulated in terms of a Volterra series (6) in the time domain or as a set of Generalised Frequency Response Functions (GFRF) (7) in the frequency domain. In these relationships, the terms h_1 and H_1 represent the linear impulse response function and the Linear frequency Response Function respectively. These terms describe the linear amplification of the waves. The terms h_2, H_2, and h_3, H_3 describe the nonlinear coupling occurring between 3 and 4 wave processes. In space plasmas higher order terms may often be neglected [10].

$$y(t) = \int_{-\infty}^{\infty} h_1(\tau_1)x(t-\tau_1) + \int\int h_2(\tau_1,\tau_2)x(t-\tau_1)x(t-\tau_2)d\tau_1 d\tau_2$$
$$+ \int\int\int h_3(\tau_1,\tau_2,\tau_3)x(t-\tau_1)x(t-\tau_2)x(t-\tau_3)d\tau_1 d\tau_2 d\tau_3 + \ldots \quad (6)$$

$$Y_f = H_1(f)X_f + \Sigma_{f_1+f_2}H_2(f_1,f_2)X_{f_1}X_{f_2} + \Sigma_{f_1+f_2+f_3}H_3(f_1,f_2,f_3)X_{f_1}X_{f_2}X_{f_3} + \ldots \quad (7)$$

In the frequency domain, identification is carried out by firstly dividing the input and output data sets into a number of subintervals. For each subinterval the frequency spectra $Y(f_k)$ and $X(f_k)$ are computed. These spectral components are then related by the GFRF (7). For reliable results, frequency domain identification requires a long, stationary data series. For this reason, this technique is not applicable for the study of shocks in space plasma. As an alternative, time domain identification is employed since it requires shorter input data series. Time domain identification of a nonlinear system is an iterative process involving the fitting of a function, say a polynomial of the type (8), term by term until the residuals ξ_k are unpredictable from the input. The frequency response function is then be calculated analytically from the coefficients of the polynomial.

$$y_k = \Sigma_{i=1}^{\infty} a_i x_{k-i} + \Sigma_{i=1}^{\infty} b_i y_{k-i} + \Sigma_{i=1}^{\infty}\Sigma_{j=1}^{\infty} c_{ij} x_{k-i} y_{k-j}$$
$$+ \Sigma_{i=1}^{\infty}\Sigma_{j=1}^{\infty} a_{ij}^2 x_{k-i} y_{k-j} + \Sigma_{i=1}^{\infty}\Sigma_{j=1}^{\infty} b_{ij}^2 y_{k-i} y_{k-j} + \cdots + \xi_k \quad (8)$$

As an example of the use of system identification methods, we will apply a methodology based on the Nonlinear AutoRegressive Moving Average with eXogenuous inputs model (NARMAX) [11, 12] to observations of a quasi-parallel shock crossing by the AMPTE satellites UKS (input) and IRM (output). The NARMAX model is a generalisation of the Volterra series and is derived from a subsection of a set of input/output measurements. The model is then verified using a different subset to assess its performance. Figure 2 shows two examples of shocklets observed by the AMPTE satellites on October 30th, 1984. The right hand panel typifies the case when a large amplitude whistler wave train is observed whilst the left hand panel shows an event whose wave train is smaller in amplitude. In both cases, the solid line represents the By component measured by IRM and the dashed line is the model prediction. In the left panel, the dash-dotted line represents the contribution of the linear terms and the dotted the nonlinear. Clearly, the linear term is in excellent agreement with the overall model and measurements except for the period at the interface between the shocklet and wave train. This panel shows that the contribution of the nonlinear terms is only important at this interface which represents the steepening of the leading edge of the shock front. Steepening can be described in terms of nonlinear wave coupling in which coupling between 3-

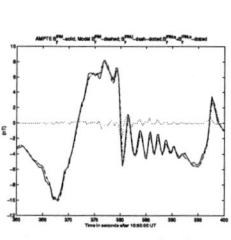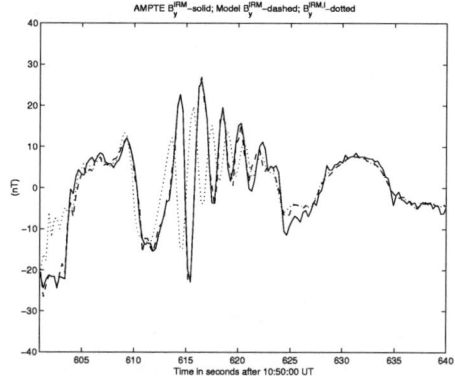

FIGURE 2. Two examples of shocklets and their associated whistler wave trains. The left hand panel shows a case when the whistler wave train had a small amplitude whilst the panel on the right shows an example with a high amplitude whistler wave train. (From Coca et al. [13] ©(2001) American Geophysical Union. Reproduced by permission of American Geophysical Union)

and 4-waves processes lead to transfer energy to shorter scales. Once the scale size of the leading edge is equal to the spatial scale of whistler wave, steepening ceases and the energy is transferred to the whistler wave train. Figure 2 is direct experimental evidence for this theoretical model. Coca et al. [13] also showed that the contribution of the quadratic nonlinearities was less important than that of the cubic term indicating that 4-wave processes are more important than 3-wave processes. This result agrees with the model of wave steepening proposed by Sagdeev and Galeev [14]. The other well known effect of so called nonlinear dispersion is illustrated in the right hand panel of Fig 2. Nonlinear dispersion takes place when the velocity of wave propagation depends upon the wave amplitude. Often nonlinear dispersion takes place due to the usual hydrodynamic nonlinearity $v\partial v/\partial x$. The right hand panel of Fig. 2 shows the effects of a large amplitude wavetrain. In this case the contribution of the linear term (dotted line) shows a phase shift when compared to the overall model output (dashed line) and observations (solid line). As a result, the wave phase speed calculated using the linear term would be incorrect. The nonlinear components provide the phase correction, demonstrating the effect of nonlinear dispersion.

SUMMARY

We have reviewed methods commonly used to study composition of observed plasma turbulence and nonlinear processes occurring within plasma turbulence. Application of these methods to multipoint measurements enables identification of a quantitative model for the observed plasma turbulence directly from experimental data. We have shown how the application of these methods may be used for shock studies to validate various theoretical models.

ACKNOWLEDGMENTS

The authors acknowledge the financial support of PPARC.

REFERENCES

1. M. A. Balikhin, and M. E. Gedalin, "Comparative analysis of different methods for distinguishing temporal and spatial variations," in *Proc. of START Conf., Aussois, France*, 1993, vol. ESA WPP 047, pp. 183–187.
2. M. A. Balikhin, L. J. C. Woolliscroft, H. S. C. Alleyne, M. Dunlop, and M. A. Gedalin, *Annales Geophysicae*, **15**, 143 (1997).
3. M. A. Balikhin, O. A. Pokhotelov, S. N. Walker, E. Amata, M. Andre, M. Dunlop, and H. S. K. Alleyne, *Geophys. Res. Lett.*, **30**, 15-1 (2003).
4. M. A. Balikhin, T. Dudok de Wit, L. J. C. Woolliscroft, S. N. Walker, H. Alleyne, V. Krasnosel'skikh, W. A. C. Mier–Jedrzejowicz, and W. Baumjohann, *Geophys. Res. Lett.*, **24**, 787–790 (1997).
5. M. A. Balikhin, S. N. Walker, T. Dudok de Witt, H. S. Alleyne, L. J. C. Woolliscroft, W. Mier-Jedrzejowicz, and W. Baumjohann, *Advances in Space Research*, **20**, 729–734 (1997).
6. H. K. Wong, and M. L. Goldstein, *J. Geophys. Res. A*, **93**, 4110–4114 (1988).
7. D. S. Orlowski, and C. T. Russell, *Advances in Space Research*, **16**, 137–141 (1995).
8. D. Tidman, and T. Northrop, *J. Geophys. Res. A*, **73**, 1543–1553 (1968).
9. V. Krasnosel'skikh, *Sov. Phys. Jetp*, **62**, 282 (1985).
10. V. E. Zakharov, S. L. Musher, and A. M. Rubenchick, *Phys. Rep.*, **129**, 285 (1985).
11. I. Leontaritis, and S. A. Billings, *Int. J. Control*, **41**, 309–328 (1985).
12. I. Leontaritis, and S. A. Billings, *Int. J. Control*, **41**, 329–344 (1985).
13. D. Coca, M. A. Balikhin, S. A. Billings, H. S. C. K. Alleyne, and M. Dunlop, *J. Geophys. Res. A*, **106**, 25005 (2001).
14. R. Z. Sagdeev, and A. A. Galeev, *Nonlinear Plasma Theory*, Benjamin, White Plains, N.Y., 1969.

Observations of Turbulence near Interplanetary Travelling Shocks

R. Kallenbach*, K. Bamert†, M. Hilchenbach** and C.W. Smith‡

*International Space Science Institute, Hallerstrasse 6, 3012 Bern, Switzerland
†Institut für Experimentelle und Angewandte Physik, University of Kiel, Leibnizstrasse 19, Kiel, 24098 Germany
**Max-Planck-Institut für Sonnensystemforschung, POB 20, 37191 Katlenburg-Lindau, Germany
‡Institute for Earth, Oceans and Space, Univ. of New Hampshire, Morse Hall, Durham, NH 03824

Abstract. The observations on magnetic field fluctuations and suprathermal ion spectra near the shocks driven by the coronal mass ejections during the time periods of the Bastille Day 2000 event and the Halloween 2003 events are summarized.

INTRODUCTION

The formation of a shock wave requires dissipation of the upstream bulk energy into the downstream medium. At a collisionless interplanetary shock the upstream bulk energy is presumably dissipated into turbulent waves which in turn heat the downstream plasma and transfer energy to suprathermal ions and electrons. Two processes dominate the amplification of magnetic field fluctuations near the shocks driven by interplanetary coronal mass ejections (CMEs): 1) Alfvén wave amplification in the upstream solar wind plasma by energetic ions [1], and 2) further amplification of the Alfvén waves in course of their passage through the shock layer [2]. These processes are accompanied by non-linear cascading and wave mixing which result in power-law turbulence spectra. We summarize the results from observations on the spatial evolution of energetic ion spectra and their relation to the turbulent wave spectra and discuss them in the context of theoretical models.

UPSTREAM ALFVÉN WAVE GENERATION

Figure 1 shows an overview of data from the two subsystems STOF (Suprathermal Time of Flight spectrometer) and PM (proton monitor) of CELIAS (Charge, Element, and Isotope Analysis System) on board SOHO (Solar and Heliospheric Observatory) and the magnetometer on board ACE (Advanced Composition Explorer), both located at the Lagrangian point L1 near Earth's orbit, during the time period of the Bastille Day 2000 event. The main shock of this event is driven by ejecta that pass the heliosphere with a speed of up to 1100 km/s. They run into solar wind at a speed of 600 km/s. The upstream Alfvén Mach number is about 10, the plasma-β close to unity, the compression ratio is about 4, and the shock normal angle to the ambient magnetic field is larger than 45°.

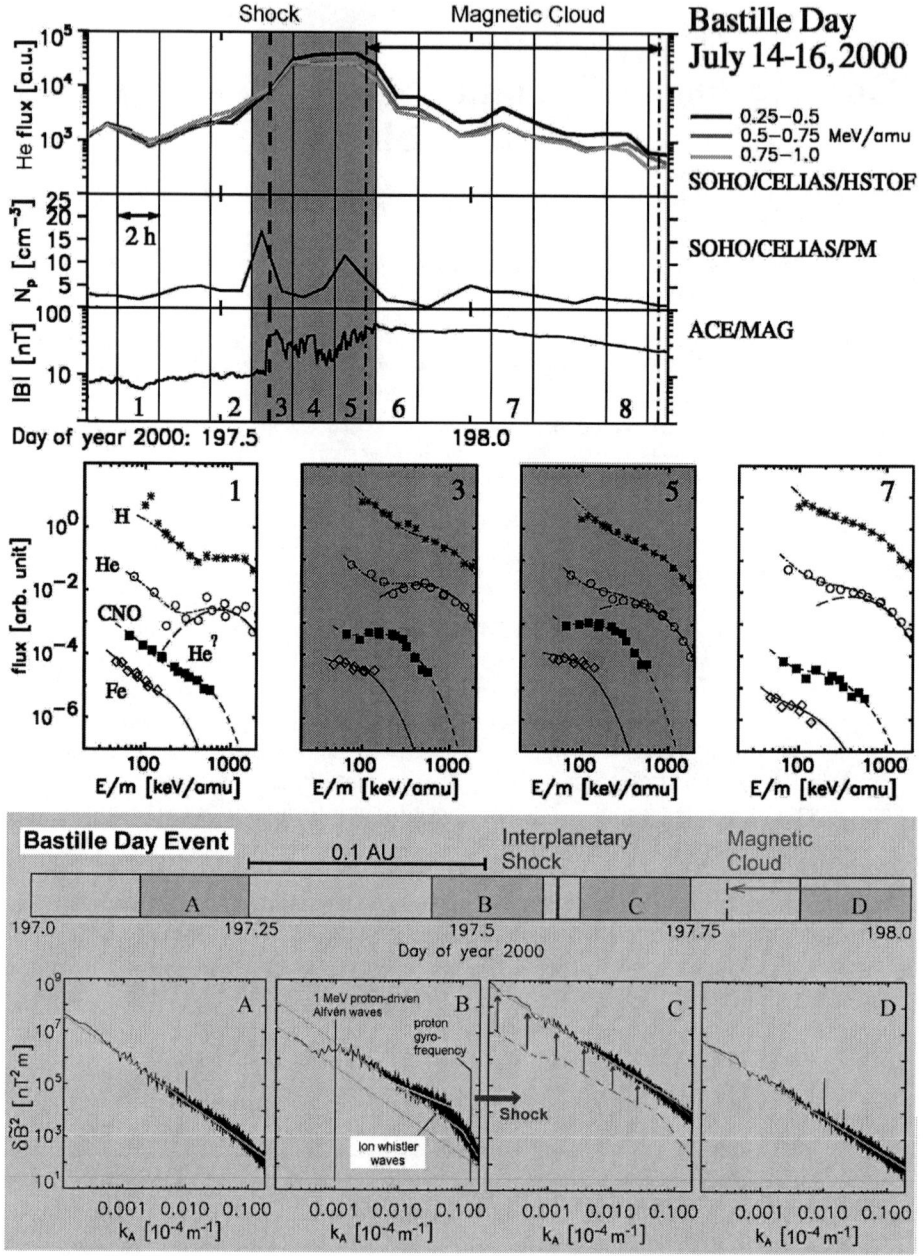

FIGURE 1. Overview of spacecraft data associated with the Bastille Day event.

For these reasons the shock can be considered as quasi-parallel when comparing data to models on diffusive acceleration [1, 3] and to the model on amplification of magnetic field fluctuations in course of their passage through the shock layer [2].

In the upstream region of this shock the suprathermal ion flux has a pronounced spatial gradient for all ion species similar to that shown for suprathermal He ions. The suprathermal ion flux, however, does not peak right at the shock layer, but rather at about 0.05 AU downstream of the shock in the region of strong magnetic turbulence. The ejecta follow this turbulence region in the form of a magnetic cloud.

Far upstream of the shock (period A), the magnetic field fluctuations are at a level of about 2×10^5 nT^2 m at the wavenumber $k_{1\,\text{MeV}}$ at which Alfvén waves resonate with protons at 1 MeV energy with $\mu = 1$ ($\mu = \cos\alpha$ with α the pitch angle). The power density spectrum scales as a power law with index 1.6 ± 0.1, similar to expectations for Kolmogorov or Kraichnan scaling. In the region within about 0.04 AU upstream of the shock (period B) there is a range of k-numbers where the fluctuations are enhanced by a factor of about 30 above the level of the ambient solar wind fluctuations. Energetic proton spectra in the range 60 keV to a few MeV and the magnetic field fluctuations in the corresponding range of gyro-resonant k-numbers match the quasi-linear theory (QLT) by Marty Lee [1] on self-consistent hydromagnetic wave excitation and ion acceleration [3]. At energies below 60 keV the proton flux is much higher than predicted by the QLT. The wave spectra at $k > 2k_{1\,\text{MeV}}$ are in the form of a Kolmogorov power law over more than one decade in k, which requires the time-scale for cascading to be shorter than the timescale for wave amplification by protons.

Panel number 1 of the ion flux spectra of H, He, CNO, and Fe in Figure 1 shows a distinct signature in the proton spectrum. There is a flat range at energies between 200 keV and 2 MeV. This is in fact the energy range, where the QLT fits the proton and wave data best [1, 3]. We therefore consider this signature to represent the driving of Alfvén waves by the protons. There is also a flat range in the He spectra marked with "He?". However, by the mass-resolving capabilitiy of the STOF/HSTOF sensor, this signature has been identified as most likely being due to a contribution of ^{3}He rather than predominantly being a signature of wave-particle interaction.

Figure 2 shows the data associated with the two main shocks during the Halloween 2003 CME events. The first shock (IPS#1) again has a compression ratio of about 4 as the Bastille Day shock, and has roughly $\beta \approx 1$ and can be considered as a quasi-parallel shock. The upstream region shows a distinct spatial gradient in energetic ion flux, suggesting that anisotropies that drive Alfvén waves could build up [4]. Indeed, the proton spectra in the region around 0.15 AU upstream of IPS#1 have the flat range beginning at about 200 keV but in this case only ranging up to 400 keV. The grey-shaded area in the wave spectrum at 0.15 AU upstream of IPS#1 show the enhancement of wave power above the ambient upstream solar wind fluctuations.

Between the passage of IPS#1 and IPS#2 at SOHO the huge X28 flare occured at the Sun. This may be the reason that interplanetary space upstream of IPS#2 is filled with suprathermal ions as shown in the lower panels of Figure 2. Of course, the shock-accelerated ions resulting from the preceding IPS#1 may also contribute to the ion flux upstream of IPS#2. As the suprathermal ion flux upstream of IPS#2 has already been high, no distinct flux gradient has been established, suggesting that no anisotropies have built up to drive Alfvén waves. Indeed the proton spectra in the region of about 0.1 AU

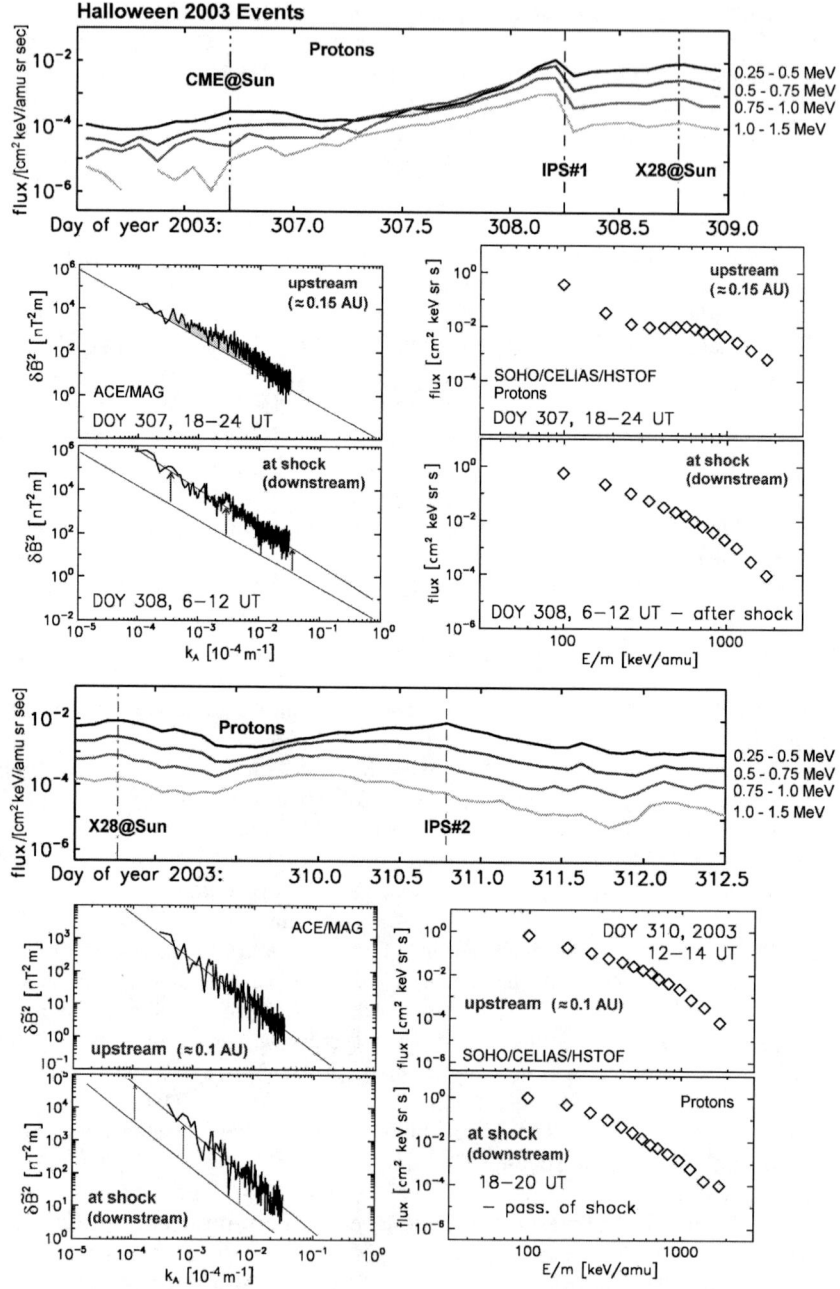

FIGURE 2. Spacecraft data associated with the Halloween 2003 CME events.

upstream of IPS#2 do not show the flat range above about 200 keV that indicates Alfvén wave driving as for the upstream region of IPS#1. Nowhere in the upstream region of IPS#2 this signature has clearly been identified. Consistent with the proton data, the wave spectra upstream of IPS#2 do not show any significant signature of Alfvén wave amplification by protons.

A special signature is visible in the upstream wave spectra of the Bastille Day event (Period B). It has been marked by "ion whistler waves." It is marked that way based on theoretical considerations. The QLT can be derived from the warm plasma dispersion tensor $\tilde{\varepsilon}$, which results in a wave growth rate $\gamma = -\varepsilon_{\text{Im}} \left[\partial \varepsilon_{\text{Re}} / \partial \omega \right]^{-1}$. In the QLT by Lee [1] pure Alfvén waves and the resonance condition $kv_{\parallel} = \Omega_p$ is assumed for the Alfvén wave amplification by protons. In this resonance condition, $v_{\parallel} \approx v$ is the proton speed parallel to the ambient magnetic field, which is in average not much different from the total proton speed v. If the Alfvén wave frequency $\omega = kv_A$ (v_A: Alfvén speed) approaches the proton angular gyro-frequency Ω_p, the correct resonance condition is $\omega + kv = \Omega_p$, and the left-handed Alfvén wave develops to a superposition of Alfvén wave and proton cyclotron wave called ion whistler wave. Its dispersion relation is $\omega = kv_A \left(\sqrt{1 + k^2 c^2 / (4\omega_{pp}^2)} - kc/\omega_{pp} \right)$ with ω_{pp} the proton plasma frequency. The corresponding wave growth rate by anisotropies can be derived to be

$$\gamma_{\text{IW}} = \gamma_{\text{L83}} \left[1 + \frac{k^2 c^2}{4\omega_{pp}^2} \right]^{-1} \frac{F_p \left([\Omega_p - \omega] / |k| \right) - F_p \left(-[\Omega_p - \omega] / |k| \right)}{F_p \left(\Omega_p / |k| \right) - F_p \left(-\Omega_p / |k| \right)}, \quad (1)$$

where the wave growth rate derived by Lee in 1983 [1] is denoted by γ_{L83} and the growth rate generalized for ion whistler waves is γ_{IW}. The quantity F_p is the integral over the radial proton speed distribution function at positive (first term) and negative pitch angle (second term with the minus sign in the approximate resonance condition). It appears that waves of frequency ω close to Ω_p resonate with significantly lower speeds $[\Omega_p - \omega] / |k|$ than in the case of the QLT by Lee [1], where the resonance condition is simply $v = \Omega_p / |k|$. Therefore, at given $|k|$, ion whistler waves interact with a higher phase space density of protons than the QLT by Lee [1] predicts. This can lead to higher wave amplification and explain the additional wave power marked for period B as "ion whistler waves".

ALFVÉN WAVE AMPLIFICATION AT THE SHOCK LAYER

At the layer of the main shock of the Bastille Day event (Figure 1, period C) the magnetic field fluctuations increase by at least a factor of 10. This number varies with wavenumber k because the injected upstream wave spectrum is not a complete power law, but the downstream wave spectrum is a pure power law, suggesting that the wave power is redistributed rapidly by cascading. As a reference we take the wavenumber $k_{100\text{keV}}$. A similar increase is observed for both the Halloween shocks IPS#1 and IPS#2, although IPS#2 has a much lower compression ratio (Figure 3). The data agree fairly well with the model by Vainio and Schlickeiser [2].

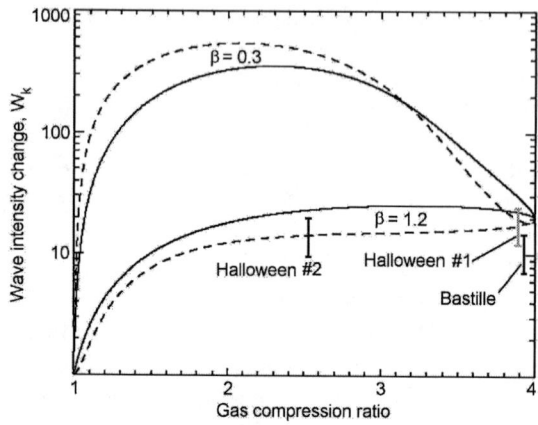

FIGURE 3. Amplification of magnetic field fluctuations at the interplanetary shocks associated with the CME-driven interplanetary shocks described in the text compared to the model by Vainio and Schlickeiser [2]. The dashed and solid lines denote the model results for cross helicities $H_{cl} = +1$ and -1, respectively. The wave amplification has been evaluated for the k-number that resonates with 100 keV protons at $\mu = 1$.

CONCLUSIONS

The quasi-linear theory (QLT) on self-consistent hydromagnetic wave excitation and ion acceleration at interplanetary travelling shocks [1, 4] is consistent with the spacecraft data for proton energies between about 200 keV and 2 MeV and corresponding Alfvén wave numbers. However, an extension of the QLT needs to include the change of the resonance condition and the dispersion relation at lower proton energies and at wave frequencies where left-handed Alfvén waves develop to ion whistler waves. This has been done above. Recently, the QLT has been extended to account more quantitatively for the variation of the pitch-angle of the resonating protons [5]. Another extension of the theory needs to include the cascading which ultimately reflects the dissipation of upstream wave energy. This point has been mentioned already by Gordon et al. [4].

ACKNOWLEDGMENTS

This work is supported in parts by INTAS grant WP 270 and by the Swiss National Science Foundation.

REFERENCES

1. Lee, M. A., *J. Geophys. Res.*, **88**, 6109–6119 (1983).
2. Vainio, R., and Schlickeiser, R., *Astron. Astrophys.*, **343**, 303–311 (1999).
3. Bamert, K., Kallenbach, R., Ness, N. F., Smith, C. W., Terasawa, T., Wimmer-Schweingruber, R. F., Hilchenbach, M., and Klecker, B., *Astrophys. J.*, **601**, L99–L102 (2004).
4. Gordon, B. E., Lee, M. A., Möbius, E., and Trattner, K., *J. Geophys. Res.*, **104**, 28'263–28'277 (1999).
5. Ng, C. K., Reames, D. V., and Tylka, A. J., *Geophys. Res. Lett.*, **26**, 2145–2148 (1999).

Nonresonant Alfvén waves driven by cosmic rays

Don Melrose

School of Physics, University of Sydney, NSW 2006, Australia

Abstract. Nonresonant growth of Alfvén waves due to streaming cosmic rays is considered, emphasizing the relation between resonant and nonresonant growth and the polarization of the growing waves. The suggested application of this mechanism to the scattering of higher energy cosmic rays in diffusive shock acceleration is discussed critically.

INTRODUCTION

Diffusive shock acceleration (DSA), proposed independently by a number of authors in the late 1970s [1], is the accepted theory for the acceleration of energetic particles in a wide variety of contexts in space and astrophysical plasmas. DSA requires efficient scattering of the particles, which is attributed to resonant scattering by Alfvén waves for ions and relativistic electrons. There is a one-to-one correlation between the parallel momentum, $p_\parallel$, of the particle and the parallel wave number, $k_\parallel \propto B/p_\parallel$, of the resonant Alfvén wave, where B is the ambient magnetic field. A gradient in the CR density upstream of the shock can cause the resonant waves to grow with a growth rate that is a function of $k_\parallel \propto B/p_\parallel$ [2]. The growth rate decreases with increasing $p_\parallel$, and above some characteristic $p_\parallel$, there are too few CRs to generate their own resonant waves. Nevertheless, it is usually postulated that the scattering occurs at the maximum possible rate, such that the diffusive mean free path is equal to the gyroradius. This implicitly requires resonant waves be present and it is usually simply postulated that they are, for example, as part of a turbulent spectrum generated independently of the CRs.

DSA can be very efficient in accelerating cosmic rays (CRs), invalidating the usual test-particle approach used in deriving the theory. For example, the CR pressure can become large enough to modify the properties of the shock [3], e.g., the recent review by [4]. In principle, the nonlinear modifications also affect the growth of waves, by allowing a nonresonant form of the instability to grow faster than the resonant form [6, 5]. Bell [5] argued that nonlinear development of the nonresonant form of the instability, driven by the current associated with the CRs, can result in another nonlinear modification: strong amplification of the magnetic field, B [7].

In this paper I explore the properties of the nonresonant instability with three specific objectives: (1) to clarify the relation between the nonresonant and resonant forms of the instability, (2) to determine the polarization of the growing waves, which has previously been assumed strictly circular [6, 5], and (3) to consider the suggestion that nonresonant wave growth coupled with the suggested increase in B might overcome the problem with the scattering of the highest energy CRs by reducing their gyroradii [5].

CP 781, *The Physics of Collisionless Shocks*, edited by G. Li, G. P. Zank, and C. T. Russell
© 2005 American Institute of Physics 0-7354-0268-X/05/$22.50

WAVE EQUATION

Starting from a general form for the dielectric tensor for a magnetized plasma with arbitrary distributions of particles of various species [2], I make the following approximations.

1. The waves are assumed to have a sufficiently low frequency that the frequency, ω, can be set to zero in the resonance condition $\omega - s\Omega - k_\| p_\| = 0$, with $\Omega = |q|B/\gamma m$ the relativistic gyrofrequency of a particle with charge q, mass m and Lorentz factor $\gamma = (1 - v_\perp^2/c^2 - v_\|^2/c^2)^{-1/2}$.
2. The small gyroradius approximation, $k_\perp v_\perp / \Omega \ll 1$ is assumed.
3. One gyroharmonics $s = \pm 1$ are retained.
4. The contribution to the dielectric tensor from the background plasma is assumed diagonal, with components $K_{11} = K_{22} = c^2/v_A^2$, $K_{12} = -K_{21} = i(c^2/v_A^2)(\omega/\Omega_i)$ $K_{33} \to \infty$, where v_A is the Alfvén speed and Ω_i is the ion gyrofrequency. The dispersion relations in the absence of the CRs are $\omega^2 = k_\|^2 v_A^2$ and $\omega^2 = (k_\|^2 + k_\perp^2)v_A^2$ for Alfvén and magnetoacoustic waves, respectively. These waves are linearly polarized in the approximation $\omega \to 0$, with the waves being circularly polarized for a small range of angles $\theta = \arctan(k_\perp/k_\|) \sim \omega/\Omega_i$ about the magnetic field when the condition $\omega \to 0$ is relaxed.

The streaming distribution of cosmic rays is modeled by the distribution function

$$f(p,\alpha) = f_0(p) + \cos\alpha f_1(p), \qquad f_1(p) = \frac{3v_{\rm CR}}{v} f_0(p), \tag{1}$$

where $v_{\rm CR}$ is the streaming velocity and v is the speed of the particle ($p = \gamma m v$). For a power-law distribution one has (e.g. page 22 of volume II of [2])

$$f_0(p) = K_{\rm CR}\left(\frac{p}{p_1}\right)^{-b}\begin{cases} 1 & \text{for } p_1 < p < p_2, \\ 0 & \text{otherwise.} \end{cases}, \qquad n_{\rm CR} = \frac{4\pi K_{\rm CR} p_1^3}{b-3}\left[1 - \left(\frac{p_1}{p_2}\right)^b\right], \tag{2}$$

where $n_{\rm CR}$ is the number density of cosmic rays.

Let the contribution of the streaming term in (1) to the dielectric tensor be denoted $\Delta K_{ij}(\omega,\mathbf{k})$. With the foregoing approximations the general result for the dielectric tensor for one species of CRs (e.g. page 41 of volume I of [2]) reduces to

$$\Delta K_{ij}(\omega,\mathbf{k}) = -\eta k_\| \frac{c^2}{v_A^2} \frac{3J_{\rm CR} B}{2\rho\omega^2} \frac{b-3}{x_1^{b-3}} \int_{x_2}^{x_1} dx\, x^{b-3} \sum_{s=\pm 1} F_{ij}(s) g_s(x), \tag{3}$$

with $x_n = |q|B/|k_\||p_n$, with $n = 1, 2$, and where $J_{\rm CR} = qn_{\rm CR}v_{\rm CR}$ is the CR current and ρ is the mass density of the background plasma. The tensor $F_{ij}(s)$ is

$$F_{ij}(s) = \begin{pmatrix} 1 & is\eta & 0 \\ -is\eta & 1 & 0 \\ 0 & 0 & 0 \end{pmatrix}, \tag{4}$$

where $\eta = q/|q|$ is the sign of the charge, and the integral over pitch angle leads to the function

$$g_s(x) = s\left[(1-x^2)\ln\left|\frac{x+1}{x-1}\right| + 2x\right] + i\pi(1-x^2)H(1-x^2), \qquad (5)$$

with $s = \pm 1$, $x = |q|B/k_\parallel p$, where $H(x)$ is the step function. The sum over s implies that only the imaginary (nonresonant) part of $g_s(x)$ contributes to $\Delta K_{11} = \Delta K_{22}$, and that only the real (resonant) part of $g_s(x)$ contributes to $\Delta K_{12} = -\Delta K_{21}$. (The strict cancelation of the resonant and nonresonant contributions to these terms, respectively, is an artifact of the approximations made.)

The wave equation then reduces to the form

$$\left|\begin{array}{cc} c^2/v_A^2 + \Delta K_{11} - n^2\cos^2\theta & \Delta K_{12} \\ \Delta K_{21} & c^2/v_A^2 + \Delta K_{22} - n^2 \end{array}\right| = 0, \qquad (6)$$

with ΔK_{ij} given by (3) and with $n^2 = |\mathbf{k}|^2 c^2/\omega^2$. Choosing ω as the dependent variable, and $|\mathbf{k}|$ and θ as the independent variables, the dispersion equation (6) becomes

$$\omega^4 - \omega^2[|\mathbf{k}|^2 v_A^2(1+\cos^2\theta) - 2i\xi] + |\mathbf{k}|^4 v_A^4\cos^2\theta - i\xi|\mathbf{k}|^2 v_A^2(1+\cos^2\theta) + 4(\xi^2+\chi^2) = 0, \qquad (7)$$

where $\Delta K_{ij} \propto 1/\omega^2$ is taken into account by writing $\xi = -i\Delta K_{11}\omega^2 v_A^2/c^2$, $\chi = -i\Delta K_{12}\omega^2 v_A^2/c^2$. The resonant and nonresonant contributions of the CRs are included through ξ and χ, respectively. The solutions of (7) are

$$\omega^2 = \omega_\pm^2 = \tfrac{1}{2}[|\mathbf{k}|^2 v_A^2(1+\cos^2\theta) - 2i\xi] \pm \tfrac{1}{2}\left\{(|\mathbf{k}|^2 v_A^2\sin^2\theta)^2 + 4\chi^2\right\}^{1/2}. \qquad (8)$$

In the absence of the CRs, $\xi, \chi \to 0$, these correspond to the magnetoacoustic and Alfvén modes respectively.

RESONANT AND NONRESONANT INSTABILITIES

The resonant instability is usually treated by neglecting the reactive effect of the CRs on the waves ($\chi \to 0$ here) and including the dissipative part by allowing the frequency to have a small imaginary part: $\omega \to \omega + i\Im\omega$. Growth corresponds to $\Im\omega > 0$. For the dispersion relations (8) one finds $\Im\omega = -\xi/2\omega$. Explicit evaluation gives

$$\Im\omega = \frac{3\pi(b-3)}{2b(b-2)}\frac{k_\parallel|J_{CR}|B}{\rho\omega}\begin{cases} x_1[b-(b-2)x_1^2] & \text{for } x_1 < 1, \\ 2x_1^{-(b-3)} & \text{for } x_1 > 1, \end{cases} \qquad (9)$$

where $x_2 \ll x_1 = |q|B/k_\parallel p_1$ is assumed. The result (9) applies to both Alfvén waves, with $\omega = |k_\parallel|v_A$, and to magnetoacoustic waves, with $\omega = |\mathbf{k}|v_A$. It also applies in the limit of small angles where the polarizations becomes circular. The condition $x_1 > 1$, corresponds to $|k_\parallel| < |q|B/p_1$ or $|k_\parallel|r_{g1} < 1$ where r_g, r_{g1} is the gyroradius for a particle with momentum p, p_1. Thus, the result applies to wave numbers $k_\parallel$ in the range where the resonance condition $|k_\parallel|r_g = 1$ can be satisfied for particles $p > p_1$ in the distribution

(2). Eq. (9) reproduces a known expression for CR-driven resonant wave growth (e.g., page 23 of volume II of [2]).

The nonresonant instability is found by retaining only the hermitian part of the response tensor for the CRs, and identifying the conditions under which the solutions for ω are intrinsically complex. When dissipation is neglected ($\xi \to 0$), the solutions (8) for ω^2 are always real, and the instability corresponds to $\omega^2 < 0$, when the growing wave has a purely imaginary frequency. This is possible only for the Alfvén mode, which becomes intrinsically growing for

$$\chi^2 > |\mathbf{k}|^4 v_A^4 \cos^2\theta. \tag{10}$$

The neglect of the dissipative part requires $\chi^2 \gg \xi^2$, which excludes the region $x_1 \lesssim 1$, where the opposite inequality applies. For $x_1 \gtrsim 1$, the non-resonant part of (3) implies

$$\chi = -\frac{3k_\| J_{\mathrm{CR}} B}{\rho} \frac{b-3}{x_1^{b-3}} \int_{x_2}^{x_1} dx\, x^{b-3} \left[(1-x^2)\ln\left|\frac{1+x}{1-x}\right| + 2x\right] \approx -\frac{2k_\| J_{\mathrm{CR}} B}{\rho}, \tag{11}$$

with $x_{1,2} = |q|B/|k_\| |p_{1,2}$, and where the approximate form applies for $x_1 \gg 1, x_2$. The condition (10) then requires

$$|J_{\mathrm{CR}}| > \frac{|\mathbf{k}|B}{2\mu_0} \gg \frac{|q|B^2}{2\mu_0 p_1}, \tag{12}$$

where the final inequality applies for $x_1 = |k_\| |r_{g1} \gg 1$ and $|\cos\theta| \approx 1$.

In the existing treatments of the nonresonant instability [6, 5] it is assumed that the waves are circularly polarized. The foregoing analysis allows one to determine the polarization. The polarization of the Alfvén mode is

$$\mathbf{e}_- = (\cos\phi, -i\sin\phi, 0), \qquad \phi = \tfrac{1}{2}\arctan[2\chi/|\mathbf{k}|^2 v_A^2 \sin^2\theta]. \tag{13}$$

In the limit of parallel propagation, $\sin\theta \to 0$, (13) implies $\phi = \pi/4$, which corresponds to circular polarization. At larger angles the polarization is elliptical. In view of the threshold condition (10), the axial ratio of the polarization ellipse for the growing waves never becomes large (the waves are never close to linearly polarized). The handedness of the growing waves is opposite to the sense of gyration of the CRs [6, 5].

DISCUSSION AND CONCLUSIONS

The treatment of the nonresonant instability in this paper confirms and extends the earlier treatments [6, 5]. Specifically, nonresonant growth applies at wavenumbers larger than that corresponding to gyroresonance with the lowest energy CR ($k_\| r_{g1} > 1$); the growing waves are on the Alfvén branch, whose frequency becomes imaginary when the condition (12) is satisfied; the growing waves are elliptically polarized in general, with the handedness determined by the CR current [6, 5]; the resonant and nonresonant forms of the instability pass over into each other when the parameters ξ^2 and χ^2 that describe dissipative and reactive effects of the CRs are roughly equal.

The threshold condition for nonresonant growth is approximately equivalent to the condition for the wave equation derived by Bell [5], his equation (15), to imply wave growth. This requires $|J_{CR}| \gtrsim |\mathbf{k}|B/2\mu_0$, which is similar to the relation, $|\delta J| = |\mathbf{k}|\delta B/\mu_0$, between the perturbations in the current and the magnetic field in a low-frequency wave. For semi-quantitative purposes, the condition (12) may be expressed in terms of the energy density, $U_{CR} \sim n_{CR}p_1 c$, in CRs in the form $U_{CR}v_{CR}/c \gg B^2/2\mu_0$, which requires that the energy flux in the CRs greatly exceed the energy density in the background magnetic field times c.

From a formal viewpoint the CR-driven instability of Alfvén waves has an unusual feature: the resonant and nonresonant contributions to the growth are determined by different terms in the dielectric tensor. The resonant growth is driven by diagonal terms (parameter ξ) that are non-gyrotropic, so that oppositely charged particles contribute in the same way. The nonresonant growth is driven by off-diagonal terms (parameter χ) that are gyrotropic, and may be described in terms of a current that depends explicitly on the sign of the charge. As a consequence, whereas the resonant growth occurs for linearly polarized waves (and causes circularly polarized waves to grow equally independent of their handedness), the nonresonant instability applies only to elliptically polarized waves with a specific handedness.

While it is plausible that nonlinear effects of CRs might lead to strong amplification of B upstream of a supernova shock [7], the suggestion [5] that the instability causing this is a nonlinear version of the nonresonant instability is speculative. The linear theory applies to a system with a uniform background magnetic field, and to waves with $k_\parallel$ larger than the inverse of the gyroradius of the lowest energy CR, whereas the nonlinear theory applies specifically to a field tangled on the same scale as the gyroradius. The question arises as to whether the linear nonresonant instability might be related to the firehose instability, which is an MHD instability that develops in the Alfvén mode when the pressure parallel to $\mathbf{B}$ is greater than the pressure perpendicular to $\mathbf{B}$. The growth rate for the firehose instability for $\theta = 0$ is $\Im\omega = |k_\parallel (P_\parallel - P_\perp - B^2/\mu_0)|/\rho$, and the growth rate for the nonresonant instability well above its threshold is $\Im\omega = |\chi|^{1/2}$, with $\chi = 2|k_\parallel J_{CR} B|/\rho$ according to (11) This suggests that the two instabilities are not closely related. The nonresonant instability does not appear to be related to any well-known instability, apart from its resonant counterpart.

ACKNOWLEDGMENTS

I thank Qinghuan Luo for helpful discussions.

REFERENCES

1. A. R. Bell,1978, MNRAS, 182, 147& 443; W. I. Axford, E. Leer and G. Skadron, 1977, Proc. 15th Int. Cosmic Ray Conf. (Plovdiv), 11, 132; G. F. Krymsky, 1977, Sov. Phys. Dokl. 23, 327; R. D. Blandford and J. P. Ostriker, 1978, ApJ, 221, L29
2. Melrose, D.B., 1980, *Plasma Astrophysics Volume I & II*, Gordon & Breach, New York

3. J. F. McKenzie and H. J. Völk, 1981, Proc. 17th Int. Cosmic Ray Conf., Paris **9**, 242; W. I. Axford, E. Leer and and J. F. McKenzie, 1982, A&A, 111, 317
4. M. A. Malkov and L. O'C. Drury, 2001, Rep. Prog. Phys., 64, 429
5. A. R. Bell, 2004, MNRAS, 353, 550
6. A. Achterberg, 1983, A&A, 119, 274
7. A. R. Bell and S. G. Lucek, 2001, MNRAS, 321, 443

Hamiltonian Approach to Nonlinear Travelling Whistler Waves

G.M. Webb*, J.F. McKenzie*,†, E. Dubinin** and K. Sauer**

*Institute of Geophysics and Planetary Physics, University of California Riverside, Riverside CA 92521, USA
†School of Physics, University of KwaZulu-Natal (Howard College), Durban, 4041, South Africa
**Max Planck Institute for Solar System Research, Katlenburg-Lindau, 37191, Germany

Abstract. A Hamiltonian formulation of nonlinear, parallel propagating, travelling whistler waves is discussed. The model is based on the equations of two-fluid electron-proton plasmas. In the cold gas limit, the complete system of equations reduces to two coupled differential equations for the transverse electron speed u and a phase variable $\phi = \phi_p - \phi_e$ representing the difference in the phases of the transverse complex velocities of the protons and the electrons. Two integrals of the equations are obtained. The Hamiltonian integral H, is used to classify the trajectories in the (ϕ, w) phase plane, where ϕ and $w = u^2$ are the canonical coordinates. Periodic, oscilliton solitary wave and compacton solutions are obtained, depending on the value of the Hamiltonian integral H and the Alfvén Mach number M of the travelling wave. The individual electron and proton phase variables ϕ_e and ϕ_p are determined in terms of ϕ and w. An alternative Hamiltonian formulation in which $\tilde{\phi} = \phi_p + \phi_e$ is the new independent variable replacing x is used to write the travelling wave solutions parametrically in terms of $\tilde{\phi}$.

Keywords: Whistlers, Solitary Waves, Oscillitons, Multi-fluid Plasmas
PACS: 94.30.Tz, 52.35.Hr

1. INTRODUCTION

Spacecraft observations of nonlinear plasma waves in the magnetosphere exhibit a variety of large amplitude solitary wave forms (Cattel et al. [1] Pickett et al. [2]). In particular, nonlinear, whistler oscillitons (Sauer et al. [3]) have been invoked to explain the 'coherent lion roars' emission in the magnetosphere (Zhang et al,[4], Baumjohann et al. [5]). Whistler oscillitons are nonlinear, travelling wave solutions of the multi-fluid plasma equations which contain an oscillating core inside a solitary wave envelope due to the different bulk gyration rates of ions and electrons about the magnetic field (Sauer et al.[6]; Dubinin et al. [7]; McKenzie et al. [8]).

Figure 1 from Sauer et al. [3], shows the transverse magnetic field components B_y and B_z for a nonlinear, travelling whistler wave in a cold plasma [panels (a) & (b)] and the Equator-S observations of Baumjohann et al. [5] and Geotail observations (Zhang et al. [4]) [panels (c) & (d)]. Note the fine scale oscillations inside the wave envelope, and the similarity of the observational wave forms and the theory.

In this paper we show that the parallel propagating, nonlinear travelling whistler waves in a cold plasma investigated by Dubinin et al. [7] and McKenzie et al. [8] are governed by a Hamiltonian dynamical system. Solutions of the system are obtained in closed form. Compacton type solitary waves, oscillitons, and periodic travelling waves

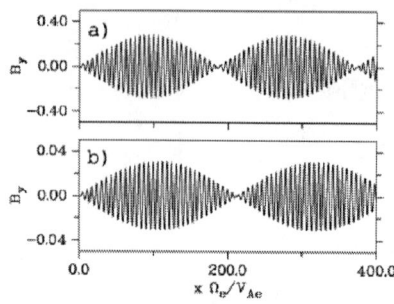
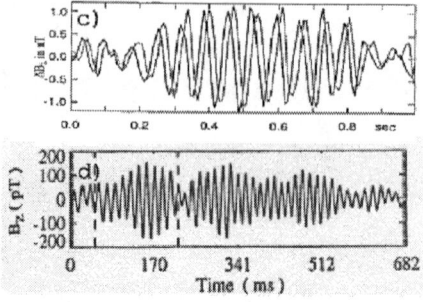

FIGURE 1. B_y and B_z for oscillitary, whistler travelling waves (panels (a) & (b)) and comparison with the Equator-S observations and Geotail observations (panels (c) & (d) : Sauer et al. [3])

are obtained, depending on the Hamiltonian integral, H, and Alfvén Mach number M, of the wave.

2. MODEL EQUATIONS

From Dubinin et al. [7], the equations describing the envelope of nonlinear, parallel propagating, travelling whistler waves in a cold, two-fluid, charge neutral electron-proton plasma can be reduced to the coupled differential equation system:

$$\frac{du_e}{dx} = M^2 u_e \sin\phi, \tag{1}$$

$$\frac{d\phi}{dx} = 2M^2(1+\cos\phi) - (1-u_e^2/\mu)^{-1/2}, \tag{2}$$

where u_e is the magnitude of the transverse electron velocity. Here $u_{j\pm} = u_{jy} \pm iu_{jz} = |u_j|\exp(\pm i\phi_j)$ ($j = e, p$) denote the transverse electron and proton fluid velocities in complex, amplitude and phase form, and the background field $\mathbf{B}_0$ is assumed to be parallel to the travelling wave velocity $\mathbf{U}$ (i.e. along the x-axis). In (1) and (2):

$$\phi = \phi_p - \phi_e, \quad x = \frac{X}{\ell_{\bar{m}}}, \quad \ell_{\bar{m}} = \frac{U}{\Omega_{\bar{m}}}, \tag{3}$$

$$\Omega_{\bar{m}} = \frac{eB_0}{\bar{m}}, \quad \bar{m} = \frac{m_p m_e}{m_p + m_e}, \quad \mu = \frac{m_p}{m_e}, \tag{4}$$

where X is the x-coordinate in the travelling wave frame, m_p and m_e are the proton and electron masses, and $\Omega_{\bar{m}}$ is the gyrofrequency corresponding to $\bar{m}$ (note that $\bar{m} \approx m_e$); $M^2 = U^2/V_{A\bar{m}}^2$ and $V_{A\bar{m}}^2 = B_0^2/(\mu_0 n_p \bar{m})$ defines the Alfvén Mach number of the wave M.

Auxiliary equations governing the evolution of ϕ_e and ϕ_p and the fine scale oscillations throughout the wave can be derived from the transverse electron and proton momentum equations, namely:

$$\frac{d\phi_e}{dx} = \frac{\mu}{(\mu+1)(1-u_e^2/\mu)^{1/2}} - M^2(1+\cos\phi), \quad (5)$$

$$\frac{d\phi_p}{dx} = M^2(1+\cos\phi) - \frac{1}{(\mu+1)(1-u_e^2/\mu)^{1/2}}. \quad (6)$$

Other equations governing the system are:

$$u_x = (1-u_e^2/\mu)^{1/2}, \quad u_p = \frac{u_e}{\mu}, \quad B_\pm \equiv B_y \pm iB_z = M_{Ae}^2(\mu u_{p\pm} + u_{e\pm}), \quad (7)$$

where M_{Ae} is the electron fluid Alfvén Mach number of the wave.

3. HAMILTON'S EQUATIONS

Equations (1)-(2) can be written in the Hamiltonian form:

$$\frac{d\phi}{dx} = \frac{\partial H_0}{\partial w} \equiv 2M^2(1+\cos\phi) - (1-w/\mu)^{-1/2}, \quad (8)$$

$$\frac{dw}{dx} = -\frac{\partial H_0}{\partial \phi} \equiv 2M^2 w \sin\phi, \quad (9)$$

where

$$w = u^2, \quad H_0 = 2M^2 w(1+\cos\phi) + 2\mu(1-w/\mu)^{1/2} \equiv 2M^2\Phi. \quad (10)$$

defines the Hamiltonian H_0 and $u \equiv u_e$. The canonical conjugate coordinates for the Hamiltonian system are (ϕ, w). The Hamiltonian is a constant of the motion. The integral (10) was obtained in the analysis of Dubinin et al. [7].

Introducing the variables:

$$\Lambda = (1-w/\mu)^{1/2}, \quad \tau = \tan(\phi/2), \quad H_0 = 2\mu H, \quad (11)$$

and using the Hamiltonian integral (10), (8) can be integrated to yield a second integral:

$$\int^\Lambda \frac{\Lambda\, d\Lambda}{\{(H-\Lambda)[2M^2(1-\Lambda^2) - (H-\Lambda)]\}^{1/2}} = -\sigma x + const., \quad (12)$$

where $\sigma = \text{sgn}(\tau) = \text{sgn}(\sin\phi)$. Note (12) is an elliptic integral.

3.1 Phase Plane Trajectories and Critical Points

The different types of solutions of the Hamiltonian system (8)-(9) can be determined from the trajectories in the (ϕ, w) phase plane, which correspond to the contours of the Hamiltonian H_0 in the phase plane. Hamiltonian critical points occur at points where

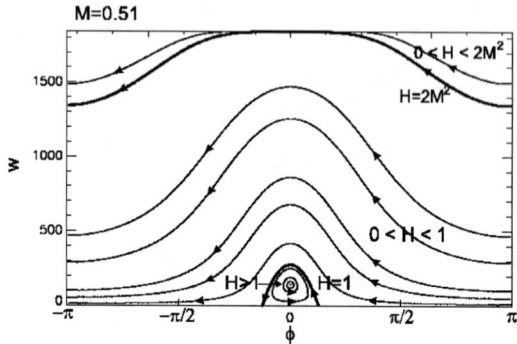

FIGURE 2. Hamiltonian contours for $H = const.$ where $H = H_0/(2\mu)$ and for the case $M = 0.51$

$H_{0\phi} = H_{0w} = 0$. There are two families of critical points. The first family occur at (ϕ_{n1}, w_{c1}):

$$w_{c1} = 0, \quad \phi_{n1} = 2n\pi \pm \phi_c, \quad \phi_c = \cos^{-1}\left(\frac{1-2M^2}{2M^2}\right), \quad H = H_{c1} = 1, \quad (13)$$

(n is an integer) and are saddles if $M > 1/2$. The second family occur at (ϕ_{n2}, w_{c2}):

$$\phi_{n2} = 2n\pi, \quad w_{c2} = \frac{\mu(M^4 - 1/16)}{M^4}, \quad H = H_{c2} = 2M^2 + \frac{1}{8M^2}, \quad (14)$$

which are centres (O-points) if $M > 1/2$. For $0 < M < 1/2$ there are no real critical points.

Phase plane trajectories for $M = 0.51$ are given in Figure 2 (the arrows correspond to increasing x). The trajectories with $2M^2 < H < 1$ correspond to compactons, (i.e. solitary waves with a finite spatial period L, in which $dw/dx = 0$ at the endpoints at $\phi = \pi$ and $\phi = -\pi$). The contour with $H = 1$ connecting the two saddle points at $w = 0$ corresponds to an oscilliton solution with an infinite spatial period (i.e. $L \to \infty$). The closed trajectories $1 < H < H_{c2}$ about the centre critical point (14) at $(0, w_{c2})$ correspond to periodic wave envelopes with a finite spatial period L.

The oscilliton solution, with $H = 1$ can be written parametrically in the form:

$$\phi = 2\tan^{-1}(\tau) = -2\tan^{-1}[\tau_c \tanh(k\tilde{\phi})]$$

$$x = \left(\frac{\mu+1}{\mu-1}\right)\tilde{\phi} - \frac{\tau_c}{M^2}\tanh(k\tilde{\phi}),$$

$$u = \frac{\sqrt{\mu}\tau_c}{2M^2}\text{sech}(k\tilde{\phi})[1 + \tau_c^2 \tanh^2(k\tilde{\phi})]^{1/2},$$

$$k = \frac{(\mu+1)\tau_c}{2(\mu-1)}, \quad \tau_c = \sqrt{4M^2 - 1}. \quad (15)$$

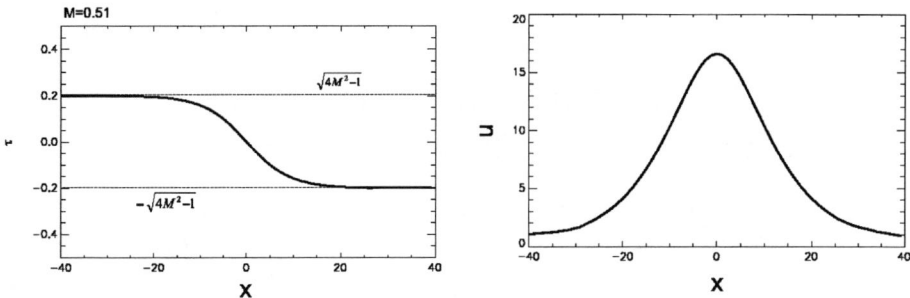

FIGURE 3. Plots of $\tau = \tan(\phi/2)$ and transverse electron fluid speed u versus x for the whistler oscilliton solution with $M = 0.51$ and $H = 1$.

where $\tilde{\phi} = \phi_e + \phi_p$. Note $\phi_p = 0.5(\tilde{\phi} + \phi)$ and $\phi_e = 0.5(\tilde{\phi} - \phi)$. Similarly, the compactons and periodic waves can be written parametrically in terms of $\tilde{\phi}$, using elliptic integrals and Jacobian elliptic functions ([9]). Figure 3 shows features of a whistler oscilliton for the case $M = 0.51$ and $H = 1$. The u-profile shows the envelope u for the oscilliton has a single hump profile similar to that of a soliton, and ϕ and τ decrease monotonically with x.

In this paper, we have provided a Hamiltonian framework to describe nonlinear, parallel propagating, travelling whistler waves in cold, multi-fluid plasmas. We anticipate that the Hamiltonian formalism, will also prove fruitful for more general cases (see Spencer and Kaufman [10], Holm and Kupershmidt [11] and Sahraoui et al. [12] for a Hamiltonian description of the multi-fluid plasma equations).

ACKNOWLEDGMENTS

The work of GMW is supported in part by NASA grant NAG13451 and by NSF grant ATM-0317509. We acknowledge stimulating discussions with G.P. Zank. The work of E.D. and K.S. is supported in part by the Max Planck Society.

REFERENCES

1. C. Cattel, et al., *J. Geophys. Res.*, **107**, 1238 (2002).
2. J. S. Pickett, et al., *Nonlin. Proc. Geophys.*, **10**, 3 (2003).
3. K. Sauer, E. Dubinin, and J. F. McKenzie, *Geophys. Res. Lett.*, **29**, 2226 (2002).
4. Y. H. Zhang, et al., *J. Geophys. Res.*, **103**, 4615 (1998).
5. W. R. Baumjohann, et al., *Ann. Geophys.*, **17**, 1528 (1999).
6. K. Sauer, E. Dubinin, and J. F. McKenzie, *Geophys. Res. Lett.*, **28**, 3589 (2001).
7. E. Dubinin, K. Sauer, and J. F. McKenzie, *J. Plasma Phys.*, **69**, 305–330 (2003).
8. J. F. McKenzie, E. Dubinin, K. Sauer, and T. B. Doyle, *J. Plasma Phys.*, **70**, 431–462 (2004).
9. G. M. Webb, J. F. McKenzie, E. Dubinin, and K. Sauer, *Nonlin. Proc. Geophys.* (2005), in press.
10. R. G. Spencer, and A. N. Kaufman, *Phys. Rev. A.*, **25**, 2437–2439 (1982).
11. D. D. Holm, and B. A. Kupershmidt, *Physica D.*, **6**, 347–363 (1983).
12. F. Sahraoui, G. Belmont, and L. Rezeau, *Physics of Plasmas*, **10**, 1325–1337 (2003).

Upstream gyrating ion events: Cluster observations and simulations

K. Sauer[1], M. Fränz[1], E. Dubinin[1], C. Mazelle[2], A.Korth[1], H.Rème[2], I.Dandouras[2], K.-H. Glaßmeier[3]

[1] Max-Planck-Institut für Sonnensystemforschung, Katlenburg-Lindau, Germany; [2] CESR/CNRS, Toulouse, France; [3] TU Braunschweig, Germany

Abstract. Localized events of low-frequency quasi-monochromatic waves in the 30s range observed by Cluster in the upstream region of Earth are analyzed. They are associated with a gyro-motion of the two ion populations consisting of the incoming solar wind protons and the back-streaming ions from the shock. A coordinate system is chosen in which one axis is parallel to the ambient magnetic field $\mathbf{B}_0$ and the other one is in the $\mathbf{v}_{sw} \times \mathbf{B}_0$ direction. The variation of the plasma parameters is compared with the result of two-fluid Hall-MHD simulations using different beam densities and velocities. Keeping a fixed (relative) beam density (e.g. $\alpha=0.005$), non-stationary 'shock-like' structures are generated if the beam velocity exceeds a certain threshold of about ten times the Alfven velocity. Below the threshold, the localized events represent stationary, nonlinear waves (oscillitons) in a beam-plasma system in which the Reynold's stresses of the plasma and beam ions are balanced by the magnetic field stress.

Keywords: Beam-excited low-frequency waves, two-fluid Hall-MHD simulations
PACS: 52.30.Ex, 52.35. Hr. 52.35.Mw, 52.35.Sb

INTRODUCTION

Large-amplitude, low-frequency, monochromatic electromagnetic waves with circular polarization are a characteristic feature of the Earth's foreshock. Their typical period is in the 20-30s range and they appear in most cases in association with gyrophase-bunched ion beams. It is thought that these gyrating ions are caused either by specular reflection of the solar wind protons at the shock and their subsequent gyration [1,2] or as a consequence of the right-hand resonant instability which may lead to a disruption of the originally field-aligned beam [3-5]. In recent papers [6,7] the latter mechanism has been considered within the frame of soliton theory looking for stationary, nonlinear solutions of the two-fluid Hall-MHD equations for a beam-plasma system. It has been shown that the simultaneous occurrence of a localized gyrating ion beam and the right-hand LF wave packet characterize a nonlinear structure which represents a soliton with superimposed spatial oscillations (oscilliton). It arises from the momentum exchange between the beam and the solar wind via the self-generated electromagnetic field. But a stationary structure only exists if the (undisturbed) beam velocity is below a certain threshold (V_{bt}) which is $V_{bt} \leq 10\ V_A$ (V_A: Alfven velocity) for beam densities $n_{bo} \leq 0.01\ n_{po}$. Since the beam velocities under foreshock conditions at Earth are mostly above this threshold, it is an interesting task to look into the Cluster data and to compare observed gyrating ion events with the results of <u>non-stationary</u>, two-fluid Hall-MHD simulations in which the beam velocity is not limited. Cluster observations and simulation results are described below. Conclusions from a comparison of both are summarized at the end.

CLUSTER OBSERVATIONS

We use measurements of the Cluster Ion Spectrograph (CIS) [10] onboard the 4 Cluster spacecraft to investigate waves in the foreshock region. CIS consists of the 2 instruments: a Composition and Distribution Function analyzer (CODIF) giving the mass per charge composition and a Hot Ion Analyzer (HIA) which does not offer mass resolution. The energy range of the sensors is ~20eV to 40keV. Each analyzer has two different sensitivities to increase the dynamic range. In solar wind instrumental mode used in this paper the high sensitivity side of the sensors measures ions from all spatial directions excluding the solar wind direction. The low sensitivity side measures the solar wind direction only.

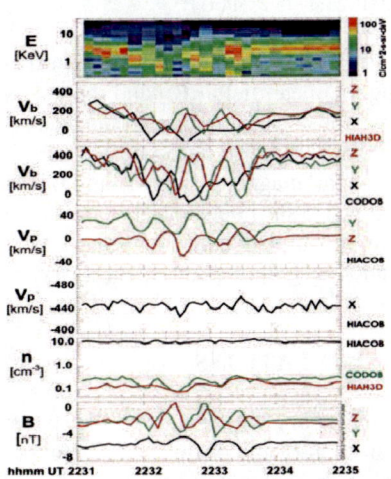

Figure 1. Observations of CIS and magnetometer on Cluster 1 for 27 January 2003 between 22:31 and 22:35 UT from top to bottom: a) differential flux of ions observed by HIA high sensitivity side as a function of energy, b) ion velocity measured by HIA high sensitivity (8s resolution), c) proton velocity measured by the CODIF sensor (4s resolution), d) y and z components of the solar wind velocity, e) x-component of the solar wind velocity, f) the densities of the solar wind observed by the HIA low sensitivity side (black) and of ions from non-solar directions observed by the CODIF sensor(green) and HIA high sensitivity side (red), g) magnetic field vector at 4s resolution. All data are given in GSE coordinates.

Fig.1 shows observations of the CIS instrument and the magnetometer on Cluster 1 recorded on 27 January 2003 between 22:31 and 22:35 UT when Cluster 1 was in the foreshock region at a distance of $19 R_E$. The parameters shown in the upper three panels exclude ions coming from the solar direction. We call this ion population the particle beam. The spectrum in the top panel of Fig.1 shows no continuous signal between 1 and 2keV. That means solar wind ions are well separated from the beam on the HIA high sensitivity side. The qualitative agreement between the ion velocity determined by the HIA high sensitivity side (panel b) and CODIF sensor (panel c) show that we can use CODIF onboard moments to determine the beam moments at 4s time resolution.

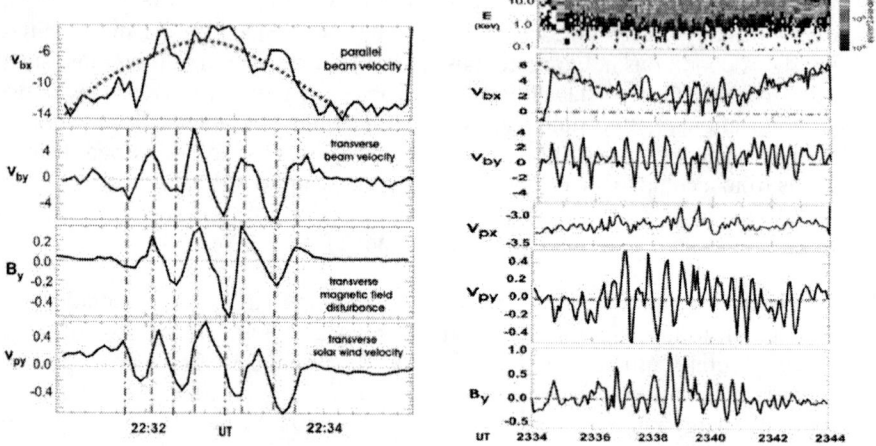

FIGURE 2. (left) Event from Fig. 1 transformed into BVM coordinates, velocities are normalized by the mean Alfven speed. The magnetic field is in units of its mean value <B>, from top to bottom: a) the field parallel beam velocity, b) a field-perpendicular component of the beam, c) the respective component of the field, and d) of the solar wind velocity. **(right)** HIA high sensitivity spectrum and respective parameters observed by Cluster 1 on 7 April 2001 23:34 to 23:44 UT.

Fig.2(left) shows the same observations as in Fig.1 rotated into a cartesian coordinate system (BVM) where the x-axis is along the mean field <B> and the y-axis is perpendicular to <B> and the mean solar wind vector <V>. Note the strong braking of the beam in field-parallel direction and the associated large gyration up to V_{by} ~5. At the same time, the field-transverse components of the magnetic field B_y and of the solar wind velocity grow up to about 0.5. The small phase shift between B_y and V_{by} at maximum indicates that most of the transverse momentum is carried by the beam. Fig.2(right) shows the respective parameters for another gyrating ion event analyzed by Mazelle et al. [8] in the same coordinate system.

TWO-FLUID HALL-MHD SIMULATIONS

As an introduction the limitation of the stationary theory with respect to the beam velocity is illustrated. Fig. 3a shows the spatial variation of the beam velocity (upper panel) together with the corresponding variation of the solar wind proton velocity (lower panel) which belong to the beam-plasma oscilliton described in earlier papers [6,7] for different beam velocities V_b at infinity and a relative beam density of $n_{bo}/n_{po}= 0.005$. With increasing beam velocity the solar wind becomes more and more accelerated up to a point where a critical value is reached which is close to the Alfven velocity V_A. For beam velocities above V_b~10, the stationary theory breaks down.

For the parameters of the gyrating ion event of Fig. 1 (n_{bo}/n_{po}=0.01, V_b=23), the LF wave dispersion of the beam-plasma system is shown in Fig. 3b. It results from the resonant interaction of the beam ($\omega=-\Omega_p+kV_b$) with the right-hand polarized mode. Vlasov theory for a warm plasma with β_p=1, T_b/T_p=5 (thin grey lines) and the cold two-fluid approach (black lines) give nearly the same results.

a)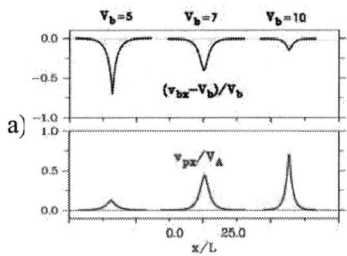
b)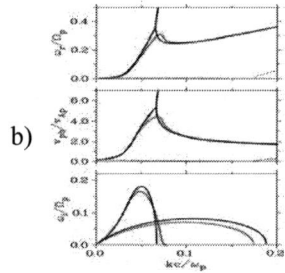

Figure 3a. Beam braking and plasma acceleration for different beam velocities V_b from the oscilliton theory [6,7].

Figure 3b. LF wave dispersion (frequency, phase velocity, growth rate) of a beam-plasma system : $n_{bo}/n_{po}=0.01$, $V_b=23$.

For one-dimensional, non-stationary simulations the two-fluid Hall MHD equations describing the interaction of the beam with the main proton plasma via the electromagnetic field [6,7] have been solved with the same parameters as in Fig. 3b. At the beginning the simulation box is homogeneously filled with the plasma and the beam which propagates parallel to the magnetic field. The left panels of Fig. 4a show the space-time evolution of the beam velocity V_{bx} and the plasma velocity V_{px} (parallel to $\mathbf{B}_o$) for T=60 up to T=100 (in units of Ω_p^{-1}). At T≈70 a strong deceleration of the beam appears which is associated with significant local plasma acceleration. This process proceeds and leads to the formation of shock-like transitions at later time. How the other plasma parameters look like just before this transition is shown in Fig. 4b. It can be seen that besides of deceleration the beam experiences a strong gyration which is indicated by the large velocity component V_{by} transverse to the magnetic field reaching about ten times the Alfven speed. The transverse solar wind velocity remains much smaller, but increases up to about the Alfven speed. At the same time the normalized transverse magnetic field component grows to ~1. Comparing these

a)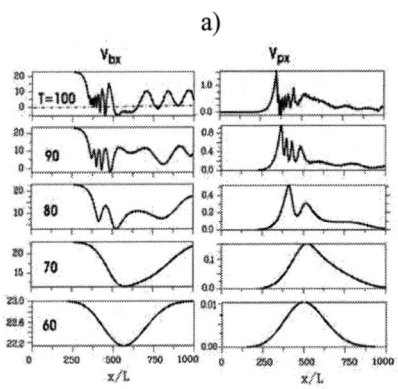
b)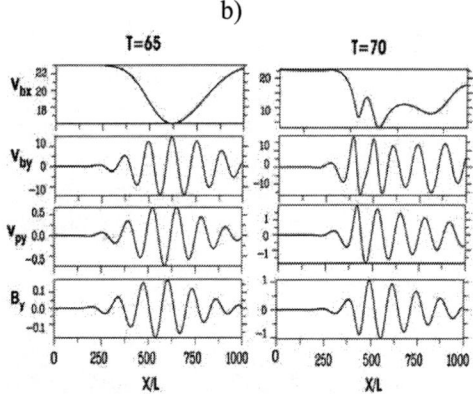

Figure 4a. Temporal evolution of both the beam velocity V_{bx} and the plasma velocity V_{px} near to the transition to a shock-like structure.

Figure 4b. Spatial profiles of the beam and plasma velocities (V_{bx}, V_{by}, Vp_y) and the transverse magnetic field B_y for two times before 'shock' transition.

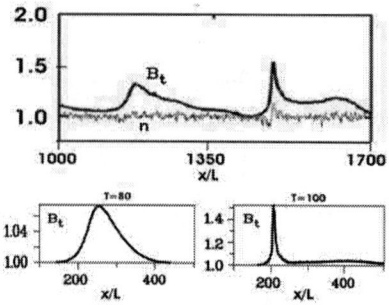

Figure 5. Transition to shock-like structures; comparison between results of hybrid code [9] shown in the upper panel and two-fluid Hall-MHD simulations (two panels below). The parameters are: $n_{bo}/n_{po}= 0.01$, $V_b=10$. Wave steepening occurs at about $T \geq 90$, and is marked by a significant enhancement of the magnetic field magnitude which is a remarkable feature in the case of parallel wave propagation.

results with the observations shown in Fig.2 (left) one finds a good agreement with respect to wave length and amplitudes. It is interesting to note that the described transition to shock-like structures have also been seen in hybrid code simulations [9].

CONCLUSIONS

The comparison of Cluster foreshock observations with the results of two-fluid Hall-MHD simulations for a beam-plasma configuration leads us to the following conclusions: In the case of small beam densities ($n_{bo}/n_{po} \leq 0.01$) the low-frequency right-handed foreshock waves in the 30s range are generated by the RH mode instability. For beam velocities below a certain threshold ($V_b \leq 10$) quasi-stationary nonlinear structures (oscillitons) may exist. The threshold is given by the condition that the plasma velocity reaches the oscilliton velocity which is close to the Alfven speed. Strong beam braking and formation of shock-like structures take place if the beam velocity exceeds the threshold at which infinite derivatives appear in the stationary theory. Adequate treatment of the shock-like structures requires multi-dimensional simulations.

ACKNOWLEDGEMENT

The authors express their gratitude to the ISSI (Bern) and thank the members of the International Team on Foreshock Ions. One of us (K.S.) thanks the IGPP of UC Riverside for the support during the 4th IGPP International Astrophysics Conference on 'Collisionless Shocks'.

REFERENCES

1. Gosling, J.T., et al., *Geophys. Res. Lett.*, **9**, 1333, 1982.
2. Gurgillio, C. Parks, G.K., and Mauk, B.H., *J. Geophys. Res.*, **88**, 9093, 1983.
3. Hoshino, M., and Teresawa, T., *J. Geophys. Res.* **90**, 57, 1985.
4. Thomsen, M.F. et al., *J. Geophys. Res.* **90**, 267, 1985.
5. Fuselier, S.A. et al., *Geophys. Res.Lett.*, **13**, 60, 1986.
6. Sauer, K., and Dubinin.E., *Geophys. Res. Lett.*, **30**, doi:10.1029/2003GL018266, 2003.
7. Dubinin.E., Sauer, K., and McKenzie, J.F., *J. Geophys. Res.*,**109**, doi:10.1029/2003JA010283, 2004.
8. Mazelle,C., et al., *Planet. Space Science.*, **51**, 785, 2003.
9. Wang, X. Y., and Lin, Y., *Phys.. Plasmas*, **10**, 3528, 2003.
10. Rème, H. et al., *Ann.Geophys.* **19**, 1303-1354, 2001.

Ion Thermalization and Wave Excitation Downstream of Earth's Bow Shock

Yong C.-M. Liu, Martin A. Lee and Harald Kucharek

*Department of Physics and Institute for the Study of Earth, Oceans and Space,
University of New Hampshire, Durham, NH 03824*

Abstract. We present a quasilinear theory for the "thermalization" of the proton and helium ions downstream of the quasi-perpendicular Earth's bow shock and for the associated ion-cyclotron waves. The residual ion temperature anisotropy, the wave polarization, and the magnetic fluctuation power and spectrum far downstream are predicted by the theory. The results are compared with the AMPTE/IRM data documented by Sckopke et al. [1]. These data are obtained at marginally supercritical shock crossings where we expect the quasilinear approximation to be valid. The agreement with the observations is generally very good if wave dispersion and the temperature of the transmitted protons are included. The helium contribution to the waves excited downstream of the shock is also estimated.

1. INTRODUCTION

Most of the time Earth's bow shock is supercritical which means some of the incoming protons in the solar wind are reflected from the shock. In a perpendicular or quasi-perpendicular shock the reflected protons are accelerated by the motional electric field before they gyrate back to the shock and pass through it. Then they gyrate in the downstream flow with a larger speed than the transmitted protons. They have a larger velocity spread in the direction perpendicular to the ambient magnetic field than in the parallel direction. This distribution with large temperature anisotropy is unstable and will excite ion-cyclotron waves; the waves in turn scatter the protons to a more stable distribution as the protons convect downstream. In a high-Mach-number shock this process evolves rapidly so that it is impossible to resolve. Sckopke et al. [1] chose several marginally supercritical shock crossings in the AMPTE/IRM data in order to investigate the evolution in detail. These events are labeled Events 1, 2, 3, 5, and 9. They find that the waves excited downstream of the shock are mainly left-hand-polarized, the magnetic fluctuation power is distributed in the frequency range 0.3Ω-0.8Ω (Ω is the proton gyrofrequency), and the spectra display a double-hump structure. Integration of the power within this frequency range is taken as a measure of magnetic fluctuation power. They also presented the ion temperature anisotropy. The downstream wave power range, peak frequency and temperature anisotropy values furthest from the shock are listed in Table 1 for all the events. The power has a larger fluctuation than that of the other quantities so that we show a characteristic range of variation. Our theoretical model is based on these species of ions: reflected protons, directly transmitted protons and helium. We predict the wave polarization and spectrum, and the residual temperature anisotropy and compare these values with the data.

2. SIMPLE THEORETICAL APPROACH

In the simple approach we neglect the contribution of the transmitted core protons and helium to the wave spectrum and assume the transmitted core protons are cold. We assume that the downstream plasma-frame speed of the reflected protons v_0 is much larger than the downstream Alfven speed V_A and we neglect wave dispersion so that all waves propagate with speed V_A. Later we relax these simplifications and assumptions. The advantage of the simple approach is that we can obtain analytical results. Assuming that all of the incoming protons have the same speed as the incoming solar wind, the velocity of reflected protons just downstream of the shock is calculated from the proton trajectory so that we obtain the initial speed v_0 and pitch angle before protons are scattered by the waves. The asymptotic proton distribution far downstream is the bispherical distribution [2]. Noting that the core protons dominate the average parallel velocity, we calculate the residual temperature anisotropy. The magnetic fluctuation power is calculated using the equations in Lee and Ip [3]; only left-hand polarized waves are generated. The peak frequency is then calculated.

The results are shown in Table 1 and the wave spectra of magnetic fluctuations for Events 1, 3, and 9 are shown in Figure 1. We immediately see that the results of the simple approach are good, but not perfect. The worst feature of the simple approach is that about 50% of the wave power is outside of the frequency range 0.3Ω-0.8Ω.

Event	$\dfrac{v_0}{V_A}$	$\left(\dfrac{\omega}{\Omega}\right)_o$	$\left(\dfrac{\omega}{\Omega}\right)_t$	P_{Lo}	P_{Lt}	$\left(\dfrac{T_\perp}{T_\parallel}\right)_o$	$\left(\dfrac{T_\perp}{T_\parallel}\right)_t$
1	2.3	.44	.64	.01-.05	.01	3.0	1.4
2	2.4	.41	.64	.01-.04	.01	2.7	1.5
3	3.1	.4	.48	.01-.04	.03	1.7	1.6
5	3.0	.5	.50	.02-.06	.02	1.9	1.7
9	4.0	.5	.37	.01-.04	.08	1.7	1.1

TABLE 1: Normalized reflected proton speed v_0/V_A; observed (subscript "o") and predicted (subscript "t") values for peak frequency (ω/Ω), wave power P_L and proton temperature anisotropy $T_\perp/T_\parallel$. Modified from [4] by permission of the American Geophysical Union.

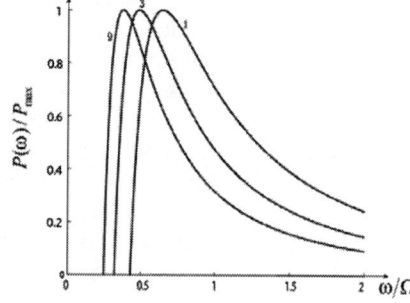

FIGURE 1: Predicted magnetic fluctuation spectrum. Reprinted from [4] by permission of the American Geophysical Union.

3. THEORY WITH WAVE DISPERSION

We now include wave dispersion in the ion-wave interaction. We use the cold plasma dispersion relation, only considering waves propagating along the magnetic field. The dispersion relation and the resonance line for a proton are shown in Figure 2. The slope of the resonance line is the parallel velocity $v_{//}$ of the resonant protons. Intersection of the two curves gives the resonant wave frequency and wavenumber. Since $v_{//}$ changes as the proton is scattered by the waves, the proton is sequentially resonant with "different" waves with different phase velocities. The protons are scattered along a path in velocity space on which their speed in the wave frame is locally conserved. The path forms a distorted sphere since the wave phase speed is changing [5]. We redo the calculation of the temperature anisotropy based on this surface. The wave power spectrum is calculated based on the energy loss of the protons during the process of redistribution [6]. The results (subscript "t") are shown in Table 2 and Figure 3. The predicted peak frequencies are now much closer to the observed values. The power spectra show that most of the power is now distributed within the frequency range 0.3Ω-0.8Ω.

Event	$\left(\dfrac{\omega}{\Omega}\right)_o$	$\left(\dfrac{\omega}{\Omega}\right)_t$	P_{Lo}	P_{Lt}	$\left(\dfrac{T_\perp}{T_{//}}\right)_o$	$\left(\dfrac{T_\perp}{T_{//}}\right)_t$	$\left(\dfrac{T_\perp}{T_{//}}\right)_{t2}$
1	.44	0.42	.01-.05	.01	3.0	1.6	2.5
2	.41	0.42	.01-.04	.01	2.7	1.6	2.6
3	.4	0.35	.01-.04	.03	1.7	1.5	1.9
5	.5	0.35	.02-.06	.02	1.9	1.5	1.9
9	.5	0.32	.01-.04	.07	1.7	1.3	1.5

TABLE 2: Observed and predicted quantities including effects of wave dispersion. Modified from [4] by permission of the American Geophysical Union.

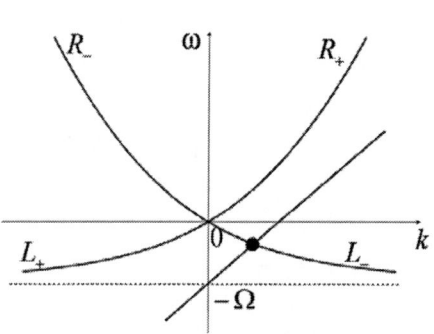

FIGURE 2: Dispersion relation $\omega(k)$ and resonance line $kv_{//} - \Omega$. Reprinted from [4] by permission of the American Geophysical Union.

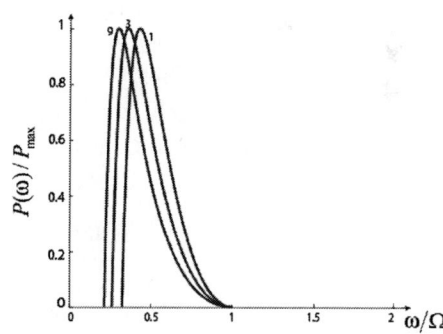

FIGURE 3: Predicted wave spectra for Event 1, 3 and 9 including wave dispersion. Reprinted from [4] by permission of the American Geophysical Union.

4. CORE PROTONS

The "core" protons, which are directly transmitted through the shock, also have a large temperature anisotropy since their thermal spread in the shock normal direction is increased when they cross the shock while the spread in the other two directions is preserved. Assuming that the upstream distribution is Maxwellian, we calculate the spatially-averaged distribution just downstream and allow it to be redistributed onto dispersive bispherical shells [7]. The contours of the distribution function just downstream and the bispherical shells are shown in Figure 4. We calculate the composite temperature anisotropy far downstream which includes the thermal spread of both reflected and core protons. The results are shown in the last column of Table 2. We also calculate the power spectrum based on the energy loss of the core protons, and add it to the spectrum excited by the reflected protons. Figure 5 shows the total spectrum predicted for Event 1. The total spectrum displays a double-hump structure with the secondary peak excited by the core protons. Using linear analysis Brinca et al. [8] also concluded that a secondary peak is excited by the core protons.

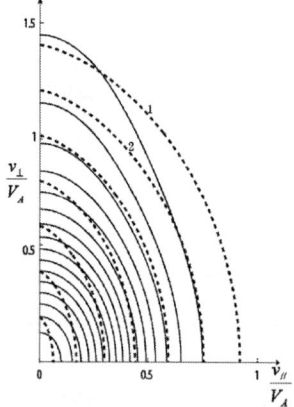 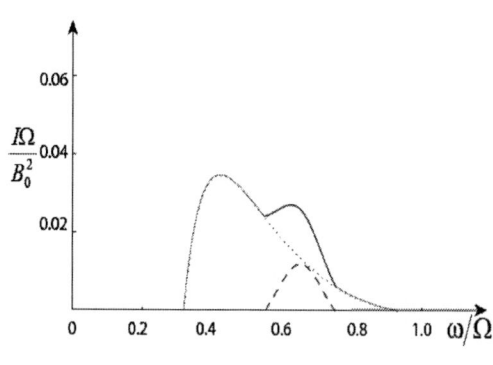

FIGURE 4: Contours of the initial core proton distribution function (solid curves). and bispherical shells (dotted curves). Reprinted from [4] by permission of the American Geophysical Union.

FIGURE 5: Total power spectrum (solid line) and the wave spectrum excited by reflected (dotted line) and core (dashed line) protons for Event 1. . Reprinted from [4] by permission of the American Geophysical Union.

5. HELIUM CONTRIBUTION

All helium ions are transmitted at the shock and gyrate downstream to form an anisotropic unstable distribution. They also excite waves downstream of the shock. The dispersion relation and resonance line imply that He^{2+} with small $v_{//}$ is resonant with waves propagating in both directions. To complicate matters, they damp resonant waves traveling in one direction and excite resonant waves traveling in the other direction. Noting that the resonant waves are also excited by protons, helium will follow a more complicated path when scattered by the waves. We estimate the contribution of helium to the excited wave spectrum by considering two extreme cases: helium dominates or protons dominate. In the helium dominant case, the wave excitation is controlled by the helium. In the proton dominant case, protons control wave excitation and helium ions behave as test particles. The resulting helium distribution in each case cannot be shown due to space limitation. The wave power excited by helium in each event is estimated based on the average of the power calculated in the two extreme cases. The results are shown in Table 3.

Event	$\frac{v_{He}}{V_A}$	P_{L1}	P_{L2}	P_{total}	P_{Lo}
1	0.4	0.003	0.0006	0.012	.01-.05
2	0.45	0.003	0.0006	0.012	.01-.04
3	0.4	0.006	0.0006	0.03	.01-.04
5	0.4	0.006	0.0006	0.03	.02-.06
9	0.35	0.003	0.0006	0.072	.01-.04

TABLE 3: The plasma-frame speed of helium downstream normalized by the Alfven speed, the predicted helium-excited power for the helium-dominant case P_{L1}, for the proton-dominant case P_{L2}, and the wave power due to helium and the reflected protons P_{total}. Modified from [4] by permission of the American Geophysical Union.

6. CONCLUSIONS

Based on the comparison of predictions and observations, we conclude:
1. Quasilinear theory works remarkably well.
2. Wave dispersion is important in the prediction of the excited wave spectrum.
3. The helium contribution appears to be small but the prediction should be improved.

We anticipate that analysis of the Cluster data will reveal new features of the ion-wave interaction downstream of Earth's bow shock.

ACKNOWLEDGMENTS

The authors are grateful for stimulating and useful discussions with P. Isenberg, E. Möbius, and B. Vasquez. This work was supported, in part, by NSF grant ATM-0091527 and NASA Sun-Earth-Connection Theory Program Grant NAG 5-11797.

REFERENCES

1. Sckopke, N. et. al., *J. Geophys. Res.*, **95**, 6337-6352 (1990).
2. Galeev, A. A. and R. V. Sagdeev, *Astrophys, Space Sci.*, **144**, 427-438 (1988).
3. Lee, M. A. and W. –H. Ip, *J. Geophys. Res.*, **92**, 11,041-11,052. (1987)
4. Liu, C.-M. Y., M. A. Lee and H. Kucharek, in print, *J. Geophys. Res.*, (2005)
5. Isenberg, P. A. and M. A. Lee *J. Geophys. Res.*, **101**, 11,055-11,066 (1996).
6. Huddleston, D. E. and A. D. Johnstone, *J. Geophys. Res.*, **97**, 12,217-12,230 (1992)
7. Isenberg, P. A. "A kinetic shell description of the ion cyclotron anisotropy instability" in *Solar Wind Ten*, edited by M. Velli, R. Bruno, and F. Malara, AIP Conference. Proceedings 679, Melville, NY, pp. 493-496.
8. Brinca, A. L., et al., *J. Geophys. Res.*, **95**, 6331-6335 (1990).

SESSION 4

PARTICLE ACCELERATION AT SHOCKS

Surfing acceleration of ions at relativistic, oblique shocks

Defne Üçer* and Vitali D. Shapiro*

*University of California, San Diego, Physics Department, California, 92093, USA

Abstract. It was shown in [1] that one of the limits to the shock surfing acceleration of ions at quasiperpendicular shocks is due to the obliqueness of the shock wave. In this paper, a critical obliqueness is found for the relativistic shock wave, below which the energy gain of the particle due to the shock surfing acceleration, is not limited by the obliqueness of the shock wave.

Shock surfing acceleration, which was proposed by Sagdeev [2] has proven to be an efficient acceleration mechanism especially for slow pick-up ions at the perpendicular shock waves [1, 5].

At the *strictly* perpendicular shock wave, the surfing ion's motion can be considered as a combination of bouncing motion normal to the shock front and acceleration in the plane of the shock front. Bouncing motion is due to fact that the ion is trapped in front of the shock between two turning points. Ion is reflected at the shock front due to the shock potential which is a characteristic of the quasi-perpendicular shock waves. Second turning point is due to the upstream magnetic field; Lorentz force pushes the ion back to the shock front. The ion is accelerated under the action of the convective electric field of the plasma flow as long as it is trapped between these two turning points.

The constraints on the acceleration can be summarized as follows:

1. The ion escapes from the acceleration region if the Lorentz force which pushes particle toward the shock front exceeds the electric force normal to the shock front.

$$E_x < \frac{v_y}{c} B_z$$

It is crucial to have E_x/B_z ratios close to one in order to keep the relativistic particle in the acceleration region. Such large E_x/B_z ratios are expected to be formed at high Mach number shocks, and in this letter it is assumed that the electric field normal to the shock front is sufficiently large such that the ion acceleration is not affected by this constraint. Self consistent numerical simulations at high Mach numbers need to be done to determine the real E_x/B_z ratios at the front as well as the accurate structure of the shock.

2. The ion escapes from the acceleration region if the ion's bounce kinetic energy exceeds the potential barrier at the shock front.

For the nonrelativistic case, the bounce energy of the ion increases during acceleration and this brings a serious constraint on the surfing acceleration. It has been shown in [4] that for a *relativistic* particle, bounce kinetic energy decreases during acceleration which allows the elimination of the second constraint. It is found in this earlier paper that if the surfing particle stays in the acceleration region until it reaches a critical energy, its bounce kinetic energy will never exceed potential barrier at the shock front. It is also shown in this paper that the ions that are slow with respect to the shock wave are more likely to reach the critical energy, once more verifying that the shock surfing acceleration is more effective for the ions that are slow in the shock frame.

3. At the oblique shocks, another limitation to shock surfing appears due to the presence of magnetic field component in the direction of the shock normal [1]. This component of the magnetic field causes cycloidal motion of the ion in the shock plane. Due to this motion, particle is untrapped from the acceleration region after a certain time.

Before we get into the equations let us clarify the structure of the shock and explain the last constraint qualitatively. Figure 1 shows the main fields in *the shock frame* and the part of the surfing ion's trajectory on the shock plane. The bouncing motion of the ion in x direction is not shown.

The shock is moving in $+x$ direction with a velocity $\mathbf{v_s} = u\,\hat{x}$, that is why the bulk plasma flow in the shock frame has a velocity $-u\,\hat{x}$. Since quasiperpendicular shocks are mainly propagating perpendicular to the ambient magnetic field there is a convective electric field at the shock frame:

$$\mathbf{E} = \frac{1}{c}\mathbf{v_s} \times \mathbf{B_0} = -\frac{u}{c}B_{z0}\hat{y} \qquad (1)$$

where $\mathbf{B_0}$ is the *upstream* magnetic field and $B_{y0} = 0$. The convective electric field is responsible for the acceleration of ion on the plane of the shock front (yz plane).

In panel (a) of figure 1, magnetic field is strictly perpendicular to the shock propagation direction. The bounce motion is ignored, therefore the ion trajectory on yz plane is a straight line in $-y$ direction. Ion is accelerating under the action of the convective electric field.

In panel (b) of figure 1 the shock is oblique, thus the magnetic field has a component normal to the shock front (x direction). It is seen in this figure that the x component of the magnetic field leads to cycloidal motion of the ion in the yz plane. When the ion velocity in y direction changes sign, the Lorentz force which is crucial for particle trapping changes direction, the particle is no longer returned to the shock front by this force, and the ion escapes from the acceleration region. In the figure, ion trajectory is shown until $v_y = 0$ when the ion can no longer be kept in the acceleration region due to the absence of upstream turning point.

There are a couple of more assumptions that we make in our analysis which are worth mentioning at this point. One of them is that the shock has a size and lifetime larger than any space and time scale of acceleration. In reality, although the surfing acceleration is quite fast, it is limited by the finite age and size of the shock. Our other assumption is that the width of the shock ramp is small compared to the ion gyroradius. Since the amplitude of bounce oscillations in front of the shock is of the order of ion gyroradius, this assumption leads to the picture of a particle bouncing between two turning points -shock potential and upstream Lorentz force- where the shock potential does not affect

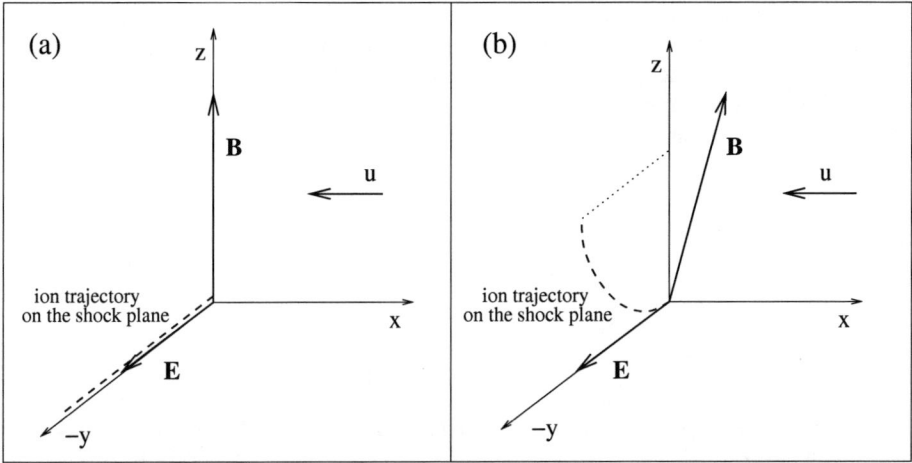

FIGURE 1. Shock geometry and surfing particle trajectory on the shock plane (yz plane). The bounce motion of the ion is not shown in this figure. a) Strictly perpendicular shock wave. Dashed line shows that the motion in the shock plane is acceleration in $-y$ direction. b) Oblique shock wave. Dashed line shows the cycloidal ion motion in the shock plane due to x component of the magnetic field.

the upstream turning point. Observations reveal such thin shock structures, where the width of the shock ramp is of the order of several electron skin depths [3]. The width of the shock ramp is also closely related to the cross-shock electric field at the shock front which was discussed as the constraint 1 earlier.

The effect of obliqueness on acceleration can be understood by *averaging the equations of motion of the ion over the period of bounce motion normal to the shock front* (x direction) following the analysis at [1]. Then for the non-relativistic case, one can obtain the following equations for ion dynamics in the shock plane:

$$\frac{dv_y}{dt} = -\frac{e}{m_i c}(uB_{z0} - v_z B_x) - \frac{e}{m_i c} v_x B_z \qquad (2)$$

$$\frac{dv_z}{dt} = -\frac{e}{m_i c}(v_y B_x - v_x B_y) \qquad (3)$$

where B_x is the component of magnetic field along the shock normal and B_y is the non-coplanar component of the magnetic field. After averaging the equations over time, the terms proportional to v_x vanish due to the periodicity of the bounce motion. The first term on the r.h.s. of equation (2) is the convective electric field which results with particle acceleration along the shock front.

It is important to emphasize that the normal component of the magnetic field (B_x) is constant through the shock front due to one dimensionality of the shock structure and $\nabla \cdot \mathbf{B} = 0$. Although B_z is not constant, only the upstream magnetic field B_{z0} plays a role in the relevant equations. This is why, variation of the magnetic field across the shock front is not taken into account in the present analysis.

The non-coplanar component of the magnetic field B_y does not have an effect on our governing equations because of the averaging. However, existence of this component could lead to a modification of the constraint 1, which was discussed earlier in the introduction. It is such that the normal component of the Lorentz force gains an additional term proportional to B_y which leads to the erosion of the electrostatic barrier at the shock front. However since the non-coplanar component of the magnetic field is small in comparison with the main field B_z, this modification is insignificant.

The equations (2) and (3) can be reduced to the following set of equations:

$$\frac{d^2 v_y}{dt^2} = -\Omega_x^2 v_y \tag{4}$$

$$\frac{d^2 v_z}{dt^2} = -\Omega_x^2 v_z + \Omega_x^2 u \frac{B_{z0}}{B_x} \tag{5}$$

where $\Omega_x = eB_x/m_i c$. Solution to these equations with initial condition $v_y(t=0) = v_z(t=0) = 0$ is:

$$v_y = -u \frac{B_{z0}}{B_x} \sin \Omega_x t \tag{6}$$

$$v_z = u \frac{B_{z0}}{B_x}(1 - \cos \Omega_x t) \tag{7}$$

As qualitatively explained earlier, at $t = \pi/\Omega_x$, the velocity component v_y changes sign and becomes positive, the upstream Lorentz force changes sign as well and the ion is no longer trapped in the acceleration region. Even if the other limitations that were discussed earlier do not apply, the acceleration is terminated after a certain time. This is a serious problem for shock surfing acceleration since there are no strictly perpendicular shock waves in nature.

Now, let us consider the relativistic case. In this case, the system of equations (2) and (3) are modified such that:

$$\frac{dp_y}{dt} = -e\frac{u}{c}\gamma_u B_{z0} + \Omega_x \frac{p_z}{\gamma} \tag{8}$$

$$\frac{dp_z}{dt} = -\Omega_x \frac{p_y}{\gamma} \tag{9}$$

$$\frac{d\gamma}{dt} = -\gamma_u \frac{u}{c} \Omega_z \frac{p_y}{\gamma m_i c} \tag{10}$$

where $\gamma = (1 - v^2/c^2)^{-1/2}$ is the usual Lorentz factor of the particle, $\gamma_u = (1 - u^2/c^2)^{-1/2}$ and $\Omega_z = eB_{z0}/m_i c$. Equations (8), (9) and (10) are obtained by averaging over the period of bounce motion similar to the nonrelativistic equations (2) and (3). An equation for the energy gain of the particle due to the convective electric field (10) is added to the usual equations of particle motion in the shock plane (8) and (9). Equations (8),(9) and (10) can easily be integrated by introduction of a new time variable τ such that:

$$\gamma(t)\frac{d}{dt} = \frac{d}{d\tau} \quad \text{or} \quad \tau = \int_0^t \frac{dt'}{\gamma(t')} \tag{11}$$

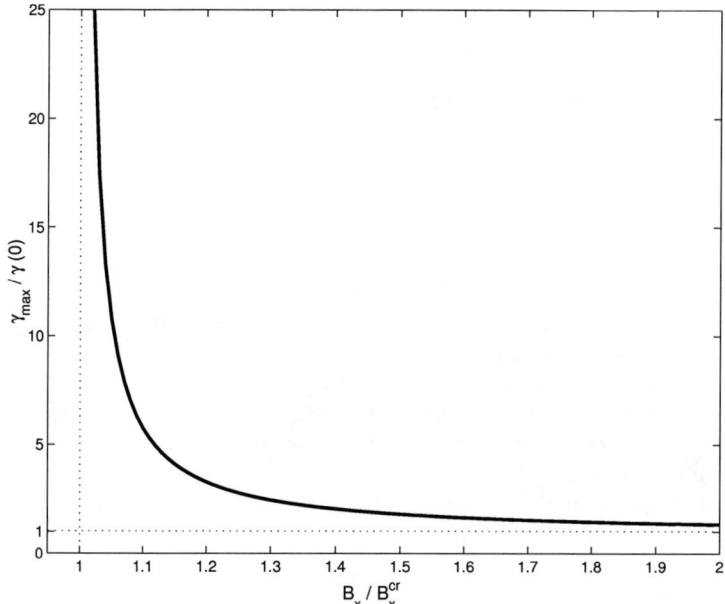

FIGURE 2. The efficiency of shock surfing acceleration as a function of obliqueness B_x/B_x^{cr}

Then differentiating equation (8) with respect to time and substituting equations (9) and (10), one can obtain:

$$\frac{d^2 p_y}{d\tau^2} + \Omega^2 p_y = 0 \qquad \text{where} \qquad \Omega^2 = \Omega_x^2 - \Omega_z^2 \frac{u^2}{c^2} \gamma_u^2 \qquad (12)$$

In the case of large obliqueness $\Omega^2 > 0$, the equation (12) yields a periodic solution for p_y. Using equation (8) and the initial conditions $p_y(0) = 0$, and $p_z(0) = 0$, one can obtain the following solution to p_y:

$$p_y(\tau) = -\gamma_u m_i u \frac{\Omega_z}{\Omega} \gamma(0) \sin\Omega\tau \qquad (13)$$

Also solutions for $p_z(\tau)$ and $\gamma(\tau)$ can be obtained as:

$$p_z(\tau) = \gamma_u m_i u \frac{\Omega_x \Omega_z}{\Omega^2} \gamma(0)(1 - \cos\Omega\tau) \qquad (14)$$

$$\gamma(\tau) = \gamma(0) \left(\frac{\Omega_x^2}{\Omega^2} - \gamma_u^2 \frac{u^2}{c^2} \frac{\Omega_z^2}{\Omega^2} \cos\Omega\tau \right) \qquad (15)$$

As discussed earlier, escape of the ion from the acceleration region occurs at time $\Omega\tau = \pi$, when p_y changes sign and the ion is no longer returned to the shock front by the Lorentz's force. Particle energy achieved at this time is then:

$$\gamma_{max} = \gamma(0)\left(1 + 2\gamma_u^2 \frac{u^2}{c^2}\frac{\Omega_z^2}{\Omega^2}\right) = \gamma(0)\left(1 + 2\frac{\gamma_u^2 u^2 B_{z0}^2/c^2}{B_x^2 - \gamma_u^2 u^2 B_{z0}^2/c^2}\right) \quad (16)$$

Now let us define a critical obliqueness:

$$B_x^{cr} = \gamma_u u B_{z0}/c \quad (17)$$

and plot the ratio of $\gamma_{max}/\gamma(0)$, which determines the efficiency of acceleration, as a function of the obliqueness of the shock B_x/B_x^{cr}. Figure 2 shows this relation.

The larger the B_x, the smaller the time of acceleration ($t = \pi/\Omega$) and the smaller the energy gain. At the critical obliqueness Ω changes sign and there is a singularity in equation (16). At $\Omega = 0$, the limit due to the obliqueness of the shock wave vanishes. If the acceleration is not terminated by any other reason, the ion is accelerated indefinitely, thus the gain is infinite.

The gain remains infinite below the critical obliqueness where $\Omega^2 < 0$. In this case the solution of equations (8) and (12) can be written as:

$$p_y = -\gamma_u m_i u \frac{\Omega_z}{\nu}\gamma(0)\sinh\nu\tau \quad (18)$$

$$p_z = -\gamma_u m_i u \frac{\Omega_x \Omega_z}{\nu^2}\gamma(0)(1 - \cosh\nu\tau) \quad (19)$$

where $\nu^2 = -\Omega^2 > 0$. As seen in these equations, below the critical obliqueness, the y component of ion momentum remains negative and normal component of the Lorentz force always pushes the ion toward the shock. Thus below the critical obliqueness the ion remains in the acceleration region indefinitely, if no other mechanism terminates the acceleration.

As a conclusion, the constraints on the shock surfing acceleration, which were overviewed earlier in this letter, are reduced significantly with the inclusion of the relativistic effects. If the electric field is sufficiently large at the shock front, then the shock surfing acceleration at quasiperpendicular collisionless shock waves, which are below critical obliqueness, can be theoretically unlimited since the two constraints 2. and 3. discussed earlier are eliminated.

REFERENCES

1. Lee, M.A., V.D. Shapiro, R.Z.Sagdeev *J.Geophys.Res*, **101**, 4777, 1996.
2. Sagdeev R.Z., *Reviews of Plasma Physics*,ed. by M.A.Leontovich, **4**, Consultants Bur., New York, p.23, 1966.
3. Scudder,J.D., A. Mangeney, C. Lecombe, C.C. Harley, T.L. Aggson, *J.Geophys.Res.*, **91**, 11053, 1986.
4. Ucer D., V.D. Shapiro, *Phys. Rev. Lett.*, **87**(7), 07500, 2001.
5. Zank G.P., H.I. Pauls, L.H. Cairns, G.M. Webb, *J.Geophys.Res.*, **101**, 457 , 1996.

Simulated 2D vs. 3D Shock Waves: Implications for Particle Acceleration

Frank C. Jones

*NASA/Goddard Space Flight Center
Greenbelt, MD 20771*

Abstract. We have given a rigorous derivation of a theorem showing that charged particles in an arbitrary electromagnetic field with at least one ignorable spatial coordinate remain forever tied to a given magnetic-field line. Such a situation contrasts the significant motions normal to the magnetic field that are expected in most real three-dimensional systems. While the significance of the theorem was not widely appreciated until recently, it has important consequences for a number of problems and is of particular relevance for the acceleration of cosmic rays by shocks.

Keywords: particle orbits, hamiltonian mechanics, ignorable coordinates, conservation theorems
PACS: 95.30.Qd,52.20.Dq

INTRODUCTION

In 1966 Jokipii [1] investigated cross field diffusion in the slab model of magnetic turbulence and found that in this model cross field diffusion was produced solely by field line wandering; the particles remained stuck to the field lines. The notion that particles could be stuck to magnetic field lines in certain situations was not a new one. In the same year Stern [2] published a review of all situations in which the concept of frozen field lines or particles stuck to field lines could have meaning. This would be a useful concept in the case of (a) a Perfectly conduction fluid (Ideal MHD)or a Guiding center plasma (adiabatic theory) with only drift motion of the particles.

In a paper by Jokipii *et al.* [3] the authors asserted a proof that particles were "fixed to field lines" in some sense if the situation involved less than three dimensions.

Unfortunately they included in the proof the sentances;

"Now define the frame moving with a velocity U_i as that in which the y-component of the Lorentz force is zero. *In the idealized hydromagnetic approximation this is the frame moving with the fluid and with which the magnetic field is convected*"(Emphasis added) This allowed many readers to assume that the theorem was only an approximation and did not apply to computer simulations that used exact equations of motion. It, therefore could be ignored. This continued even after Giacalone and Jokipii [4] published numerical simulations that demonstrated (See Fig. 1) the truth of the proposition.

To help remedy this situation we published [5] a paper that demonstrated rigorously and in detail the proof and precise meaning of this theorem.

Problem: Potentials do *not* have the same symmetry as the fields; but the Hamiltonian and the conserved canonical momentum are defined in terms of the potentials. An example: A uniform electric field has translational symmetry in three directions but the potential must have a gradient if the field is to have a non-zero value. The solution is to

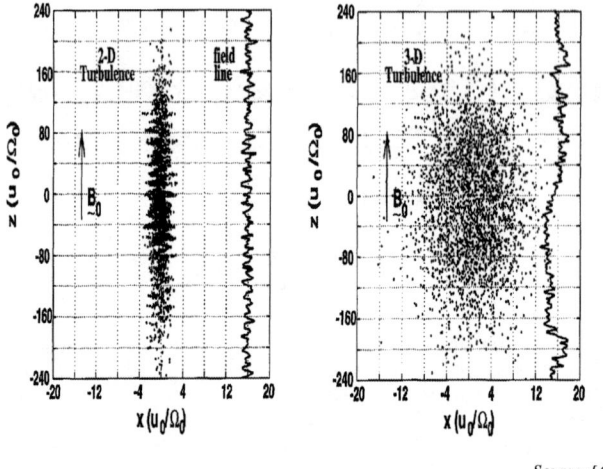

Source: [4]

FIGURE 1. Numerical results of Giacalone and Jokipii [4]

find a gauge in which the coordinate that is ignorable in the field is also ignorable in the potentials. Such a gauge exists; we were able to prove this by explicit construction. It turns out that this gauge is unique up to an overall scalar constant (in space and time). Therefore, any change in time of the potentials reflects real physical changes in the fields - no gauge relabeling motion.

RESULTS

It can be shown that the vector potential component in the y direction, A_2, is constant in the y direction (if y is chosen as the ignorable coordinate) and along a field line so it can be used to label a field line. This allows one to define the motion of a field line (surface) without invoking ideal MHD or adiabatic approximations *BUT* in the cases where these conditions exist this motion *IS* the frozen field motion. It should be pointed out that motion of a particle along the ignorable coordinate is allowed; since nothing varies in this direction such morion is of no interest. In fact when we refer to a field line we will henceforth mean the surface generated by moving the field line along the ignorable coordinate, gradient and curvature drift motion is of this type. The only situation in which this theorem does not apply, it gives essentially a null answer, is that in which the field lines themselves lie along the y direction.

We have from a basic result of Hamiltonian theory that the canonical momentum $p_y + \frac{e}{c} A_y = $ const., where p_y is the mechanical momentum. Therefore, since $-p \leq p_y \leq p$, where p is the scalar mechanical momentum, we have

$$(A_y)_{\text{Max}} - (A_y)_{\text{Min}} \leq \frac{2cp}{e},$$

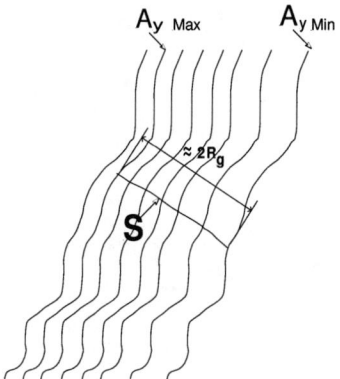

FIGURE 2. Field Lines showing the path S, its length and labeling the field line with the minimum and maximum values of A_y.

To obtain an estimate of the instantaneous spatial scale of this separation consider a curve, S, in the x_1, x_3 plane, from the field line with the minimum value of A_y to the one with the maximum value, that is everywhere perpendicular to the field $\vec{B}'(x_1, x_3) \equiv \hat{e}_1 B_1(x_1, x_3) + \hat{e}_3 B_3(x_1, x_3)$ where $d\vec{s} \cdot \vec{B}' = 0$ along the curve S (see Fig. 2).

We have then

$$(A_y)_{\text{Max}} - (A_y)_{\text{Min}}$$
$$= \int_S dx_1 \frac{\partial(A_y)}{\partial x_1} + \int_S dx_3 \frac{\partial(A_y)}{\partial x_3}$$
$$= -\int_S (d\vec{s} \times \vec{B}')_y \geq \int_S ds\, |B'|$$

Since the absolute value of this integral

$$\leq \frac{2cp}{e}$$

we may divide by $\frac{cp}{e}$ to obtain

$$\int_S \frac{ds}{r'_g} \leq 2$$

where r' is the instantaneous, local value of the particle's gyroradius in the field $\vec{B}'$. This is not unexpected; it primarily says that a particle will gyrate about field lines with an orbit whose extent is the diameter of its (approximately) circular orbit. See again Fig. 2.

As long as the canonical momentum remains fixed a particles mechanical momentum p can change and hence the values of $(A_y)_{\text{Min}}$ and $(A_y)_{\text{Max}}$ but it cannot wander away an arbitrary distance - hence *NO CROSS FIELD DIFFUSION*

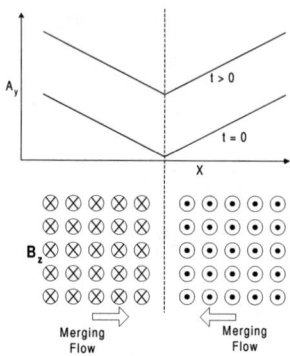

FIGURE 3. Merging field lines, the different curves represent the value of A_y at different times. The dotted circles are field lines out of the plane of the figure and the crossed ones are field lines into the plane of the picture

Field Line Merging

This theorem does allow a particle's energy to change and hence particle acceleration can take place in reduced diminsionality situations. Consider field anihilation. The fundamental equation remains

$$p_y + \frac{e}{c}A_y = \text{const}$$

So if the available values of A_y change (see Fig. 3) the value of p_y must change with it.

$$\begin{aligned}\frac{e}{c}\frac{\partial A_y}{\partial t} &= \frac{e}{c}U_x \cdot \frac{\partial A_y}{\partial x} \\ &= \frac{e}{c}U_x \cdot B \\ &= eE_y \\ &= \frac{\partial p_y}{\partial t}\end{aligned}$$

So we see that the particle's energy increases due to the electric field that exists in the neutral sheet of field line merging and anihilation. This result was known in 1988 by Ambrosiano *et al.* [6].

Shock Surfing

There are also situations in which a given field line is not anihilated but something prevents a particle that is bound to it from following its motion. A particular case is

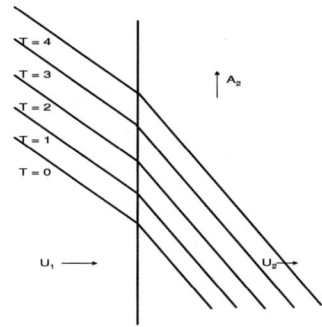

FIGURE 4. Given value of A2 as a function of time (NOT Field Lines!)

when a retarding or Cross-shock potential exists that prevends or at least slows down a positively charged particle from crossing the shock and following the field lines to which it is "attached". This is illustrated by Fig. 4. In this case the particle will move parallel to the shock and be energized, as in the previous example, by the $\vec{U} \times \vec{B}$ electric field and hence its gyroradius will increase by

$$\frac{\partial r_g}{\partial t} = \frac{c}{eB} \cdot \frac{\partial p_y}{\partial t} = U_x$$

. So r_g lengthens at just the rate that its initital value of A_y moves away. Hence when the particle finally passes through the shock it will still gyrate around the field lines that it previously encircled

REFERENCES

1. Jokipii, J.R., Apj 164, 480, (1966)
2. Stern, D.P., Sp. Sci. Rev. 6, 147, (1966)
3. Jokipii, J.R., Kóta, J., & Giacalone, J., J.Geophys. Res. 20, 1759, (1993)
4. Giacalone, J. & Jokipii, J.R., ApJ 430, L137, (1994)
5. Jones, F.C., Jokipii, J.R., & Baring, M.G., Apj 509, 238 (1998)
6. Ambrosiano, J. J.,Matthaeus, W. H., & Goldstein, M. L., J.Geophys. Res. 93, 14383 (1988)

Particle acceleration at collisionless shocks: An overview

G.P. Zank, Gang Li, G.M. Webb, J.A. le Roux, V. Florinski, X. Ao, and W.K.M. Rice

Institute of Geophysics and Planetary Physics
University of California, Riverside, CA 92521, USA.

Abstract. An overview of shock acceleration is presented, focusing primarily on interplanetary shocks and the termination shock as examples. An extended discussion of recent advances in modeling real solar energetic particle (SEP) and energetic storm particle (ESP) events is presented. When the energy of accelerated particles becomes very large, their back reaction on the flow can result in a shock that is significantly mediated, and as an example, we consider some results for the termination shock.

INTRODUCTION

Particle acceleration is ubiquitous at shock waves, occurring on scales ranging from supernova remnants to interplanetary shocks and cometary bow shocks. The mechanism thought to be responsible for the almost universally observed power-law spectra is diffusive shock acceleration [1-4]. The observed cosmic ray spectrum, until about 10^{14} eV/nuc., can be explained by particle acceleration at a supernova remnant shock [e.g., 5]. Diffusive shock acceleration is also generally thought to account for gradual SEP events [e.g., 6], but the correspondence between the simple predictions of theory and observations is often not compelling [e.g., 7] and more elaborate time-dependent models have had to be developed [8-12]. We discuss the basic elements of diffusive shock acceleration in terms of particle acceleration at coronal mass ejection (CME) driven shocks or interplanetary shocks.

Near the shock front, Alfven waves are responsible for particle scattering. The particle distribution f, and wave energy density A are coupled together through [2],

$$\frac{\partial f}{\partial t} + u \frac{\partial f}{\partial r} - \frac{p}{3} \frac{\partial u}{\partial r} \frac{\partial f}{\partial r} = \frac{\partial}{\partial t}\left(\kappa \frac{\partial f}{\partial r}\right); \qquad \frac{\partial A}{\partial t} + u \frac{\partial A}{\partial r} = \Gamma A - \gamma A;$$

$$\kappa(p) = \frac{\kappa_0}{A(k)} \frac{B_0}{B} \frac{(p/p_0)^2}{\sqrt{(m_p c/p_0)^2 + (p/p_0)^2}}; \qquad \kappa_0 = \frac{4}{3\pi} r_{g0} c = \frac{4}{3\pi} \frac{p_0 c}{eB_0}, \qquad (1.1)$$

where $\Gamma \propto \partial f/\partial r$ and γ are the growth and damping rates of the wave energy density, κ the spatial diffusion coefficient, κ_0 the Bohm limit, u the flow speed, B the magnetic field, and p the particle momentum. The particle transport equation and the wave energy density equation are coupled through the spatial diffusion tensor and the wave driving term Γ, and these equations must be solved simultaneously to determine both the accelerated particle spectrum (responsible for generating the waves) and the wave energy density (responsible for scattering the particles) self-consistently. However, the wave intensity driven by the accelerated particle gradient at a strong shock can often exceed the background mean magnetic field intensity, which violates the quasi-linear assumption implicit in equations (1.1). In this case, the Bohm limit is typically applied i.e., κ_0 is used when the wave energy density per logarithmic bandwidth exceeds the local solar wind magnetic energy density [8]. In [9] and [10], the formula in [13] is used to evaluate the wave intensity from (1.1). Steady state solutions to (1.1) show that the accelerated particle intensities are constant downstream of the shock and exponentially decaying upstream of the shock. The scale length of the decay is determined by the momentum dependent diffusion coefficient (steady state solution) (see Figure 1). Downstream of the shock, particles trapped by the turbulence convected from upstream of the shock experience convection, cooling as the flow expands back to its ambient state, and of course diffusion. Owing to imperfect trapping upstream and downstream of the shock, particles can escape upstream from the shock into the undisturbed solar wind. The escaping particles stream along the IMF, experiencing some scattering, and are observed at 1 AU as gradual SEP events.

One cannot assume that CME-driven shocks or interplanetary shocks are stationary since they have a variable, finite propagation time to 1 AU and weaken as they expand into the supersonic solar wind. A time-dependent model of diffusive shock acceleration at expanding interplanetary shocks is introduced in [8], modeling the solar wind flow and expanding shock self-consistently by solving the radial 2D gas dynamic or MHD equations numerically. At a given time, the accelerated particle spectrum was assumed to be the steady-state spectrum appropriate to local conditions,

$$f(t_k;t_k,p) = \frac{Q(t_k)\eta R^2(t_k)}{p_0^3} \frac{3-q(t_k)}{\left[p_{max}(t_k)/p_{inj}(t_k)\right]^{3-q(t_k)} - 1} \left(\frac{p_0}{p_{inj}(t_k)}\right)^{3-q(t_k)} \left(\frac{p}{p_0}\right)^{-q(t_k)}$$

$$\times \{H(p-p_{inj}(t_k)) - H(p-p_{max}(t_k))\}$$
(1.2)

where t_k is the kth time increment, $H(*)$ is the Heaviside step function, $p_{inj/max}$ denotes the injection/maximum momentum of the spectrum, R denotes the radius of the shock, $q(t_i) = 3s_i/(s_i-1)$ the accelerated particle spectral index, s_i the local shock compression ratio, and Q the local injection rate per unit area. In the absence of a widely accepted injection model, a simple prescription that the injection momentum is some fraction of the downstream thermal energy per particle is adopted in [8]. To compute the maximum local momentum is more laborious and essentially follows from equating the dynamical timescale of the shock to the acceleration timescale (i.e., the time available for a particle to be accelerated can be no longer than the lifetime of the shock)

$$\frac{R(t)}{\dot{R}(t)} \approx \frac{q(t)}{u_1^2} \int_{p_{inj}}^{p_{max}} \kappa(p') d(\ln(p')) .$$
(1.3)

On solving this equation for p_{max}, we obtain

$$p_{max} = \left\{ \left[\frac{M^2(t)+3}{5M^2(t)+3} \frac{R(t)\dot{R}(t)}{q(t)\kappa_0} \frac{B}{B_0} + \sqrt{\left(\frac{m_p c}{p_0}\right)^2 + \left(\frac{p_{inj}}{p_0}\right)^2} \right]^2 - \left(\frac{m_p c}{p_0}\right)^2 \right\}^{1/2} , \quad (1.4)$$

where M is the shock Mach number. The maximum momentum to which a particle can be accelerated is a balance of three factors: age through $R(t)$, shock strength through $\dot{R}(t)$ and $M(t)$, and the magnetic field strength $B \propto 1/r$, r heliocentric distance. Both the weakening of the shock and the decrease in magnetic field strength with increasing heliocentric distance, at least in the solar wind, ensure that the size/age of the shock does not dominate in determining the maximum energy, as has frequently been assumed. This is illustrated in Figure 2.

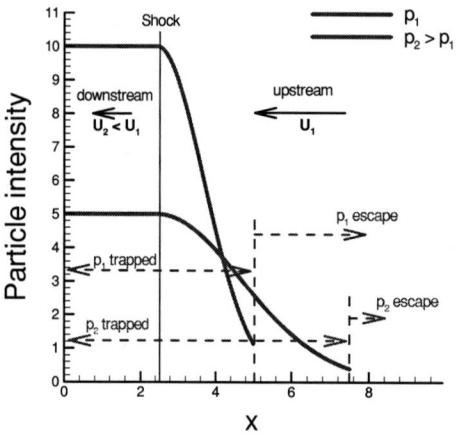
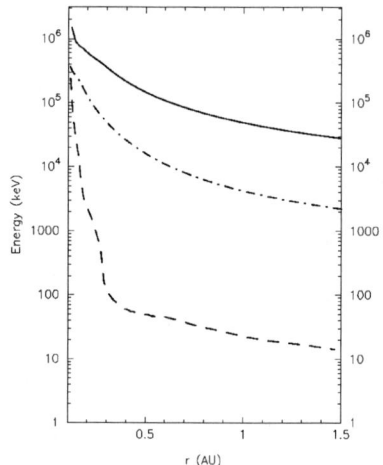

FIGURE 1: Schematic of the accelerated particle distribution at an interplanetary shock for two momenta, illustrating the momentum dependent scale length of the exponential decay ahead of the shock, and the corresponding trapping and escape or leakage of particles.

FIGURE 2: The maximum particle momentum accelerated at the shock front as a function of radial distance for a strong, intermediate, and weak shock respectively.

By following the evolution of the shock and using the local shock parameters to determine the local accelerated particle spectrum and the above prescriptions to obtain the injection and maximum momenta, we can then follow the time dependent acceleration of particles at an interplanetary shock. This has been done in a series of papers [8-11, 14], who coupled the above acceleration model to a transport model (not discussed here). The evolution of the spatial diffusion coefficient as a function of radial distance for a strong quasi-parallel shock is illustrated in Fig. 3, showing the strong time dependence of the particle acceleration timescale. The corresponding downstream spectra at a given solar wind location at different times after the shock has passed by are shown in Fig. 4. Note the rollover at high energies, which corresponds to particle escape from the post-shock flow, and the invariant spectra [e.g., 6].

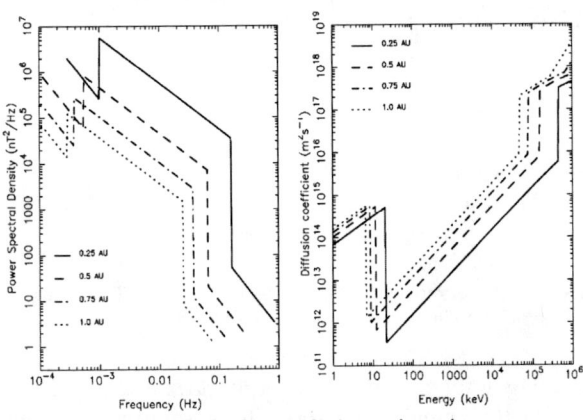

FIGURE 3: The evolution of the excited wave intensity at a strong quasi-parallel shock (left) and the corresponding diffusion coefficient (right) at different radial distances.

Observed at 1 AU, well upstream of the approaching interplanetary shock, the highest energy escaping particles arrive first, exhibiting a pronounced power law spectrum. At later times, the lower energy particles arrive and the spectrum begins to unroll. However, at later times, the shock is no longer accelerating particles to energies as high as those achieved close to the sun and the highest energy part of the spectrum is not replenished, leading to a pronounced double power law spectrum. The evolution of the spectrum observed at 1 AU is illustrated in Fig. 5. These represent cumulative spectra for five time intervals and the shock arrival time at 1 AU occurs T=1.3 days after shock initiation. Each panel corresponds to a different particle scattering mean free path in the interplanetary medium, and the right panel shows the spectra at very early times after the shock was initiated. Clearly, the spectra exhibit a power law feature, and a broken power law is apparent at later times, especially for larger mfp's (λ_0 = 1.6AU); e.g., for energies of 20 MeV for the time interval t = 4/5-1T since

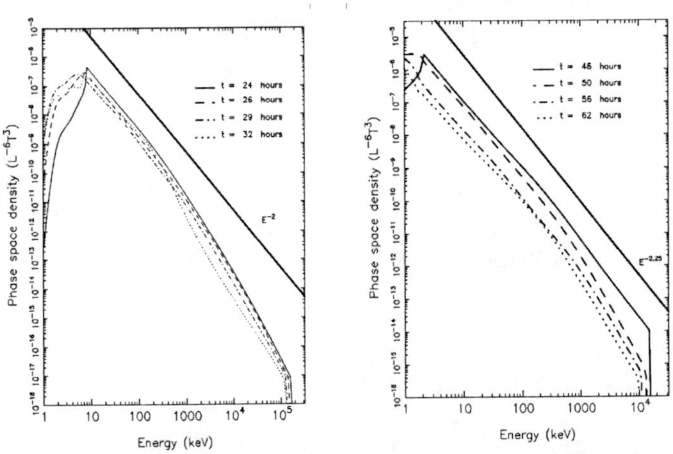

FIGURE 4: Spectra observed at 0.8 AU at four times after a strong (left) and an intermediate (right) strength shock has passed the observer. The heavy lines are reference power laws to guide the eye.

particle acceleration at the shock no longer occurs to these energies. The total or cumulative spectrum at 1AU can be obtained by integrating all the escaping particles over the time from shock initiation to the arrival of the shock at 1AU. Cumulative or

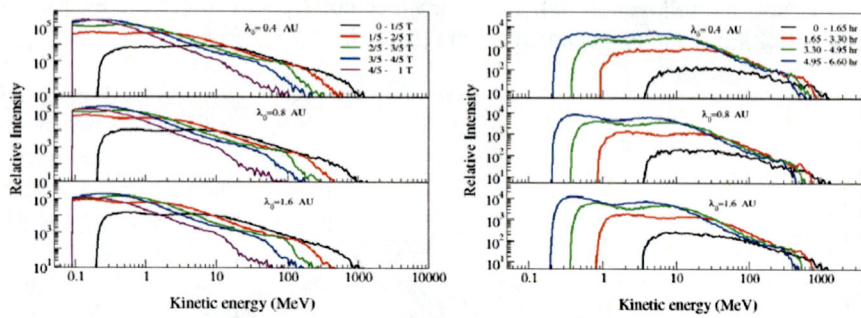

FIGURE 5: Particle spectra measured at 1 AU prior to the arrival at of a strong shock at T = 1.3 days after initiation. Three values for the scattering mfp in the solar wind are considered. The right panel corresponds to very early times.

event integrated spectra are shown in Fig. 6 for a strong and a weak shock example. Note the relatively pronounced roll-over in the cumulative strong shock spectrum and the rather flat power-law spectrum in the weak shock case. Fig. 6 illustrates how apparently

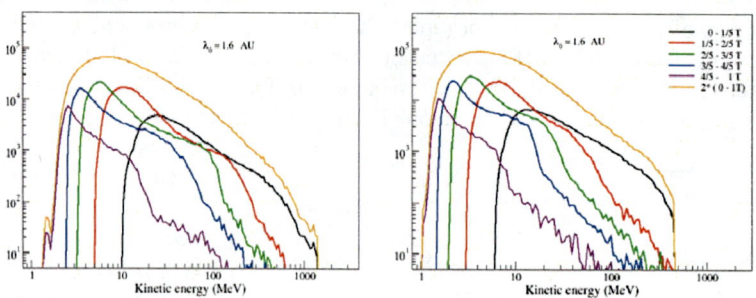

FIGURE 6: Total or cumulative spectrum at 1AU, integrated over the time from shock initiation to the arrival of the shock at 1AU.

"non-diffusive shock acceleration" features can emerge from a time-dependent diffusive shock acceleration model.

ACCELERATION OF HEAVY IONS

The preceding overview considered the acceleration of protons exclusively, but in situ observations of SEP events have for many years focused on the acceleration of heavy ions, one reason being to distinguish gradual and impulsive SEP events. Theoretically, it is necessary to understand proton acceleration at interplanetary shocks since their greater numbers means that they control the intensity and range of the turbulence spectrum responsible for scattering both protons and heavy ions. By contrast, heavy ions behave

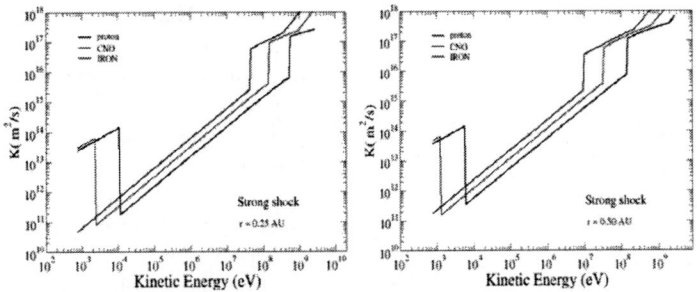

FIGURE 7: Diffusion coefficients at two heliocentric locations for protons (black), and CNO (red) and Fe (blue) ions. Note the shift towards lower energies according to Q/A at the maximum energy.

essentially as test particles and their low abundance means that they excite virtually no scattering turbulent spectrum. Based on this consideration, [11] apply the time dependent particle acceleration models of [8-10] to the acceleration and transport of heavy ions, specifically CNO and Fe, at CME-driven shocks. The effect of a proton induced turbulence field on heavy ions is manifested through the resonance condition and the particle mfp in the solar wind,

$$k = \frac{\gamma m_p \Omega}{\mu p}; \quad \Omega = \frac{(Q/A)eB}{\gamma m_p c}; \quad \lambda_p = \lambda_0 \left(\frac{pc}{1 GeV}\right)^{1/3} \left(\frac{A}{Q}\right)^{1/3} \left(\frac{r}{1 AU}\right)^{2/3} \quad (1.5)$$

which then determines the maximum energies for different mass ions and particle transport – both factors that distinguish heavy ion acceleration and transport from the proton counterpart. The diffusion coefficient for protons, CNO, and Fe ions is plotted in Fig. 7, showing that the maximum energy of heavy ions shifts towards lower energies as a result of the Q/A dependence in the cyclotron resonance condition (1.5). The maximum particle energy for a given heavy specie can then be computed as before on the basis of (1.3). The time dependence of the maximum energy for different species is illustrated in Fig. 8. Particles having higher energies, which are accelerated at earlier times and then trapped in the shock complex, will "see" a sudden change of κ. The maximum energy/nucleon for CNO is higher than iron since the former has a larger Q/A, and thus a smaller κ. In the case of the strong shock, the Bohm approximation is used throughout the simulation but only at early times in the weak shock case.

In [11], spectra at 1AU are plotted for nine different time periods corresponding to (i-1)/10T to i/10T with i = 1 to 9, where T is the shock transit time as before. The injected number of CNO and Fe ions is normalized to that of protons for comparison purpose. The spectra are plotted in Fig. 9 for a strong shock example. Initially, the spectra do not resemble power laws but, as more low energy particles reach 1 AU, the spectra evolve towards a power law spectrum. The power laws show clear breaks at higher energies, this a consequence of particle escape from the imperfect trapping region about the shock (see also Li et al., these proceedings). The $(Q/A)^2$ dependence of the highest energy particles exhibited in Fig.7 is evident in Fig. 9, although a little less clear because the observed spectrum at 1 AU is formed by particles escaping the shock complex over a range of times (unlike the single "snapshot" of Fig. 7). Comparatively, more low energy CNO and Fe particles than protons of the same energy per nucleon arrive at 1 AU at early times

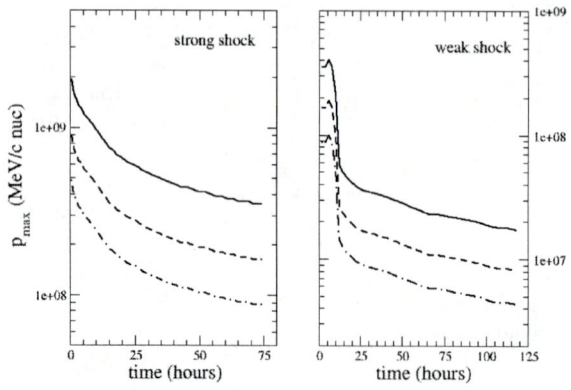

FIGURE 8: Maximum particle energies accelerated at a CME-driven shock for protons (solid curve), CNO ions (dashed), and Fe ions (dash-dot) as a function of time for a strong (left) and weak shock (right) shock. The Bohm approximation is used throughout the strong shock simulation but only initially in weak shock case.

(Fig. 9). This is because heavy ions have longer mfps at the same energy per nucleon, and, because protons have a shorter mfp, they are more easily caught and reabsorbed (reaccelerated) during their transport by the fast moving shock. Fig. 9 also reveals that the spectral indices for protons, CNO and Fe ions are different because of their different transport characteristics i.e., because the mfps have a $(Q/A)^{1/3}$ dependence, and there are more (relatively) low energy per nucleon Fe ions than CNO ions than protons, making the spectrum for Fe softer than CNO and softer than that of the protons. This is well illustrated in the second to fifth subpanels of Fig. 9.

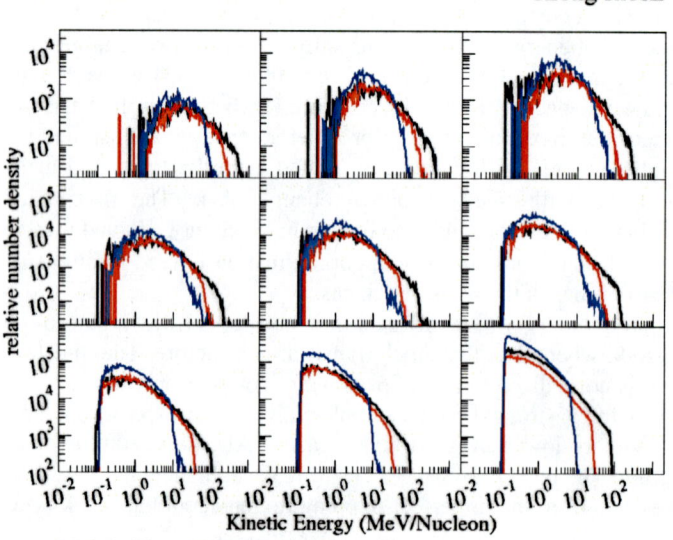

FIGURE 9: The time evolution of accelerated particle spectra observed at 1 AU for t = 0- 1/9T, t = 1/9-2/9T,...t = 9/10 – 1T, where T ~ 1.58 days or 50 hours (relative number density vs kinetic energy in MeV/nuc. Black curves refer to protons, red to CNO, and blue to Fe.

Many of the features of Fig. 9 appear to be in close accord with related observations [A. Tylka, private communication] and a closer analysis of theory and observations is clearly warranted. Finally, the integrated spectra at 1 AU for strong and weak shocks can be plotted (Fig. 10). For the strong shock, protons, CNO, and Fe exhibit similar spectral slopes at low energies, with Fe perhaps slightly softer. A roll-over or broken power law-like feature at high energies with an approximately dependence is also apparent. The weak shock case, by contrast, exhibits a quite different spectral shape with a strongly concave shape rather than the convex form of the strong shock case. By way of concluding this section, we show in Fig. 11 the Fe to O ratio for the strong and weak shock cases, normalized to coronal values, for different energies as a function of time. The ratio at early times easily exceeds 1, and is energy dependent. At later times, the lower energy ratio can drop well below 1. As discussed in [11], the ratios can be ascribed to differences in Fe and CNO propagation and trapping characteristics.

In summary, we have developed a fully time-dependent model of shock wave propagation (1- and 2-D), which includes local particle injection, Fermi acceleration at the shock, and non-diffusive transport in the interplanetary medium. Qualitatively, the modeling does very well in describing many characteristics of observed SEP events; this includes spectra, intensity profiles, and anisotropies. However, given the immense complexity of individual events, it remains challenging to model particular events. We can similarly model heavy ion acceleration and transport in gradual events, even understanding differences in Fe / O ratios. Furthermore, attempts to modeling mixed events to explore the consequences of a pre-accelerated particle population (from flares, for example)has been taken [12], which has also tried to relate mixed SEP events to the timing of flare – CME events.

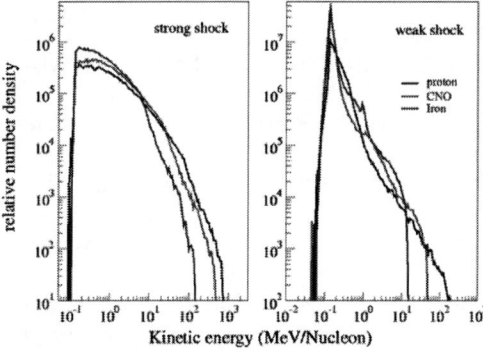
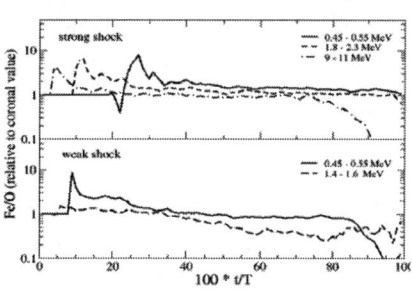

FIGURE 10: Event integrated particle spectra for a strong (corresponding to Fig. 9) and weak shock case.

FIGURE 11: Ratio of Fe to CNO ions, normalized to coronal values, as a function of time for different energy particles. T is the shock transit time.

COSMIC RAY MEDIATED TERMINATION SHOCK STRUCTURE

Although not generally thought to be a factor in interplanetary shocks, energetic particles accelerated at the termination shock (anomalous cosmic rays), for example, are expected to be sufficiently energetic that they react back onto the background flow and mediate the structure of the shock itself. Since energetic particles diffuse up- and

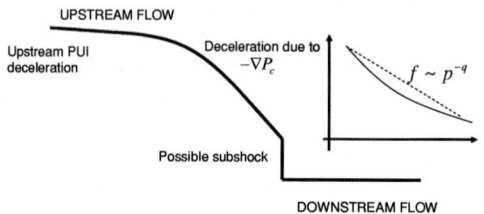

FIGURE 12: Cartoon of a cosmic ray mediated termination shock, showing the deceleration of the background flow by interstellar pickup ions, the formation of the shock precursor by the upstream cosmic ray pressure gradient, the subshock, and the downstream state. The inset shows the effect that shock mediation has on the usual power law particle spectrum accelerated at an unmediated shock (exaggerated for emphasis).

downstream, a pressure gradient develops ahead of the (termination) shock, acting to decelerate the incident flow prior to the shock itself. The overall structure therefore consists of a shock precursor and a subshock, illustrated in Fig. 12. Since the compression ratio of the shock effectively varies for particles of different energy, the accelerated particle spectrum acquires a convexity that is absent from the standard power law solution for a step discontinuity. The length scale of the precursor is evidently given by the diffusive scale of upstream particles $L_{diff} \sim \kappa/u_1$, where u_1 is the upstream flow speed. Computing what L_{diff} should be for the termination shock precursor is not trivial and estimates range from $L_{diff} = 0.1 - 5$ AU for 10 MeV protons and pickup ion decelerated solar wind. Because the shock is expected to be highly perpendicular, particle scattering and acceleration is by the random walk of magnetic field lines and therefore favors "fast" particles, which leaves an injection problem.

An interesting implication of the cosmic gradient is that it can subject the termination shock to Rayleigh-Taylor instabilities (in a generalized sense where the role of gravity is taken by the pressure gradient in the precursor; [15]) and overturning instabilities [16]. In the former case, the shock is unstable to backward propagating short wavelength (magneto)acoustic waves, which drive density, velocity and magnetic field fluctuations in the precursor. In the latter case, the shock is unstable to fluctuations propagating along shock front, these driving primarily coupled velocity and magnetic field perturbations. The linear theory for these instabilities has been developed and simulations show that the cosmic ray mediated shock can be completely destabilized. The presence of large amplitude fluctuations driven by a cosmic ray gradient introduces an interesting bootstrapping to a low frequency instability driven by pickup ions. In this case, the physical picture is one in which charge exchange in a compressed fluid element leads to overpressure and subsequent unstable expansion [17]. Consequently, the termination shock, or indeed any shock mediated by energetic particles, may be a highly turbulent structure. This raises the interesting possibility, discussed by [17, 18] and le Roux et al., 2005 (in preparation), that particle transport in a fluid turbulent medium may need to be described by a modified form of the cosmic ray transport equation (1.1), which includes corrections to the spatial diffusion coefficient and the convection term and includes a new second-order acceleration term. The precise effect that these new terms have on the accelerated particle spectrum and the acceleration time scale remain to be evaluated.

Thus, in summary, three primary results emerge from the turbulent shock model. (1) Cosmic ray gradient instabilities exist in a cosmic ray mediated termination shock. (2) A low-frequency instability driven by pickup ions will be bootstrapped to the CR gradient instabilities. Points 1 and 2 suggest a termination shock that is highly turbulent with large velocity, density, and magnetic field fluctuations. It is possible that these can steepen into "shocklets" leading to particle trapping, the formation of randomly distributed beams, etc. (3) Finally, to describe particle acceleration at a highly turbulent shock, we have developed a new form of transport equation from the usual diffusion equation that incorporates a large-amplitude random velocity field.

ACKNOWLEDGMENTS

This work is supported partially by NASA grants NAG5-10932, NAG5-11621, NAG5-12903, NAG5-11604 and NSF grant ATM-0317509.

REFERENCES

1. Axford, W. I., E. Leer, and G. Skadron, Proc. 15th Int. Cosmic Ray Conf. (Plovdiv), 11, 132, 1977.
2. Bell, A. R., Mon. Not. Roy. Astron. Soc., 182, 147-156, 443-455, 1978a,b.
3. Blandford, R. D., and J. P. Ostriker, Astrophys. J., 221, 269-280, 1978.
4. Krymsky, G. F., Doklady Akad. Nauk. SSSR, 234, 1306, 1977.
5. Völk, H. J., L. A. Zank, and G. P. Zank, Astron. Astrophys., 198, 274-282, 1988
6. Reames, D. V., Space Sci. Rev., 90, 413-491, 1999.
7. Desai, M. I., G. M. Mason, J. R. Dwyer, J. E. Mazur, R. E. Gold, S. M. Krimigis, C. W. Smith and R. M. Skoug, Astrophys. J., 588, 1149, 2003.
8. Zank, G. P., W. K. M. Rice, and C. C. Wu, J. Geophys. Res. (Space), 105, 25079-25095, 2000.
9. Li, G., G. P. Zank and W. K. M. Rice, J. Geophys. Res., 108, 1082, doi:10.1029/2002JA009666, 2003.
10. Rice, W. K. M., Zank, G. P. and G. Li, J. Geophys. Res.108, 1369 doi:10.1029/2002JA009756, 2003.
11. Li, G., G. P. Zank and W. K. M. Rice, J. Geophys. Res., in press, 2005.
12. Li, G. and G. P. Zank, Geophys. Res. Lett., 32, L02101, doi:10.1029/2004GL021250., 2005.
13. Gordon, B. E, A. M. Lee, E. Möbius and K. J. Trattner, J. Geophys. Res., 104, 28,263, 1999.
14. Li, G., G. P. Zank, M. Desai, G. M. Mason and W. Rice, AGU mono graph: Particle Acceleration in Astrophysical Plasmas in Geospace and Beyond, in press. 2004.
15. Zank G. P., and J.F. McKenzie, J. Plasma Physics, 37, 347—361, 1987.
16. Zank, G.P., W.I. Axford, and J.F. McKenzie, Astron. Astrophys., 233, 275—284, 1990.
17. Zank, G.P., X. Ao, W.I. Axford, V. Florinski, G. Li, J.A. le Roux, and G.M. Webb, in Proceedings of the 3rd Annual IGPP Astrophysics Conference Physics of the Outer Heliosphere, Ed.'s V. Florinski, N.V. Pogorelov, and G.P. Zank, 329 – 340, 2004.
18. Li, G., G.P. Zank, J.A. le Roux, G.M. Webb, and W.I. Axford, Particle acceleration at a turbulent shock, G., in Proceedings of the 3rd Annual IGPP Astrophysics Conference Physics of the Outer Heliosphere, Ed.'s V. Florinski, N.V. Pogorelov, and G.P. Zank, pp. 347 – 352, 2004.

The Energetic Storm Particle Event on 2003 October 24: A Test of Diffusive Shock Acceleration Theory

D. Lario*, R. B. Decker*, G. C. Ho*, Q. Hu[†], C. W. Smith[¶], M. I. Desai[‡], A.-F. Viñas[§]

*The Johns Hopkins University, Applied Physics Laboratory, Laurel, MD 20723, USA
[†]Institute of Geophysics and Planetary Physics, University of California, Riverside, CA 92521, USA
[¶]Space Science Center and Department of Physics, University of New Hampshire, Durham, NH 03824
[‡]Department of Physics, University of Maryland, College Park, MD 20742, USA
[§]Laboratory for Solar and Space Physics, Goddard Space Flight Center, NASA, Greenbelt, MD 20771

Abstract. A strong interplanetary shock accompanied by low-energy (<300 keV) ion intensity enhancements was observed by the ACE spacecraft at 14:48 UT on 2003 October 24. Energetic particle data collected by the EPAM instrument onboard ACE show that the time profiles of the 47-321 keV ion intensities, the ion anisotropies observed upstream and downstream of the shock, and the downstream energy spectra of the 47-321 keV ions agree with the predictions of the quasi-linear theory of diffusive shock-acceleration. This shock was also accompanied by an intense level of magnetic field fluctuations as observed during an extended (6 hours) fore-shock region by ACE/MAG. Although many of the signatures of this event are consistent with the predictions of the diffusive shock-acceleration theory acting on a low-energy particle population, analyses of the elemental abundances of heavy ions from ^{4}He to Fe measured by ACE/ULEIS indicate that the abundances of the accelerated ions are poorly correlated with those measured in the solar wind. This result is at considerable odds with the notion that the seed population for the shock acceleration mechanism is dominated by solar wind ions. By examining the relative behavior of the heavy ion spectra, we also investigate the rigidity-dependent nature of the acceleration mechanism operating in this event.

Keywords: Shock waves, Particle acceleration.
PACS: 96.50.-e, 96.50.Pw, 96.50.Fm

INTRODUCTION

The study of energetic particle signatures associated with the passage of transient fast forward interplanetary shocks has revealed a rich variety of both shock structures and types of energetic storm particle (ESP) events [1, 2, 3, 4]. The quasi-linear theory of diffusive shock acceleration (DSA) developed by Lee [5; see also 6] predicts the asymptotic steady state approached by energetic ion distributions and low-frequency wave spectra upstream and downstream of interplanetary traveling shocks. Several examples of ESP events analyzed in the literature show partial agreement between theoretical predictions and observations [e.g., 7, 8]. However, only one event (the ESP event observed by ISEE-3 on 1978 November 12) has been compared with the complete set of predictions [9]. Numerous points of agreement were found between theory and observation of both energetic particles and magnetic field fluctuations [9]; however, several points of disagreement such as the ion flow anisotropies [10] and the magnetic field frequency spectra [6] were also noted [9].

When a shock moves past the observer, it carries with it the history of events occurring during its transit time from the Sun to the observer. The dynamic evolution of both the shock and the medium through which it propagates contribute to shape the properties of the observed ESP events. The asymptotic steady-state energy spectrum up to some maximum energy (E_{max}), as predicted by DSA theory, can only be attained if (1) there is enough time to reach this steady-state during the transit time of the shock, and (2) the characteristics of both the shock and the energetic particle population do not change substantially as the shock approaches the observer. The in-situ observation of shocks and energetic particle distributions provide the only meaningful way to check both the attainment of the steady-state spectrum for $E<E_{max}$ and other predictions of the quasi-linear DSA theory. In this paper we present one of the ESP events observed by ACE during solar cycle 23 [4] that shows several points of agreement with DSA theory.

DATA ANALYSIS

A fast forward interplanetary shock moved past the ACE spacecraft at 14:48 UT on 2003 October 24 (day of the year 297). By solving the complete set of Rankine-Hugoniot (R-H) relations we obtain the following shock parameters: shock speed along the normal direction $Vs=198\pm21$ km s^{-1}, angle between the upstream magnetic field and the shock normal $\theta_{Bn}=57°\pm6°$, magnetic field compression ratio $r_b=2.1\pm0.5$, density compression ratio $r_n=2.3\pm0.7$, and Alfvenic Mach number $M_A=3.1\pm0.5$. Since the resolution of the R-H relations involves a least-squares fitting technique with r_n as a floating parameter, we have also computed the upstream to downstream proton solar wind density ratio measured before and after the shock passage $H=3.4\pm0.7$ that is within the ±1-sigma error bar of r_n.

Figure 1 shows from top to bottom the ion (mainly proton) intensities as measured by the LEMS120 telescope of the EPAM instrument [11], the proton solar wind speed, density and temperature measured by the SWEPAM instrument [12], and the magnetic field magnitude (B) measured by the MAG instrument on board ACE [13]. The bottom panel shows the rms value of $\vec{B}$ (dBrms) using the high-resolution measurements (3 vectors/second) and computed as $\left[\sum_{i=1}^{3}<(B_i-<B_i>)^2>\right]^{1/2}$ where B_i is each component of $\vec{B}$ and $<>$ is the average over the indicated time interval. The fluxes of 47-68 keV, 68-115 keV and 115-195 keV ions increased roughly exponentially at least 20 minutes ahead of the shock, maximized 6 minutes after the shock and remained approximately constant downstream of the shock. In contrast to most of the transient shocks observed by ACE, this specific shock shows an increase in the dBrms values for at least 15 minutes before the shock passage. This shock also presented enhanced magnetic field fluctuations in its downstream region, as commonly observed in most of the fast forward transient shocks observed by ACE.

Figure 2 shows the evolution of the spectral index γ obtained by fitting the 47-321 keV ion fluxes to a power-law $E^{-\gamma}$ function and the value predicted by the quasi-linear DSA theory (e.g., $\gamma=(H+2)/(2H-2)$; solid horizontal line). The spectral index γ increases 21 minutes before the shock passage and reaches a constant value 6 minutes after the shock with values contained within the range of uncertainty of the value predicted by DSA theory (dashed lines).

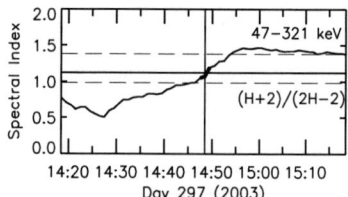

FIGURE 2. Spectral index γ before and after the shock passage.

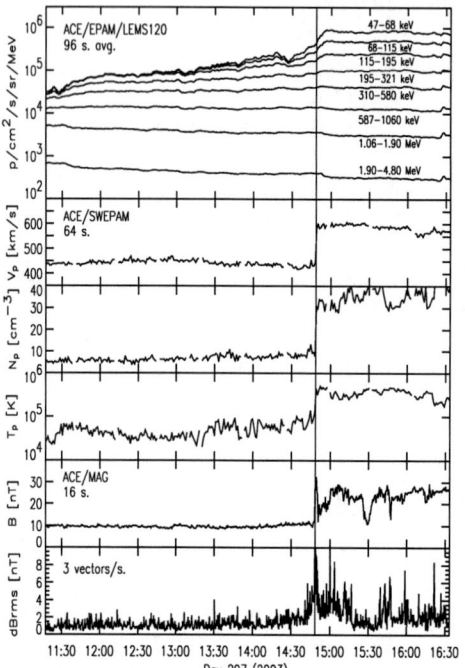

FIGURE 1. Energetic ion fluxes, solar wind and magnetic field observations upstream and downstream of the shock.

Figure 3 shows the ion intensity and the parallel streaming anisotropy computed in the solar wind frame for different energy channels (the definitions of the parallel anisotropy and the details of the procedure used to derive it are given in [14]). Figure 3 shows that ion anisotropies decay or stay relatively constant before the shock

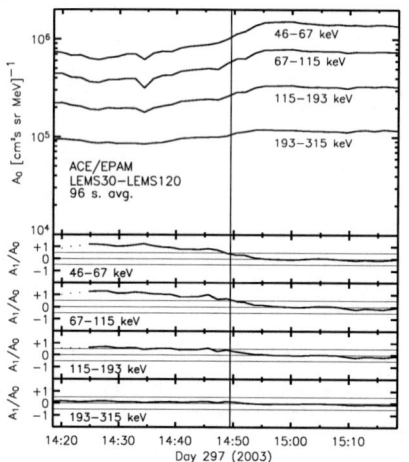

FIGURE 3. Energetic ion flux (top panel) and parallel streaming anisotropy in the solar wind frame and along the magnetic field (four bottom panels).

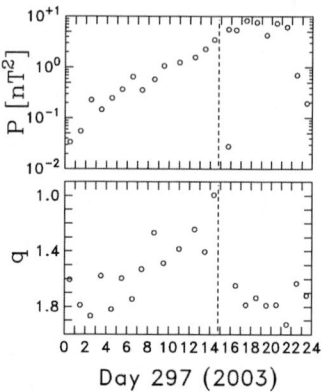

FIGURE 4. Hourly averages of the power spectra of the magnetic field fluctuations in the range frequencies 0.01-0.1 Hz (measured in the spacecraft frame of reference) and the power – law index q obtained from fitting f^q to the power spectra in the same range of frequencies.

passage (vertical line) and remain close to zero in the downstream region of the shock (once the fluxes reach their highest values). The value of the streaming anisotropy upstream of the shock decreases with the particle energy. The behavior of the flow anisotropies for this event is directly comparable to those shown in Fig. 2 of [9].

The intensity increase observed just before the shock passage allows us to estimate the diffusion coefficient of the low-energy particles. Following the method described in [15] and fitting an exponential function to the last upstream intensity increase, we obtain $\kappa_{\parallel}=2.01\times10^{18}$ cm^2 s^{-1} for protons of 88 keV that corresponds to a parallel mean free path $\lambda_{\parallel}=0.0009$ AU with an energy dependence $\kappa_{\parallel}\propto E^{-0.30}$ that according to the quasi-linear theory of particle scattering is equivalent to a spectral index $q=2.4$ of the magnetic power spectrum.

Figure 4 shows the magnetic power of the field fluctuations integrated over the interval 0.01-0.1 Hz on an hour-by-hour basis approaching the shock (top panel) and the power-law index (q) over the same range (bottom panel). This frequency range approximately corresponds to waves that resonate with 1-100 keV protons (see Fig. 4 in [9]). The peculiarity of this shock is the increasing level of field fluctuations with a flattening of the power spectrum as the shock approaches the observer. Comparison of the power magnetic field spectrum with the energy particle spectrum will be performed in a future work. The value of q deduced from the quasi-exponential increase ($q=2.4$) differs substantially from the q shown in Figure 4.

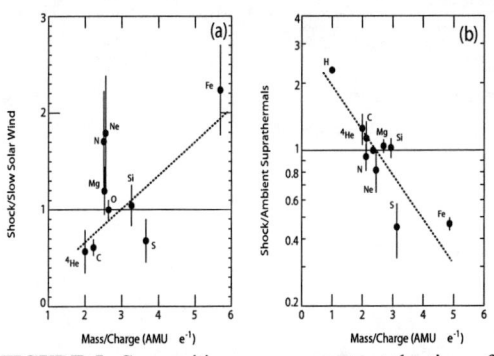

FIGURE 5. Composition measurements at the time of the shock compared to (a) slow solar wind and (b) ambient energetic particle populations vs. particle mass to charge ratio. Abundances are normalized to O.

Finally, Figure 5 shows the correspondence between the elemental abundances measured by the ULEIS instrument on board ACE [16] around the passage of the shock (from 10:24 to 21:20 UT on day 297) and those typical of the slow solar wind (left panel) and those measured in an interval upstream of the shock (from 18:06 UT on day 296 to 06:20 UT on day 297) (right panel). Analysis of the energy spectra of C, O and Fe and the energy-dependent behavior of Fe/O and C/O ratios (not shown here) reveals that the Fe/O ratio decreases with energy and C/O remains constant, as also observed by [17]. This indicates that the ions are accelerated by a mechanism in which the higher rigidity ions (i.e., larger M/Q values) are accelerated less efficiently than the lower rigidity ions. Figure 5a shows that the shock abundances are poorly correlated with the solar wind values, with Fe/O being significantly higher at the shock and C/O being significantly lower. Thus, it is clear that this shock did not accelerate particles out of the solar wind thermal distribution. However, a comparison with the ambient population measured upstream of the shock (Figure 5b), indicates a much improved correlation with the shock abundances, with M/Q dependent behavior that is consistent with the energy spectra. This indicates that this shock accelerated ions from a suprathermal population that was present in the interplanetary medium and not directly from the solar wind.

SUMMARY

We have presented a preliminary analysis of one of the ESP events observed by ACE during solar cycle 23 that shows the most consistency with the predictions of the quasi-linear DSA theory. Point-by-point comparisons of theory and individual events, such as the one performed in this study and in, e.g., [7], [8] and [9] have always revealed significant differences between model predictions and observations. Quantification of these differences for the event on 2003 October 24 will be performed in a future study. It is worth noting that theoretical models generally deal with ideal scenarios and averaged quantities. Therefore, a more fruitful approach would be the use of statistical data to construct averaged quantities (e.g., superposed epoch analyses of time-intensity profiles and shock parameters) that can be compared directly with ensemble-averaged model predictions. However, statistical approaches require assemblage of a sufficient number of shock events and the percentage of ESP events observed by ACE that show time-intensity profiles and energy spectra consistent with theory predictions is quite small [3, 4]. The study of the averaged properties of the few events that show both exponential intensity increases upstream to the shock, flat intensity profiles downstream, and energy spectra that soften as the shock approaches the observer (classified as classic ESP events in [3] and [4]) is in progress.

ACKNOWLEDGMENTS

We acknowledge the use of ACE Level 2 data and thank the ACE Science Center for providing these data. This work was performed under the support of NASA (grant NAG5-13487).

REFERENCES

1. Tsurutani, B. T., and Lin, R. P., *J. Geophys. Res.* **90**, 1-11 (1985).
2. van Nes, P., et al., *J. Geophys. Res.* **89**, 2122-2132 (1984).
3. Lario, D., et al., "ACE observations of energetic particles associated with transient interplanetary shocks" in *Solar Wind Ten: Proceedings of the 10th International Solar Wind Conference*, edited by M. Velli et al., AIP Conference Proceedings 679, American Institute of Physics, 2003, pp. 640-643.
4. Ho, G. C., et al., "Transient shocks and associated energetic particle distributions observed by ACE during solar cycle 23", submitted to *J. Geophys. Res.*, doi: 10.1029/2005JA011155 (2005).
5. Lee, M. A., *J. Geophys. Res.* **88**, 6109-6119 (1983).
6. Gordon, B. E., et al., *J. Geophys. Res.* **104**, 28263-28277 (1999).
7. Sanderson, T. R., et al., *J. Geophys. Res.* **90**, 3973-3980 (1985).
8. Bamert, K., et al., *Astrophys. J.* **601**, L99-L102 (2004).
9. Kennel, C. F., et al., *J. Geophys. Res.* **91**, 11917-11928 (1986).
10. Smith, C. W., *J. Geophys. Res.* **94**, 5474-5478 (1989).
11. Gold, R. E., et al., *Space Sci. Rev.* **86**, 541-562 (1998).
12. McComas, D.J., et al., *Space Sci. Rev.* **86**, 563-612 (1998).
13. Smith, C. W., et al., *Space Sci. Rev.* **86**, 613-632 (1998).
14. Lario, D., et al., *J. Geophys. Res.* **109**, A01107, doi:10.1029/2003JA010071 (2004).
15. van Nes, P., et al., *Adv. Space Res.* **4**, 315-318 (1984).
16. Mason, G. M., et al., *Space Sci. Rev.* **86**, 409-448 (1998).
17. Desai, M.I., et al., *Astrophys. J.* **611**, 1156-1174 (2004).

The Role of Quasi-Perpendicular Shocks in Solar Energetic Particle Events

Allan J. Tylka

E.O. Hulburt Center for Space Research, Naval Research Laboratory, Washington, DC 20375 USA

Abstract. Solar energetic particles (SEPs) are an important venue for testing and refining our understanding of acceleration processes that are ubiquitous in astrophysical plasmas. Large SEP events occur at a rate of about 10 per year during solar maximum. The dominant accelerators in these events are believed to be shocks driven by fast coronal mass ejections (CMEs). A particular challenge has been the dramatic event-to-event variability in composition and spectral characteristics at energies above a few tens of MeV per nucleon. I discuss recent efforts to understand this variability in terms of the interplay of two factors: seed populations, typically comprising at least suprathermals from flares and suprathermals from the corona or solar wind; and shock geometry, which generally begins as quasi-perpendicular near the Sun but evolves toward quasi-parallel as the shock moves outward.

Keywords: solar energetic particles, shocks, coronal mass ejections, solar flares
PACS: 96.40.Fg, 96.50.Pw, 96.60.Rd, 96.60.Wh

THE PROBLEM OF HIGH-ENERGY SEP VARIABILITY

Figure 1 (from [1]) illustrates the problem of high-energy SEP variability. The figure shows proton spectra from two events [2, 3], both produced by fast (>1800 km/s) CMEs on the west limb of the Sun. The two events differ in the so-called "knee" energy at which their spectra steepen. Spectral knees are of great physical interest, in that they reflect processes that ultimately limit the particle energization[1].

Heavy ions are test particles that identify seed populations and probe the velocity- and rigidity-dependent effects that govern acceleration and transport. The events in Fig. 1a also differ in another way: the event with low knee-energy has an Fe/O ratio that drops precipitously with increasing energy, falling to only ~1% of the coronal value at ~50 MeV/nuc [3]. In the event with high knee-energy, on the other hand, Fe/O rises with energy and attains roughly five-times the coronal value [4].

Figure 1b (from [5]) pursues this clue to origins of high-energy SEP variability. The figure shows a correlation between key characteristics of the 42 largest SEP events of 1997-2004. All of these events are associated with large, fast (>800 km/s) CMEs. The event selection required a total *GOES* proton fluence exceeding 2 x 10^5 p/cm^2-sr above 30 MeV. This energy biases the selection toward particles accelerated

---

[1] Spectral knees also have a huge impact on the events' potential radiation hazard. Behind 10 g/cm^2 of shielding – a significant amount, sufficient to stop protons with energies below ~100 MeV – dose rates from these two spectra are ~0.05 and ~4 rem/*hour*. For comparison, the average exposure in the US is ~0.4 rem/*year*; the legal limit for terrestrial radiation workers is 5 rem/year; and the limit for astronauts is 50 rem/year.

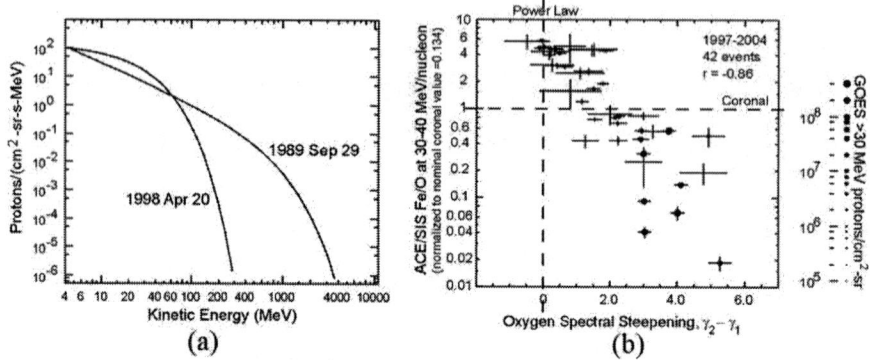

Figure 1: (a) Proton spectra from two events with different knee energies. (b) Correlation plot of event-integrated Fe/O (normalized to the nominal coronal value [29]) vs. the steepening of the oxygen spectrum, as described in the text. Symbol size indicates the fluence of >30 MeV protons, as given in the legend at the right.

near the Sun. The y-axis shows Fe/O from *ACE* at 30-40 MeV/nuc. The x-axis quantifies the event's spectral character. Specifically, we fit the event-integrated oxygen spectrum in each event to two independent power-laws, $E^{-\gamma_1}$ covering 3-10 MeV/nuc (from *Wind*) and $E^{-\gamma_2}$ covering 30-100 MeV/nuc (from *ACE*). The difference $\gamma_2-\gamma_1$ provides a measure of spectral steepening.

The spectral shapes in Fig. 1b range from power-laws (with $\gamma_2-\gamma_1 \sim 0$) to nearly exponentials (with $\gamma_2-\gamma_1 \gg 0$). The Fe/O ratios vary by nearly three orders of magnitude. The >30 MeV proton fluences also vary by three orders of magnitude, with strong Fe enrichments absent, or at least rare, among the largest events.

It is important to realize that there are no selection biases in Fig. 1b. That is, events were chosen on the basis of proton fluence, without reference to heavy-ion characteristics. The clear anti-correlation in Fig. 1b (with r = -0.86, corresponding to a random probability < 10^{-5}) indicates that there are common factors behind the composition and spectral variability. This anti-correlation is an important constraint for modeling efforts: if we can discover the factors behind the high-energy Fe/O variability, we can also understand the origins of spectral knees and, at least in part, variability in event size at high energies. Conversely, explanations that deal with these features in a piece-meal fashion are unlikely to be adequate.

Figure 2 (from [5]) shows two events that illustrate the extremes of variability in Fig. 1b. Both events arise from very fast CMEs near the west-limb. The events have similar Fe/O ratios below ~10 MeV/nuc. But at higher energies the events differ. Exponential rollovers are clear in the spectra of the Fe-poor event. In the event with enhanced high-energy Fe/O, however, the spectra are power-laws above ~10 MeV/nuc, with Fe having a harder spectrum than O.

In terms of the physics of SEP production, this variability goes to the heart of our understanding of the relative roles of flares and CME-driven shocks in accelerating

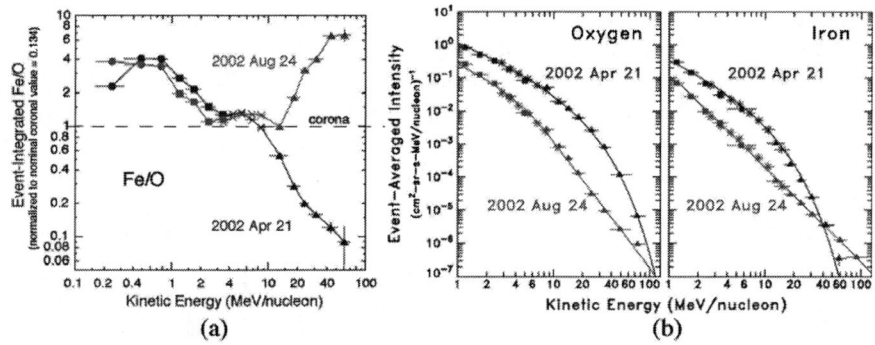

Figure 2: (a) Fe/O vs. energy from instruments on *ACE* and *Wind* in two events; (b) the event-integrated oxygen and iron spectra for these two events.

SEPs to high energy. Understanding this variability is also critical for efforts aimed at predicting potential SEP hazards for space-system operation and design[2].

FACTORS IN HIGH-ENERGY SEP VARIABILITY

The origin of suppressed Fe/O at high energies is readily described [3]. The differential energy spectrum of shock-accelerated particles has roughly the form $F(E) \sim E^{-\gamma} exp(-E/E_0)$ [8]. For some events, E_0 has a relatively low value, and we see the exponential rollover below ~100 MeV/nuc. E_0 also scales, at least in some events, with the ion's charge-to-mass (Q/A) ratio [3]. Thus, if a shock operates on a seed population dominated by solar-wind-like-suprathermals (with Q/A~0.4 for O while Q/A~0.2 for Fe), the Fe spectrum rolls over more steeply than the oxygen spectrum. In fact, Tylka et al. [3, 5, 9] exploited species-dependent exponential rollovers to determine charge states that agreed well with directly measured values from *SAMPEX* and *ACE*. In some events, the mean charges of the various elements are all consistent with a common temperature of ~1-2 MK, typical of the corona and solar wind.

However, shocks also accelerate suprathermals from flares. These flare suprathermals can be remnants from previous flare activity [10, 11, 9]; they might also come from the flare that accompanies the CME launch, if the field topology is favorable [12,13]. These flare suprathermals bring distinctive characteristics to the seed population, including Fe/O ~10 times the coronal value, Fe ions with ionic charge $Q_{Fe} \geq 16$, and ~1000-fold enhancements in $^3He/^4He$ and trans-Fe ions [14-16]. Surveys of $^3He/^4He$ above ~200 keV/nuc indicate that flare remnants are common at solar

---

[2] Events with large proton fluences are important for solar panel degradation, sensor backgrounds, and total-dose effects. The smaller events are comparatively unimportant in those regards. However, the harder spectra and heavy-ion enrichments in the smaller events make them a prime concern for single-event-effects in electronics. In fact, the Fe-rich events in Fig. 1b comprise a potentially distinct class of radiation hazard: although these events produce only ~12% of the >100 MeV solar protons in 1997-2004, they contributed ~60% of the solar Fe observed by *ACE* above 120 MeV/nuc. Heavy-ions (especially Fe) may also pose a qualitatively distinct radiation hazard for astronauts [6,7].

maximum [17-20]. The flare suprathermals are also sufficiently numerous [21] to be a significant contributor to the seed population for CME-driven shocks.

One might be tempted to think that a variable combination of solar-wind and flare-suprathermals in the shock's seed population might in itself be sufficient to explain high-energy SEP variability. But, as shown in Fig. 1b, the variability in high-energy Fe/O is closely tied to variability in spectral shape. The spectral variability indicates that some additional factor(s) must also be coming into play[3].

A recent hypothesis [5] suggests that the variability above a few tens of MeV/nuc is affected by the interplay of two factors – (1) shock geometry and a (2) compound seed population, comprising at least solar-wind (or coronal) suprathermals [22; henceforth "SW"] and suprathermals from flares. The ^{3}He/^{4}He surveys [17-20], compared to the average SW ^{3}He/^{4}He value, suggest that flare suprathermals become more important at higher seed energies, as sketched in Fig. 3a (from [5]). Under most conditions, efficient acceleration requires a higher initial speed at quasi-perpendicular shocks than at quasi-parallel shocks [23, 24]. Thus, as also sketched in Fig. 3a, flare suprathermals are more likely to dominate over SW suprathermals as the seed particles for quasi-perp shocks. Moreover, quasi-perp shocks have acceleration time-scales that can be orders-of-magnitude faster than those of parallel shocks [23, 24], making them particularly effective in attaining high energies. As a reflection of this fact, Lee [25] derived explicit dependence $E_0 \sim [\sec\theta_{Bn}]^{(2/(2\gamma-1))}$ in $F(E) \sim E^{-\gamma}\exp(-E/E_0)$, where θ_{Bn} is the angle between the shock normal and the upstream magnetic field.

One can therefore envision the following scenario. Non-radial expansion of CMEs generally produces extensive regions of quasi-perp shocks on their flanks, at least out to altitudes of a few solar-radii. (This behavior is seen in MHD simulations [26], and the quasi-perp regions on the flanks of CMEs have previously been identified as a potential source of metric type-II radio emission [27].) Altitudes of a few solar-radii are also those at which the highest-energy particles are produced, e.g. [28].)

But as the shock moves out from the Sun, θ_{Bn} generally changes toward quasi-parallel. As θ_{Bn} decreases, the nature of the accessible seed population would also change (as sketched in Fig. 3a), and the spectra would also soften at high energies. The net effect of this evolution is to allow the unique compositional characteristics of flare-suprathermals to be preferentially reflected among the higher-energy SEPs, causing Fe/O, <Q_{Fe}>, and ^{3}He/^{4}He to increase with energy. Along other field lines, the quasi-perp phase is absent [26] and we see exponential rollovers within the energy ranges of our instruments, as well as charge states characteristic of SW suprathermals.

A HEURISTIC CALCULATION

Figures 3b and 3c [29] show results from a heuristic calculation embodying this scenario. In these calculations, the differential spectrum of ion species X is parametrized as $F_X(E, \theta_{Bn})=C_x E^{-\gamma}\exp(-E/E_{0X})$, where $E_{0X}=E_0(Q_X/A_X)(\sec\theta_{Bn})^{2/(2\gamma-1)}$. The power-law index γ is explicitly the same for all species, and the θ_{Bn}-dependence arises through the e-folding energy, E_{0X} [25]. The species dependence also resides in

---

[3] One might also entertain the notion of a *direct* flare component, rather than flare particles processed through a shock. But as detailed in [5], timing studies, longitude distributions, and spectral characteristics disfavor a direct flare component.

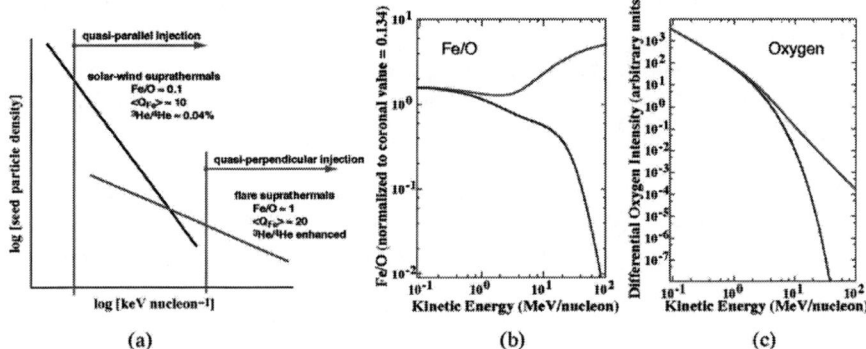

Figure 3 (a) Schematic representation of the suprathermal seed population, comprising both solar-wind and flare-accelerated ions. The flare ions are more likely to be apparent in quasi-perp shocks, for which the injection threshold is higher. (b) Fe/O vs. energy after averaging shock-accelerated particle spectra over two different ranges of θ_{Bn}, as described in the text. (c) corresponding oxygen spectra.

the e-folding energy, through proportionality to Q_X/A_X. To approximate evolution in θ_{Bn}, this spectrum is averaged over $\xi = \cos\theta_{Bn}$. The flare component is accessible to the shock at all ξ values. But to mimic the threshold effects in Fig. 3a, we introduce a weighting factor of ξ for the SW component, so that it makes increasingly less contribution as the shock approaches perpendicular. To complete the calculation, we assume nominal composition [30] and charge state distributions [31, 32] for the SW and flare seed particles. (See Figure 3a.) Figures 3b and 3c show two calculations of Fe/O vs. energy and the oxygen spectrum for $\gamma=1.5$ and $E_0=3$ MeV/nuc. In both calculations, flares provide 10% of the seed oxygen ions (and correspondingly, 47% of the Fe seeds). However, in the top curves in each plot, the spectrum was averaged over the full range of $0° \leq \theta_{Bn} \leq 90°$; in the bottom curves, only over $0° \leq \theta_{Bn} \leq 60°$. This calculation also yields energy-dependent ^{3}He/^{4}He and $<Q_{Fe}>$ similar to those that have been reported [33, 34]. Moreover, by adjusting the range of θ_{Bn}-averaging and the relative size of flare- and SW components, the calculations reproduce the whole "zoo" of energy-dependent morphology in the high-energy Fe/O data.

These highly simplified calculations suffice only to motivate further study. It should be noted that these calculations neglect evolution in other shock parameters, such as the compression ratio, as well as details of the transport. These factors may be particularly important at energies below ~1 MeV/nuc, which are produced more or less continuously as shock travels out from the Sun and the observer's magnetic connection point sweeps across a broad extent of the shock front.

As discussed in [5], the shock-geometry hypothesis may also help to resolve other puzzles. For example, events with enhanced high-energy Fe/O tend to have smaller proton fluences. This qualitatively agrees with the notion of smaller accessible seed population at quasi-perp shocks. Quasi-perp shocks are also particularly good at making GeV-particles. We therefore generally expect ground level events (GLEs) to be Fe-rich. In fact, ~85% of GLEs since 1973 are Fe-rich above 50 MeV/nuc [35, 5].

The notion of near-Sun shock geometry as a critical factor in high-energy SEP variability is a new one. Some theoretical issues, such as the dependence of injection threshold on θ_{Bn}, require clarification (i.e., compare [24] and [36]). However, this hypothesis appears to offer a promising framework for understanding many complex and subtle features of the SEP data. Future efforts will compare detailed time-dependent measurements with simulations that incorporate sophisticated CME models [37-39], particle acceleration codes for shocks of arbitrary and evolving obliquity, and more realistic treatments of transport [40] and the structure of the seed population.

This work has been supported by NASA DPR S13791G and the Office of Naval Research.

REFERENCES

1. D.V. Reames, in *Acceleration and Transport of Energetic Particles Observed in the Heliosphere*, edited by R. A. Mewaldt et al., AIP Conference Proceedings 528, Melville, New York, 2000, pp. 289-300.
2. J. L. Lovell, M. Dulding, and J. E. Humble, *J. Geophys. Res.* **103**, 23,733-23,742 (1998).
3. A. J. Tylka, *J. Geophys. Res.* **106**, 25,333-25,352 (2001).
4. A.J. Tylka and W.F. Dietrich, *Radiation Measurements* **30** (3), 345-360 (1999).
5. A.J. Tylka et al., *Astrophys. J.* **625**, 474-495 (2005).
6. G. Casadesus et al., *Adv. Space Res.* **33**, 1340-1346 (2004).
7. B.M. Rabin, J.A. Joseph, and B. Shukitt-Hale, *Adv. Space Res.* **33**, 1330-1333 (2004).
8. D.C. Ellison and R. Ramaty, *Astrophys. J.* **298**, 400-408 (1985).
9. A.J.Tylka, et al., *Astrophys. J.* **558**, L59-L63 (2001).
10. G.M. Mason, J.E. Mazur, and J.R. Dwyer, *Astrophys. J.* **525**, L133-L136 (1999).
11. M.I. Desai, et al., *Astrophys. J.* **533**, L89-L92 (2001).
12. D.V. Reames, *Astrophys. J.* **571**, L63-L66 (2002).
13. G. Li and G.P. Zank, *Geophys. Res. Lett.* **32**, L02101, doi:10.1029/2004GL021250 (2005).
14. D.V. Reames, *Astrophys. J.* **540**, L111-L114 (2000).
15. D.V. Reames and C.K. Ng, *Astrophys. J.* **610**, 510-522 (2004).
16. G.M. Mason, et al., *Astrophys. J.* **606**, 555-564, (2004).
17. I.G. Richardson et al., *Astrophys. J.* **363**, L9-L12 (1990).
18. J. Laivola, J. Torsti, & L. Kocharov, Proc. 28th Int. Cosmic Ray Conf. (Tsukuba) **6**, 3233-3236 (2003).
19. M.E. Wiedenbeck et al., in *Proc. 10th Internat. Solar Wind Conf.*, edited by M. Velli et al., AIP Conference Proceedings 679, Melville, New York, 2003, pp. 652-655.
20. J. Torsti, J. Vaivola, and L. Kocharov, *Astron. & Astrophys.* **408**, L1-L4 (2003).
21. R.A. Mewaldt, R.A. et al., Proc. 28th Int. Cosmic Ray Conf. (Tsukuba) **6**, 3229-3232, (2003).
22. G. Gloeckler et al., in *Acceleration and Transport of Energetic Particles Observed in the Heliosphere*, edited by R. A. Mewaldt et al., AIP Conference Proceedings 528, Melville, New York, 2000, pp. 221-228.
23. J. R. Jokipii, *Astrophys. J.* **313**, 842-846 (1987).
24. G.M. Webb et al., *Astrophys. J.* **453**, 178-206 (1995).
25. M.A. Lee, *Astrophys. J. Suppl.* **158**, 38-67 (2005).
26. W.B. Manchester et al., *Astrophys. J.* **622**, 1225-1239 (2005).
27. R.S. Steinolfson, *Solar Phys.* **94**, 193-202 (1984).
28. J.W. Bieber et al., *Astrophys. J.* **601**, L103-L106 (2004).
29. M.A. Lee and A.J. Tylka, in preparation.
30. D.V. Reames, *Adv. Space Res.* **15**(7), 41-51 (1995).
31. R. von Steiger et al., *J. Geophys. Res.* **105**, 27,217-27,238 (2000).
32. A. Luhn et al., *Astrophys. J.* **317**, 951-955 (1987).
33. A.J. Tylka et al., *Astrophys. J.* **581**, L119-L123, (2002).
34. W.F. Dietrich and A.J. Tylka, Proc. 28th Int. Cosmic Ray Conf. (Tsukuba) **6**, 3291-3294 (2003).
35. W.F. Dietrich and C. Lopate, Proc. 26th Int. Cosmic Ray Conf. (Salt Lake City) **6**, 71-74 (1999).
36. J. Giacalone, *Astrophys. J.* **624**, 765-772 (2005).
37. I. Roussev et al., *Astrophys. J.* **605**, L73-L76 (2004).
38. G. Li, G.P. Zank, and W.K.M. Rice, *J. Geophys. Res.* **108** (A2), 1082-1103 (2003).
39. W.K.M. Rice, G.P. Zank, and G. Li, *J. Geophys. Res.* **108** (A10), 1369-1382 (2003).
40. C.K. Ng, D.V. Reames, and A.J. Tylka, *Astrophys. J.* **591**, 461-485 (2003).

Energetic Particle Transport in Strong Compressive Wave Turbulence Near Shocks

J. A. le Roux, G. P. Zank, G. Li, and G. M. Webb

Institute of Geophysics and Planetary Physics, University of California, Riverside

Abstract. Strong interplanetary coronal mass ejection driven shocks are often accompanied by high levels of low frequency compressive wave turbulence. This might require a non-linear kinetic theory approach to properly describe energetic particle transport in their vicinity. We present a non-linear diffusive kinetic theory for suprathermal particle transport and stochastic acceleration along the background magnetic field in strong compressive dynamic wave turbulence to which small-scale Alfvén waves are coupled. Our theory shows that the standard cosmic-ray transport equation must be revised for low suprathermal particle energies to accommodate fundamental changes in spatial diffusion (standard diffusion becomes turbulent diffusion) as well as modifications to particle convection, and adiabatic energy changes. In addition, a momentum diffusion term, which generates accelerated suprathermal particle spectra with a hard power law, must be added. Such effective first stage acceleration possibly leads to efficient injection of particles into second stage diffusive shock acceleration as described by standard theory.

INTRODUCTION

In energetic particle kinetic transport theories, it is customary to use to 1st approximation Alfvén waves as the MHD wave mode that causes particle scattering and parallel diffusion in the quiet solar wind, but also in the enhanced turbulence fluctuation levels upstream of interplanetary shocks in terms of a self-generated Alfvén wave description [1]. However, the magnetic fluctuations near interplanetary shocks in the solar wind are marked by strong magnitude fluctuations, interpreted as an indication of strong compressibility, and the fluctuations are isotropic or even predominantly field-aligned relative to the background magnetic field [2]. This indicates a possible significant presence of compressive wave modes such as fast mode magnetosonic waves or ion-acoustic waves at interplanetary shocks. It is usually thought that these modes are Landau-damped very efficiently, but Landau damping could be strongly suppressed given that the observed thermal electron to proton temperature ratio can be much larger than one near interplanetary shocks [3].

Within this context we developed a non-linear energetic particle kinetic theory (particle trajectories are assumed to be diffusive instead of undisturbed gyro orbits as in standard quasi-linear theory) for field-aligned diffusive transport in strong compressive wave turbulence velocity fluctuations at a turbulent shock. For details, see [4].

THE TRANSPORT EQUATION

Our theory is valid in the situation where energetic particles, microscopically diffusing along the magnetic field due to gyro-resonant interaction with weak small-scale parallel-propagating Alfvén waves, find themselves propagating through a non-uniform compressive medium filled with strong, relatively large-scale electrostatic ion-acoustic velocity fluctuations near a shock on larger scales. The undamped ion-acoustic waves are assumed to propagate along the magnetic field [5]. In the fluid frame the energetic particle transport equation is given by

$$\frac{\partial f}{\partial t} + \frac{1}{3p^2}\frac{\partial}{\partial p}(p^3 U^*)\frac{\partial f}{\partial z} - \frac{\partial}{\partial z}\left(\kappa_{eff}\frac{\partial f}{\partial z}\right) + \left(\frac{\partial U^*}{\partial z}\right)\frac{p}{3}\frac{\partial f}{\partial p} + \frac{2}{3}pU^*\frac{p}{3}\frac{\partial^2 f}{\partial z \partial p}$$
$$= \frac{1}{p^2}\frac{\partial}{\partial p}\left(p^2 D_{pp}\frac{\partial f}{\partial p}\right), \qquad (1)$$

where f is the direction-averaged particle distribution function of position z along the background magnetic field, momentum p, and time t. In addition, $U^* \propto \langle \delta U^2 \rangle_{dif}$ ($\langle \delta U^2 \rangle_{dif}$ is the average energy difference between forward and backward propagating ion-acoustic waves along the background magnetic field) is an effective large-scale convection speed, κ_{eff} is the effective parallel diffusion coefficient, and D_{pp} is the momentum diffusion coefficient.

According to equation (1), energetic particles experience coherent convection (2nd term on the left hand side of equation (1)), effective parallel diffusion (3rd term on the left), adiabatic energy changes (4th term on the left), effective changes in both position and momentum (5th term on the left), and stochastic particle acceleration (term on the right hand side of equation (1)), because particles are subject to the effects of many random compressive ion-acoustic velocity fluctuations on large scales. The convection effect, adiabatic energy changes, and the mixed derivative term disappear when ion-acoustic waves propagate with equal intensities in both forward and backward directions. Stochastic acceleration, however, always occurs, even when waves propagate only in one direction, which is different from the case of incompressible Alfvén waves [6]. The reason is compressibility. For Alfvén waves propagating in one direction, an observer in the wave frame will see no motional electric field so that particle acceleration will not occur in that frame, but an observer in the wave frame of the compressive fluctuations will see particles undergoing adiabatic energy changes due to compression or decompression effects. Effective parallel diffusion on large scales is the combined effect of microscopic diffusion averaged over the non-uniform medium and turbulent diffusion due to the strong ion-acoustic velocity fluctuations as is discussed below.

Equation (1) suggests that the standard cosmic-ray transport equation needs to be extended when applied to a strongly turbulent shock, so that convection and adiabatic energy changes due to the large-scale solar wind flow, and spatial diffusion reflect the modification by the compressive wave effects, while the mixed derivative and stochastic acceleration terms are new additions that needs to be considered.

PARALLEL DIFFUSION

According to our non-linear theory, κ_{eff} is in general determined by a 5th order polynomial equation. For the special case of equal ion-acoustic wave intensities in both propagation directions, the equation simplifies to a tractable 3rd order polynomial for which a solution can be found. The solution is complicated, however, so that we present a simplified approximate solution which clearly illustrates the physics. For strong ion-acoustic waves assuming $\langle \delta U^2 \rangle \approx V_s^2$ where $\langle \delta U^2 \rangle$ the average energy in the ion-acoustic velocity fluctuations and V_s is the phase speed, the solution is

$$\kappa_{eff} \approx \frac{4}{9}\kappa_t + \kappa, \qquad (2)$$

where $\kappa_t = \langle \delta U^2 \rangle^{1/2} l_c$ is the classical turbulent diffusion coefficient, and l_c is the correlation scale associated with the strong dynamic ion-acoustic velocity fluctuations. In equation (2), $\kappa = (1/3)v\langle \lambda \rangle$ is the microscopic parallel diffusion coefficient associated with small-scale Alfvén waves where $\langle \lambda \rangle$ is the mean free path averaged over large scales of the compressive non-uniform medium. In the derivation of equation (2) it was assumed that $\gamma = 1/t_c = \langle \delta U^2 \rangle^{1/2}/l_c$ where t_c is the characteristic time scale (correlation time) for non-linear wave interactions at the scale l_c.

In the low suprathermal energy limit, that is $v \gg V_s$, but $v \ll (l_c/\lambda)\langle \delta U^2 \rangle^{1/2}$ (which is equivalent to $\kappa/\langle \delta U^2 \rangle^{1/2} \ll l_c$ or $\kappa \ll \kappa_t$), $\kappa_{eff} \approx \kappa_t$ and turbulent diffusion by ion-acoustic waves dominates microscopic diffusion κ by Alfvén waves on large spatial scales. The particles are strongly coupled to the Alfvén waves (κ is small) which in turn, according to standard WKB theory, directly responds to the density variations of compressive velocity fluctuations. Thus the particles are approximately passively convected with the turbulent flow field via Alfvén wave mediation. The particles experience turbulent diffusive motion on long time scales assuming that strong dynamic wave turbulence effects associated with non-linear wave interactions lead to random diffusive motions of the waves in which the particles are quasi-trapped.

In the opposite limit valid for high energy suprathermal particles, that is $v \gg (l_c/\lambda)\langle \delta U^2 \rangle^{1/2}$ ($\kappa/\langle \delta U^2 \rangle^{1/2} \gg l_c$ or $\kappa \gg \kappa_t$), the particles are weakly coupled to the small-scale Alfvén waves so that compressive velocity fluctuations have little effect on the particles via Alfvén wave mediation. Turbulent diffusion is unimportant and microscopic diffusion dominates. Consequently, $\kappa_{eff} \approx \kappa$ in this limit.

While κ_t is independent of particle speed, $\kappa \propto v^{4/3}$ for standard quasi-linear theory, assuming that the suprathermal particles interact resonantly with the Kolmogorov inertial range of the magnetic field fluctuation power spectrum of Alfvén waves. Thus, at low suprathermal particle energies κ_{eff} is independent of particle speed due to the dominance of turbulent diffusion, while at high suprathermal energies $\kappa_{eff} \propto v^{4/3}$ when microscopic diffusion dominates.

However, in the absence of dynamic ion-acoustic wave turbulence effects (strong but non-interacting compressive wave turbulence), $\gamma = 1/t_c = 0$ and we have:

$$\kappa_{eff} = \left(\frac{\kappa}{\kappa_t}\right)^{1/3} \kappa_t + \kappa. \tag{3}$$

For low suprathermal energies, $\kappa_{eff} \approx (\kappa/\kappa_t)^{1/3} \kappa_t$, and turbulent diffusion occurs at a reduced level because the particles are quasi-trapped in waves that now exhibit only coherent motion. Some turbulent diffusion still occurs because κ is small, but finite, allowing particles to eventually cross into neighboring fluid elements in a random way on long time scales.

STOCHASTIC ACCELERATION

The approximate expression for the momentum diffusion coefficient D_{pp} is

$$D_{pp} = p^2 \frac{\langle \delta U^2 \rangle}{(9/4)\kappa_t + \kappa}. \tag{4}$$

In the low suprathermal energy limit, $\kappa/\langle \delta U^2 \rangle^{1/2} \ll l_c$ ($\kappa \ll \kappa_t$), the expression for D_{pp} simplifies to $D_{pp} \propto p^2 \langle \delta U^2 \rangle/\kappa_t$ showing that particles gain energy when turbulent diffusion κ_t in strong dynamic compressive wave turbulence dominates microscopic diffusion κ.

Although small, microscopic diffusion κ is nonetheless essential for diffusing particles to gain energy stochastically since it allows particles to randomly sample compressive regions (adiabatic energy gain) and rarefaction regions (adiabatic energy loss) on long time scales. Overall the particles gain energy because they spend more time in compression regions where the microscopic scattering rate is enhanced compared to rarefaction regions where particles spend less time because scattering is less efficient.

In the high energy limit $\kappa/\langle \delta U^2 \rangle^{1/2} \gg l_c$ ($\kappa \gg \kappa_t$) we find that $D_{pp} \propto p^2 \langle \delta U^2 \rangle/\kappa$ because microscopic particle diffusion across compressive fluid elements dominates turbulent diffusion of particles by the compressive fluid elements ($\kappa \gg \kappa_t$). In addition, $D_{pp} \propto v^2$ at low suprathermal energies when turbulent diffusion dominates while $D_{pp} \propto v^{2/3}$ when microscopic diffusion due to Alfvén waves dominates at high particle energies (assuming that $\kappa \propto v^{4/3}$ according to standard kinetic theory for Alfvén waves as discussed above). Thus unlike κ, D_{pp} is strongly dependent on particle speed at low suprathermal energies but only weakly dependent on particle speed at high suprathermal speeds.

In the absence of dynamical wave turbulence ($\gamma = 0$), we find that

$$D_{pp} \approx p^2 \frac{\langle \delta U^2 \rangle}{(\kappa_t/\kappa)^{1/3} \kappa_t + \kappa}. \tag{5}$$

The reduction in the effectiveness of turbulent diffusion when $\gamma = 0$ reduces the effectiveness with which particles can execute random motions across compression and rarefaction regions. The particles are quasi-trapped in fluid elements that execute mainly coherent motions so that stochastic particle acceleration becomes less effective.

DISCUSSION

Our theory, as given by equation (1), shows that the standard cosmic-ray transport equation needs to be modified at relatively low suprathermal particle energies. Strong compressive wave turbulence results in (i) turbulent diffusion which dominates standard parallel diffusion due to Alfvén waves so that parallel diffusion is independent of particle energy, (ii) modification of particle convection and adiabatic energy changes by the large-scale solar wind, and (iii) efficient stochastic acceleration so that a momentum diffusion term with coefficient D_{pp} needs to be included.

At low suprathermal energies, the D_{pp}-term in equation (1) is more important than terms containing κ_{eff} and U^* so that one can, to a first approximation, neglect them. After assuming a steady state one finds that the accelerated particle spectrum f has a hard power law so that $f \propto v^{-3}$. At high suprathermal particle energies the diffusion term with κ_{eff} dominates the D_{pp}-term. This domination becomes more pronounced with increasing particle energy. Thus we expect the particle spectrum to develop an exponential rollover at high particle speeds. This suggests that, in the enhanced compressive turbulence near strong ICME-driven shocks close to the Sun, a suprathermal accelerated particle spectrum will develop that features a hard power law at low energies followed by an exponential rollover at high energies. This result is qualitatively consistent with observations of suprathermal ion spectra near coronal mass ejection driven shocks [7]. Such efficient stochastic acceleration of suprathermal ions can possibly serve as a first stage acceleration process, resulting in efficient injection of particles into a second stage diffusive shock acceleration mechanism.

ACKNOWLEDGMENTS

The authors acknowledge financial support from NASA grant NAG5-11604 and NSF grant ATM-0112772.

REFERENCES

1. Crooker, N. U., Gosling, J. T., et al., *Space Sci. Rev.* **89**, 179-220 (1990).
2. Lee, M. A. 1983, *J. Geophys. Res.* **88**, 6109-6119 (1983).
3. Gurnett, D. A., Neubauer, F. M., and Schwenn, R. J., *J. Geophys. Res.* **84**, 541-552 (1979).
4. le Roux, J. A., Zank, G. P., Li, G., and Webb, G. M., *Astrophys. J.* **626**, in press (2005).
5. Gary, S. P., *Theory of space plasma microinstabilities*, Cambridge: Cambridge University Press, 1993, p. 101.
6. Schlickeiser, R., *Astrophys. J.*, **336**, 243-293 (1989).
7. Kallenbach, R., Bamert, K., Hilchenbach, M., and Smith, C. W., this volume (2005).

Anomalous diffusion of energetic particles: implications for diffusive particle acceleration at a quasi-perpendicular shock

Olga P. Verkhoglyadova* and Jakobus A. le Roux*

Institute of Geophysics and Planetary Physics, University of California, Riverside, CA 92521-0412, USA

Abstract. We study energetic particle transport in turbulent magnetic fields. The composite turbulence is modeled by a nonlinear structure (vortex) and a randomly fluctuating background magnetic field. Results of our numerical simulations show that an anomalous diffusion regime (subdiffusion and superdiffusion) occurs when the correlation length of the magnetic field disturbances is about the same or larger than the average particle gyroradius. Subdiffusion of cosmic rays in the vicinity of a quasi-perpendicular shock can lower the density of particles at the shock front. Particles become temporarily trapped in the turbulent magnetic fields. The particle spectrum in this case can steepen as compared to the spectrum in the classic diffusive acceleration case. We estimate a power law index for the particle spectrum for a number of cases and discuss implications for cosmic ray transport in the heliosphere.

MODEL OF TWO-DIMENSIONAL TURBULENCE IN THE HELIOSPHERE

The heliosphere is a complex system populated with large-amplitude waves, discontinuities, shocks and structures [1, 2]. Recent satellite data show that a network of flux tubes and possibly vortices are observed in the heliosphere as well [3, 4, 5]. All the above phenomena constitute composite turbulence in the plasma medium. Complex electric and magnetic fields can alternate cosmic ray motion in the heliosphere. In the paper we study numerically how cosmic ray diffusion is modified by vortex structures and a random wave field.

Recent observational data and theoretical models revealed the importance of pronounced 2D (two-dimensional) turbulence in the heliosphere [6, 7, 8, 9]. Thus, we approximate heliospheric turbulence as consisting of a 2D coherent component represented by a vortex field and a random 2D component in (XY)-plane. Electric and magnetic field disturbances in the vortex occur strictly in a 2D plane orthogonal to the background uniform magnetic field $\mathbf{B_0} = B_0 \mathbf{e_z}$. The vortex field is described by disturbances in a hydrodynamic flux function, Ψ, and z-component of the magnetic field potential, A_v, which can be expressed through the vortex magnetic field $\mathbf{B_v}$ and the plasma flow velocity $\mathbf{u_v}$:

$$\begin{aligned} \mathbf{u_v} &= \mathbf{e_z} \times \nabla \Psi, \\ \mathbf{B_v} &= \nabla A_v \times \mathbf{e_z}. \end{aligned} \quad (1)$$

In the cylindrical coordinates (r, θ, z) the vortex field can be expressed as follows:

$$\begin{aligned} \Psi(r,\theta) &= C_0 J_1(kr) \sin\theta, \; r \leq a, \\ \Psi(r,\theta) &= 0, \; r > a. \end{aligned} \quad (2)$$

where C_0, k, a are free parameters representing the amplitude of flux modulations, the vortex parameter controlling the field distribution in the vortex and an effective vortex radius, respectively, and J_1 is the first-order Bessel function of the first kind. The origin of the cylindrical coordinate system coincides with the vortex geometrical center. A stationary solution implies that $A_v = s_v \Psi$, where $s_v \neq \pm 1$ is a free parameter. Details of the derivations can be found elsewhere [10, 11, 12, 5, 13].

The random component of the turbulence, $\delta \mathbf{B_z} = \delta B_z(x,y)\, \mathbf{e_z}$, describes linear fluctuations superimposed on the background magnetic field. $\delta \mathbf{B_z}$ has a power-law spectrum defined by:

$$\delta B_z = \sum_m \sum_n A(k) \cos(mx + ny + \phi_{mn}), \; A(k) = A_0 k^{-\gamma}, \quad (3)$$

where the wavenumber $k = \sqrt{m^2 + n^2}$, $\gamma = 1.5$ and ϕ_{mn} represents a random phase [14]. We choose 10x10 harmonics in the (XY) plane.

Particle motion in the turbulent fields is described by its Lagrangian, L, and the corresponding equations of motion in the framework of classical mechanics:

$$L = \frac{\dot{r}^2 + r^2 \dot{\theta}^2 + \dot{z}^2}{2} + s\,\Psi(r,\theta)\,\dot{z} - \Psi(r,\theta) + \delta B_z\, r^2\, \dot{\theta}, \quad (4)$$

where the dot denotes a derivative over time. We used the following scaling: a particle gyroperiod for time, B_0 for magnetic field, and the vortex radius a for spatial variables. The generalized momentum along the z-direction, $P_{\|}$, is an integral of motion. The equations of motion are [15, 16]:

$$\begin{aligned} \dot{z} &= P_{\|} - s\,\Psi, \\ \frac{d}{dt}\dot{r} &= r\dot{\theta}^2 - \frac{\partial}{\partial r}\Psi(1+s^2\,\Psi) \\ &\quad + 2r(1+\delta B_z)\dot{\theta} + r^2\dot{\theta}\frac{\partial}{\partial r}\delta B_z, \\ \frac{d}{dt}r^2(\dot{\theta}+1+\delta B_z) &= -\frac{\partial}{\partial \theta}\Psi(1+s^2\,\Psi) + r^2\dot{\theta}\frac{\partial}{\partial \theta}\delta B_z. \end{aligned} \quad (5)$$

We define the diffusion coefficient in the direction of a coordinate x by

$$D_{xx} = \frac{\langle \Delta x^2 \rangle}{2\tau}, \quad (6)$$

where a particle displacement from an initial position along the x direction, Δx, and a time interval corresponding to this displacement, τ, are introduced. Then, all displacements are averaged over an ensemble of particles. In general, the diffusion coefficient has a power-law dependence on τ:

$$D_{xx} \propto \tau^{\alpha-1}, \tag{7}$$

where the power law parameter α defines a diffusion regime. Thus, $\alpha = 1$ corresponds to classical diffusion and all other cases correspond to anomalous diffusion. Subdiffusion, superdiffusion and ballistic regimes are usually introduced for $\alpha < 1, \alpha > 1$ and $\alpha = 2$, correspondingly [14, 17]. Strong interaction between a particle and a background wave turbulence can produce areas of temporary trapping or "sticky domains". The averaged macroscopic effect results in subdiffusion, or slowing down of classical diffusion (see [16, 17] and discussion therein).

NUMERICAL SIMULATION RESULTS ON ANOMALOUS PARTICLE DIFFUSION

We studied energetic particle diffusion in the turbulent fields introduced by equations(1-3) by integrating numerically the particle equations of motion (5). A radial diffusion coefficient D_{rr} transverse to the magnetic field was estimated according to formula (6). Particle dynamics depends on the ratio ε_1 of maximum magnitude of random field component $A(k)$ over the coherent vortex field magnitude B_v. Another parameter controlling particle diffusion is the ratio ε_2 between the average particle gyroradius and the magnetic field correlation length λ_c [14]. The latter is defined by

$$\lambda_c^2 = \frac{\langle (\frac{2\pi}{k})^2 B(k) \rangle}{\langle B(k) \rangle}, \tag{8}$$

where $B(k) = A(k) + \hat{F}(B_v)$ is the Fourier transformation of the total turbulent magnetic fields. Numerical simulations of particle diffusion regimes were performed for different values of ε_1 and ε_2. We obtained distinct diffusion regimes, i.e., subdiffusion, superdiffusion, ballistic regime and classical diffusion [16]. Subdiffusion arises when $\varepsilon_2 \ll 1$ or $\varepsilon_2 \simeq 1$ and is in a good agreement with the work by Otsuka and Hada [14]. Simulations were performed for correlation scales in the range of $\lambda_c \simeq 3.25 - 32.5$.

MODIFICATION OF COSMIC RAY SPECTRUM AT A QUASI-PERPENDICULAR SHOCK

Recent studies show that the spectrum of accelerated particles in the vicinity of a quasi-perpendicular shock can be strongly affected when particles experience anomalous diffusion instead of classical diffusion due to turbulence [18, 19]. The accelerated spectrum in the presence of turbulence is given by

$$f(p) \propto p^{-s}, s = s_{diff}\left(1 + \frac{1-\alpha}{\rho_c}\right). \tag{9}$$

Here, s is the modified power law index of the spectrum, s_{diff} is defined by

$$s_{diff} = \frac{3\rho_c}{\rho_c - 1} \quad (10)$$

is the power law index if particles experience classical diffusion at the shock with ρ_c the shock compression ratio. Note that $s = s_{diff}$ if $\alpha = 1$, which is the classical diffusion case (see equation (7)) when energetic particles execute a random walk in the turbulence at the shock. When particles are subdiffusing ($\alpha < 1$), $s > s_{diff}$ and the accelerated spectrum is softer which reflects less efficient diffusive shock acceleration, compared to the case of classical diffusion, while $s < s_{diff}$ implies a harder accelerated spectrum (more efficient diffusive shock acceleration).

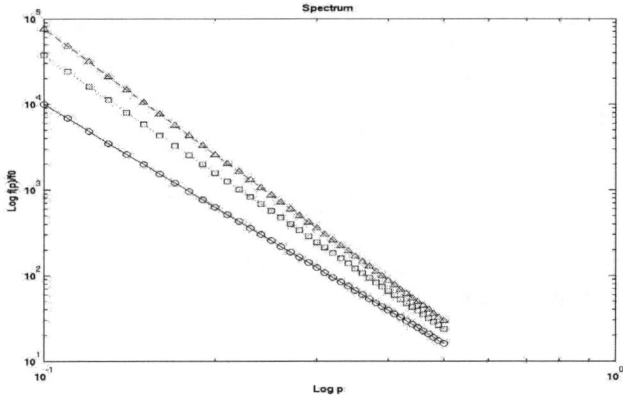

FIGURE 1. A model spectrum for energetic particles accelerated in a vicinity of a quasi-perdendicular shock. Results for the classical diffusive acceleration (circles) are compared to spectra obtained under conditions of anomalous particle diffusion in the turbulence. Spectra for the "weak" subdiffusion regime ($\varepsilon_1 \simeq 2 \cdot 10^{-3}$, $\varepsilon_2 \simeq 2$, squares) and the "strong" subdiffusive regime ($\varepsilon_1 \simeq 2 \cdot 10^{-5}$, triangles) are shown. All variables are dimensionless.

We assumed a strong shock for which $\rho_c = 4$ and $s_{diff} = 4$, and calculated a modified power law index for two subdiffusive cases. The first case of subdiffusion corresponds to $\varepsilon_1 \simeq 2 \cdot 10^{-3}$, $\varepsilon_2 \simeq 2$, $\lambda_c \simeq 3.25$ and we estimated $\alpha \simeq 0.43$. Thus the modified power law index is $s = 4.57$. For the second more pronounced subdiffusive case we estimated that $\varepsilon_1 \simeq 2 \cdot 10^{-5}$, $\varepsilon_2 \simeq 0.2$, $\lambda_c \simeq 32.5, \alpha \simeq 0.12$ and $s = 4.88$. Both cases show a departure from the power law index for the classic diffusion, s_{diff} (see Fig.1). The obtained results show that cosmic ray acceleration could be less effective at turbulent shocks. Energetic particles experiencing subdiffusion in the vicinity of a shock gain less energy.

CONCLUSIONS

Our results illustrate the importance of turbulence in the heliosphere for cosmic ray transport. We studied particle diffusion in the random-phase fluctuating static background magnetic field with a stationary 2D vortex magnetic field in the orthogonal direction. Particles can be efficiently trapped by large-scale "islands" created by the random and coherent turbulent fields within the correlation length λ_c. Anomalous diffusion regimes were studied numerically and the corresponding power law parameter α was estimated. The obtained results depend on the ratio ε_2 between the average particle gyroradius and the magnetic field correlation length. It was shown that the ratio ε_1 between the random and the vortex field magnitudes does not define the diffusion regime.

The spectrum of energetic particle diffusively accelerated in a vicinity of a quasi-perpendicular shock can be modified by strong 2D turbulence at the shock. This modification takes place if particles experience anomalous diffusion instead of classical diffusion. For $\varepsilon_2 \ll 1$, we found a steepening in the acceleration spectrum when particles are temporarily trapped by the vortex field. The acceleration is less efficient because the particles are convected away from the shock with the turbulence.

ACKNOWLEDGMENTS

This study was partially supported by the grant NSF ATM-0112772. Authors would like to acknowledge fruitful discussions with T. Hada (Kyushu University), B.T. Tsurutani (JPL/Caltech) and G. Webb (UCR).

REFERENCES

1. T.S. Horbury, & A. Balogh, *Journ. Geophys. Res.*, **106**, A8, 15929 (2001)
2. B.T. Tsurutani, C. Galvan, J.K. Arballo, et al., *Geophys. Res. Lett.*, **29**, 23 (2002)
3. N.U. Crooker, M.E. Burton, J.L. Philips, E.J. Smith & A. Balogh, *Journ. Geophys. Res.*, **101**, A2, 2467 (1996)
4. N.U. Crooker, S.W. Kahker, J.T. Gosling et al., *Journ. Geophys. Res.*, **106**, A8, 15963 (2001)
5. O.P. Verkhoglyadova, B. Dasgupta & B.T. Tsurutani, *Nonlinear Proc. in Geophys.*, **10**, 1-9 (2003)
6. J.W. Bieber, W. Wanner & W.H. Matthaeus, *Journ. Geophys. Res.*, **101**, A2, 2511 (1996)
7. W.H. Matthaeus, M.L. Goldstein, & D.A. Roberts, *Journ. Geophys. Res.*, **95**, 20673 (1990)
8. S. Ghosh, Matthaeus, W.H., Roberts, D.A. & Goldstein, M.L., *J. Geophys. Res.*, **103**, A10, 23705 (1998)
9. J.A. le Roux, G.P. Zank, & W.H. Matthaeus, *Journ. Geophys. Res.*, **107**, A7, 1138 (2002)
10. B.B. Kadomtsev & O.P. Pogutse, *Sov. Phys. JETP*, **5**, 575- 590 (1973)
11. H.R. Strauss, *Phys. Fluids*, **19**, 134 (1976)
12. D.C. Montgomery, *Physica Scripta*, **T2/3**, 83 (1982)
13. O.P. Verkhoglyadova & J.A. le Roux, *Astrophys. Journ.*, **602**, 1002-1005 (2004)
14. F. Otsuka, & T. Hada, *Space Sci. Rev.*, **107**, 499 (2003)
15. O.P. Verkhoglyadova, & J.A. le Roux, *Adv. Space Sci.* (in press) (2005)
16. O.P. Verkhoglyadova, & J.A. le Roux, *Journ. Geophys. Res.* (in press) (2005)
17. G.M. Zaslavsky, *Physics Reports*, **371**, 6, 461 (2002)
18. P. Duffy, J.G. Kirk, Y.A. Gallant, & R.O. Dendy, *Astron. Astrophys.*, **302**, L21 (1995)
19. J.G. Kirk, P. Duffy, & Y.A. Gallant, *Astron. Astrophys.*, **314**, 1010 (1996)

Simulation of SEP Acceleration and Transport at CME-driven Shocks

J. Kóta*, W.B. Manchester†, J.R. Jokipii*, D.L. de Zeeuw† and T.I. Gombosi†

*The University of Arizona, Lunar and Planetary Laboratory, Tucson, AZ 85721-0092, USA
†University of Michigan, Center for Space Environment Modeling, Ann Arbor, MI 48109-2143, USA

Abstract. Our code of solar energetic particle (SEP) acceleration and transport developed in Arizona is combined with the realistic CME simulations of Michigan, using the solar wind and magnetic field data of the Michigan CME-simulation as input to the SEP code. We suggest that, in addition to the acceleration at the shock significant acceleration may also occur in the sheet behind the shock, where magnetic field lines are compressed as they are bent around the expanding cloud. We consider field aligned motion and cast the proper Fokker-Planck equation into a non-inertial comoving frame, that follows field lines as they evolve. Illustrative simulation results are presented.

INTRODUCTION

Gosling [1] suggested that Coronal Mass Ejections (CMEs) play a central role in large solar energetic particle (SEP) events. The present paradigm [2] is that SEP events can be divided into the two basic classes of impulsive and gradual events. Gradual events, which are more prolonged in time and are more extended in longitudinal range, and pose the most severe radiation hazard in geospace, are thought to be accelerated by CME driven shocks. The precise mechanism of SEP acceleration to high energies is, however, not yet fully understood.

It is the purpose of this paper to address the acceleration of SEPs in a scenario of realistic CME simulations. We combine our numerical code of SEP acceleration and transport developed in Arizona with the realistic 3-D MHD simulation of CME evolution obtained by the BATS-R-US code at Michigan. The SEP code considers field-aligned motion of ions and follows the constantly changing structure of the shock and the magnetic field as the CME evolves. For this purpose, the field-aligned transport-equation [3] [4] [5] [6] are cast into a non-inertial Lagrangian frame, comoving with the solar wind plasma. We outline some of the important features of the CME simulation [7] and emphasize that descriptions with a single shock may be insufficient. There are more than one site of acceleration to consider: the compression of the field lines behind the shock can also accelerate particles, which can play an important role in both the injection and acceleration processes. In the forthcoming sections we discuss some implications of this acceleration in the downstream region, outline our transport equation, and present illustrative simulation results.

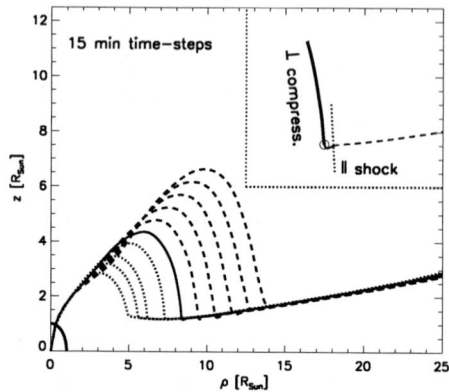

FIGURE 1. Time-evolution of one selected field line from the simulation of Manchester et al. [7]. Lines depict the meridional projection of the selected field line in 15 minute consecutive time steps. Note the draping as the CME pushes the field line upward around the magnetic cloud. The box (upper right) shows a schematic illustration around the shock: the circle indicates the location where the magnetic field strengthens.

ACCELERATION IN THE SHEET BEHIND THE SHOCK

Current models of CME propagation into interplanetary space [7] [8] [9] [10] [11] [12] indicate that the shock is not a simple quasi-parallel shock, and the downstream region immediately beyond the shock shows a remarkable structure of it own. The shock geometry must evolve in time. Recently, *Lee and Tylka* [13] emphasized that the geometry is likely to change from an initially quasi-perpendicular shock to a more or less parallel shock later. The structure of the sheet between the shock and the magnetic cloud of the CME is discussed in detail by *Manchester et al.* [12].

In this work, we select a field line from the 3-D MHD simulation of [7] that connects almost head-on to the CME. The plasma is pushed upward in latitude and shows little longitudinal deflection. Figure 1 shows the evolution of the field line in 15 minute consecutive time steps. The shock is nearly parallel, but field lines bend around and are strongly compressed by the expanding CME in a thin sheet between the shock and the magnetic cloud. In passing we note that, in this sheet, the $B \times V$ electric field, which is largely in the azimuthal direction in this case, can well exceed 10GV/degree when the shock is only a few solar radii from the Sun.

In this scenario, there are at least two important sites of SEP acceleration. The sharp changes in the density and magnetic field strength do not occur at the same place, but a finer structure can be revealed. First there is a nearly parallel shock, where the density jumps but the magnetic field does not. Second, due to the bending and compression in the sheet, there is a sharp increase of the magnetic field strength close, but distinctly behind the shock (marked by circle in the box of Fig. 1). Particles can be accelerated at both sites. We recall that acceleration can occur even without net compression. For instance, an increase in the field strength without net compression will accelerate particles of large pitch angles ($\mu \approx 0$) and decelerate particles of small pitch angles. This can result in a net statistical acceleration if scattering is not very frequent.

We note that, in fact, our SEP simulations show large quadrupole anisotropies indicating that this process may occur. This structure of the field around the CME is an almost inescapable result of the CME's pushing the field line near the Sun. At later stages this feature is likely to weaken or decay, and may not be present at 1 AU.

COMOVING FIELD-ALIGNED TRANSPORT EQUATION

The theoretical description of SEP acceleration and transport is based on either a diffusion picture [15] [16] [17] [18] [19] [20] or focussed transport along the field line [3] [4] [21] [22]. The diffusive treatment is faster to calculate but requires stronger assumptions. In this work, we employ the focussed transport equation, which has a broader range of validity, and can cover high anisotropies as well as low particle speeds, thus remains applicable in the energy region of injection. Also, acceleration at the bending of the field line would not be included in the diffusive equation [14]. We re-write the transport equation into a non-inertial frame, so that our scheme could follow the field lines carried by the solar wind. The evolution of the distribution function, $f(\mathbf{r},p,\mu,t)$, in terms of position, $\mathbf{r}$, time, t, momentum, p, and cosine of pitch-angle, μ, is described by [23] [24]

$$\left(1 - \frac{(\mathbf{Vb})w\mu}{c^2}\right)\frac{df}{dt} + w\mu\frac{\partial f}{\partial z} + \frac{(1-\mu^2)}{2}\left[w\frac{\partial \ln B}{\partial z} - \frac{2}{w}\mathbf{b}\frac{d\mathbf{V}}{dt} + \mu\frac{d\ln(n^2/B^3)}{dt}\right]\frac{\partial f}{\partial \mu} +$$
$$+ \left[-\frac{\mu \mathbf{b}}{w}\frac{d\mathbf{V}}{dt} + \mu^2\frac{d\ln(n/B)}{dt} + \frac{(1-\mu^2)}{2}\frac{d\ln B}{dt}\right]\frac{\partial f}{\partial \ln p} = \frac{\partial}{\partial \mu}\left(\frac{D_{\mu\mu}}{2}\frac{\partial f}{\partial \mu}\right) + q \quad (1)$$

The right hand side accounts for pitch-angle scattering, and injection of seed population, $q(\mathbf{r},p,t)$. p, μ, and the particle speed, w, are measured in the frame co-moving with the solar wind at a velocity, $\mathbf{V}$. The unit vector $\mathbf{b}$ points along the magnetic field, $\mathbf{B}$. c denotes the speed of light. The relativistic correction in the first term results from the use of mixed coordinates [4].

Eq (1) contains only quantities on the given field line, eliminating the need to calculate spatial variation of the solar wind velocity, $\mathbf{V}$ (i.e the full $\nabla \mathbf{V}$ tensor that appears in the standard equation [3] [5]). A numerical derivation of $\nabla \mathbf{V}$ would, almost inevitably, introduce an undesirable computational noise. The relevant components of the $\nabla \mathbf{V}$ tensor, which correspond to the compression rates, parallel and perpendicular to the magnetic field, are expressed in (1) with the rates at which the plasma density, n, and magnetic field, B, at a given fluid element change.

The most important implication of (1) is that the acceleration rate is divided into two major terms (the inertial term containing $d\mathbf{V}/dt$ is small except at the lowest energies). Compressions parallel to the field, which act at parallel shocks, will change B/n, but not B. On the other hand, compressions perpendicular to $\mathbf{B}$, which occur at perpendicular shocks, do change B, but not B/n.

We note that a similar co-moving formulation of Parker's diffusive equation [14] was discussed by *Sokolov et al* [20].

ILLUSTRATIVE NUMERICAL SIMULATIONS

The present work represents a step in our effort to couple our SEP acceleration and transport code with realistic CME simulations. Since particles cannot significantly affect the evolution of the CME itself, the coupling is uni-directional: the SEP acceleration model needs the CME simulation results as input. The coupling may be carried out at different levels depending on how frequently plasma and magnetic field are taken from the CME simulation. The most desirable way is, naturally, to update the plasma and magnetic field data from the CME simulations frequently. This is also our ultimate goal. At the same time, the frequent coupling is the computationally most demanding.

In this work, we take a simple approach and use similarity solutions as a plausible way for describing the evolution of the field and plasma around the CME. Figure 1 suggests that the CME evolves in quite self similar way. In the illustrative example presented here, we take the simulation of *Manchester et al.* [7] at a given time, and extend it as a similarity solution to serve as the time-depending background field. Then, in this background field, a seed population of superthermal ~5keV particles is injected at the shock.

Our simulation, at its present stage, treats scattering phenomenologically. We assume a pitch-angle scattering coefficient, which is enhanced in the vicinity of the shock to account for self-generated waves [25]. The form we use in this work is:

$$D_{\mu\mu} = A(r/r_{sh}) \, r^{-3/2} P^{-1/3} \frac{(1-\mu^2)w}{2\lambda_0} \qquad (2)$$

where r is radial distance in AU, P is rigidity in GV, and λ_0, which is the parallel mean free path at 1 AU, and 1 GV is chosen 0.4 AU. The factor $A(r/r_{sh})$ accounts for the enhancement of scattering due to self-generated waves near the shock at r_{sh}. $A = 1$ far from the shock and here we use an enhancement of 100 decreasing upstream exponentially on a scale of $0.1 r_{sh}$. A more physics based inclusion of these waves is a goal of future work.

During an SEP event, the Earth, or spacecraft, intersects and samples different field lines. This becomes particularly apparent after the passage of the shock, when connection to the Earth changes drastically. To obtain the full predicted time-profiles at Earth, the simulation is to be repeated for a number of different field lines, according to their different connection to the Earth.

Figure 2 shows simulated time profiles at 1 AU for different energies. These profiles represent variations on one field line and are no longer applicable for observations at Earth after the passage of the shock. The reason is that the Earth moves on a new field line, into an entirely different environment.

Figure 3 exhibits the energy spectra obtained at different radial distances in an early (left panel) and a later (right panel) phase of the event. Inspection of Fig. 3 shows that the spectra can be characterized as double power-laws with a breaking point around a few MeV. This double power-law can be the results of particle escape upstream as suggested by [16]. Another possible explanation is that the acceleration at the bending of the field line, which is effective at lower energies, becomes ineffective above the breaking point. It shall require further diagnostic work to distinguish between these alternatives.

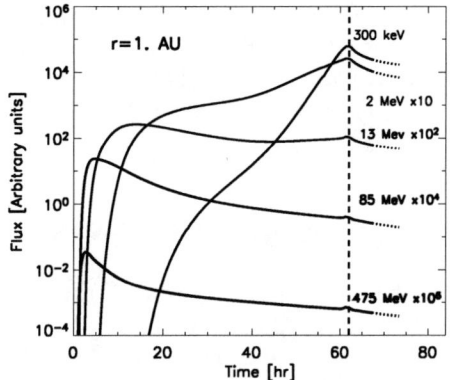

FIGURE 2. Time-profiles obtained for 0.3, 2, 13, 85 and 475 MeV proton fluxes at 1 AU in a a simulated SEP event. The vertical dashed line marks the arrival of the CME-driven shock at Earth. After the shock arrival the connection changes dramatically and the simulation along one field line is no longer applicable.

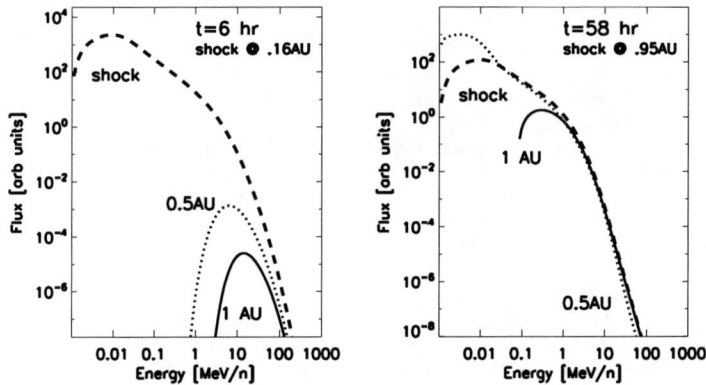

FIGURE 3. Simulated proton spectra shown at 0.5 AU (dotted) and 1 AU (solid) from the Sun at an early (6hr - left panel) and a later (58 hr - Right panel) stage of the event. The dashed lines represent the spectra at the shock, which is at r=0.16 AU and 0.9 AU, respectively.

CONCLUSIONS

As we outlined above, a most important element of CME driven shocks is that the shock evolves in time; the time variation in the location and strength and geometry of the shock need to be followed in the the calculation. We have developed a numerical code that employs a co-moving Lagrangian grid following the evolving field lines and suits this task. The Lagrangian scheme has the additional advantage of eliminating the need for computing the spatial derivatives of the solar wind velocity (i.e. the $\nabla \mathbf{V}$ tensor).

The code has been successfully applied to accelerate ions to energies above ~100MeV/nucl in a realistic CME scenario. We find that the interplay of the two acceleration sites may increase the efficiency of acceleration. This potentially important feature, which needs further exploration, cannot be properly addressed in diffusive models.

ACKNOWLEDGMENTS

The authors greatly benefited from discussions with J. Giacalone, I.I. Roussev, and I.V. Sokolov. This work has been supported by NASA grant NAG5-10884, NASA ESS Cooperative agreement NCC5-614, and by DoD MURI grant F49620-01-1-0359.

REFERENCES

1. Gosling, J.T., *J. Geophys. Res*, **98**, 18937 (1993)
2. Reames, D.V., *Revs. Geophys. Suppl.* **33**, 585 (1995)
3. Skilling, J., *Astrophys. J.*, **170**, 265–273 (1971)
4. Ruffolo, D., *Astrophys. J.*, **442**, 861–874, (1995)
5. Isenberg, P.A., *J. Geophys. Res*, **102**, 4719 (1997)
6. Kóta, J., and J.R. Jokipii, *Proc. 25th Int. Cosmic Ray Conf.*, **1**, 213 (1997)
7. Manchester, W,B. et al., *J. Geophys. Res.*, **109**, 1102 (2004)
8. Roussev, I. I., et al., *Astrophys. J.*, **605L**, 73 (2004)
9. Linker, J. A., Z. Mikic, P. Riley, and R. Lionello, *2004 AAS Conf.* (2004)
10. Osdtrcil, D., P. Riley, and X.P. Xhao, *J. Geophys. Res.*, **109**, 21160 (2004)
11. Schmidt, J.M., and J.P. Cargill, *J. Geophys. Res.*, **108**, SSH5S, (2003)
12. Manchester, W.B. IV et al., *Astrophys. J.*, **662**, 1225 (2005)
13. Lee, M.A. and A.J. Tylka, to be published
14. Parker, E.N. 1965, *Planet. Space Sci.*, **13**, 9 (1965)
15. Lee, M.A., and J.M. Ryan, *Astrophys. J.*, **303**, 829 (1986)
16. Zank, G.P., W.K.M. Rice, and C.C. Wu, *J. Geophys. Res.*, **105**, 25 079-25 095 (2000)
17. Berezhko, E.G, and S.N. Taneev, *Proc. 28th Int. Cosmic Ray Conf.,*, **6**, 3343 (2003)
18. Rice, W.K.M., G.P. Zank, and Gang Li, *J. Geophys. Res.*, **108**, SSH5R (2003)
19. Li, Gang, G.P. Zank, and W.K.M. Rice, *J. Geophys. Res.*, **108**, SSH10L (2003)
20. Sokolov, I.V. et al., *Astrophys. J.*, **616**, L171 (2004)
21. Tylka, A.J., *J. Geophys. Res.*, **106**, 25,333, (2001)
22. Ng, C.K., D.V. Reames, and A.J. Tylka, *Astrophys. J.*, **591**, 461 (2003)
23. Kóta, J., and J.R. Jokipii, In *The Physics of the Outer Heliosphere*, eds Florinski et al., AIP Conf. Proc. 719, pp 272–278
24. Kóta, J., et al., in preparation (2005)
25. Lee, M.A., *J. Geophys. Res.*, **88**, 6109 (1983)

Diffusive Acceleration of Ions at Interplanetary Shocks

Matthew G. Baring & Errol J. Summerlin

Department of Physics and Astronomy, MS-108, Rice University, P. O. Box 1892, Houston, TX 77251-1892, USA
Email: baring@rice.edu, xerex@rice.edu

Abstract. Heliospheric shocks are excellent systems for testing theories of particle acceleration in their environs. These generally fall into two classes: (1) interplanetary shocks that are linear in their ion acceleration characteristics, with the non-thermal ions serving as test particles, and (2) non-linear systems such as the Earth's bow shock and the solar wind termination shock, where the accelerated ions strongly influence the magnetohydrodynamic structure of the shock. This paper explores the modelling of diffusive acceleration at a particular interplanetary shock, with an emphasis on explaining in situ measurements of ion distribution functions. The observational data for this event was acquired on day 292 of 1991 by the Ulysses mission. The modeling is performed using a well-known kinetic Monte Carlo simulation, which has yielded good agreement with observations at several heliospheric shocks, as have other theoretical techniques, namely hybrid plasma simulations, and numerical solution of the diffusion-convection equation. In this theory/data comparison, it is demonstrated that diffusive acceleration theory can, to first order, successfully account for both the proton distribution data near the shock, and the observation of energetic protons farther upstream of this interplanetary shock than lower energy pick-up protons, using a single turbulence parameter. The principal conclusion is that diffusive acceleration of inflowing upstream ions can model this pick-up ion-rich event without the invoking any seed pre-acceleration mechanism, though this investigation does not rule out the action of such pre-acceleration.

INTRODUCTION

Evidence for efficient particle acceleration at collisionless shocks in the heliosphere abounds, including direct measurements of accelerated populations in various energy ranges at the Earth's bow shock (e.g. [1, 2]) and interplanetary shocks (for the pre-Ulysses era see, for example, [3–5]). The development of theories of shock acceleration is therefore strongly motivated, and a variety of approaches have emerged. One possible means for the generation of non-thermal particles is the Fermi mechanism, often called diffusive shock acceleration; this process forms the focus of this paper.

There are various approaches to modelling diffusive shock acceleration. Among these are hybrid and full plasma codes (e.g. [6–9]), which place an emphasis primarily on plasma structure and wave properties in the environs of shocks, and the convection-diffusion differential equation approach [10]. In addition, the kinematic Monte Carlo technique of Ellison and Jones (e.g., [11–13]) also focuses on diffusion and convection, and describes the injection and acceleration of particles from thermal energies to the highest relevant energies, addressing both spectral and hydrodynamic properties. The simulation technique makes no distinction between accelerated particles and thermal ones, using an identical phenomenological description of diffusion for both. In this work, following previous invocations, it is assumed that a particle's mean free path λ is proportional to its gyroradius r_g, i.e. $\lambda = \eta r_g$, with η =const. for all particle momenta. Upstream plasma quantities are input from observational data, and downstream quantities are determined using the full MHD Rankine-Hugoniot relations.

The Monte Carlo technique was used by Ellison et al. [14] to perform the first detailed theory/data comparison for the quasi-parallel portion of the Earth's bow shock. They compared predictions of the Monte Carlo method with particle distributions of protons, He^{++} and a C, N and O ion mix obtained by the AMPTE experiment. The agreement between model predictions and data was impressive, but required modeling in the non-linear acceleration regime, when the dynamic effects of the accelerated particles control the shock structure. A similar theory/data comparison was explored for interplanetary (IP) shocks in the work of Baring et al. [15], where impressive agreement was found between the Monte Carlo predictions and spectral data obtained by the Solar Wind Ion Composition Spectrometer (SWICS) aboard Ulysses, in the case of two shocks observed early in the Ulysses mission. Such agreement was possible only with the assumption of strong particle scattering (i.e. near the Bohm diffusion limit) in the highly oblique candidate shocks. For a third shock, detected a month later, the comparison failed with significant differences arising in the 500-800 km/sec range of the phase space distribution. Baring et al. [15] attributed this discrepancy to the omission of pick-up ions from the model: such an extra component would be expected to provide a substantial contribution to the accelerated population in this particular event.

This paper explores the role of pick-up ions in such shocks via modeling the accelerated population for the specific IP shock detected by the SWICS and HI-SCALE instruments aboard the Ulysses spacecraft at around 4.5 AU, as reported in [16]. Phase space distributions from the simulations are compared with SWICS and HI-SCALE data, yielding acceptable fits for the proton populations using standard prescriptions for the injected pick-up ion distribution. The simulation results successfully account for the observation of energetic protons farther upstream of the forward shock than lower energy pick-up protons, since a rigidity-dependent diffusion is used in the modeling.

THE ULYSSES EVENT OF DAY 292, 1991

The forward shock of a CIR encountered by Ulysses on Day 292 of 1991, is appropriate for a case study, with downstream particle distributions published in Gloeckler et al. [16]. Various plasma parameters for this shock were input for the Monte Carlo simulation, and were obtained from [16] and the data compilations of [17, 18]. The shock was quite oblique, with $\theta_{Bn1} = 50° \pm 11°$ being the angle the upstream magnetic field made with the shock normal. It was also quite weak, with a sonic Mach number of $M_s \sim 2.53$, and [16] inferred a value of $r = u_1/u_2 = 2.4 \pm 0.3$ for the velocity compression ratio. The normalization of solar wind proton distributions was established using $n_p = 2.0 \mathrm{cm}^{-3}$ as the solar wind proton density. Other parameters, such as the fluid speeds and upstream plasma temperatures, are detailed in Summerlin & Baring [19], yielding an upstream flow speed of $u_1 \approx 55$ km/s in the shock rest frame. The pick-up proton distribution input for the Monte Carlo simulation was taken from [20], a developed expression that is modeled on the seminal work of [21], and is similar in conception to pick-up ion distributions used in [22]. The pick-up ion model provides both the detailed shape and normalization of this superthermal distribution at 4.5AU; it incorporates the gravitational focusing of interstellar neutrals, the physics of their ionization as a function of distance from the sun, and adiabatic losses incurred during propagation away from the sun.

The Monte Carlo shock acceleration simulation is described in [12, 13, 15, 19]. Particles are injected upstream and allowed to convect into the shock, meanwhile diffusing in space so as to effect multiple shock crossings, and thereby gain energy through the shock drift and Fermi processes. The particles gyrate in laminar electromagnetic fields, with their trajectories being obtained by solving the Lorentz force equation in the shock rest frame, in which there is, in general, a $\mathbf{u} \times \mathbf{B}$ electric field in addition to the magnetic field. The effects of magnetic turbulence are modeled by scattering these ions in the rest frame of the local fluid flow. While the simulation can routinely model either large-angle or small-angle scattering, in this paper, large-angle scattering is employed, appropriate for the turbulent fields in IP shocks. For this phenomenological scattering, it is assumed that a particle's mean free path λ is proportional to its gyroradius r_g, i.e. $\lambda = \eta r_g$, with $\eta =$ const. for all particle momenta. Other dependences on particle rigidity can be employed, however the results are not extremely sensitive to such choices.

At every scattering, the direction of the particle's momentum vector is randomized in the local fluid frame, with the resulting effect that the gyrocenter of a particle is shifted randomly by a distance of the order of one gyroradius in the plane orthogonal to the local field. Accordingly, cross-field diffusion emerges naturally from the simulation, and is governed by a kinetic theory description [13, 23], where the ratio of the spatial diffusion coefficients parallel ($\kappa_\parallel = \lambda v/3$) and perpendicular ($\kappa_\perp$) to the mean magnetic field is given by $\kappa_\perp/\kappa_\parallel = 1/(1+\eta^2)$. Clearly then, η couples directly to the amount of cross-field diffusion, and is a measure not only of the frequency of collisions between particles and waves, but also of the level of turbulence present in the system, i.e. is an indicator of $\langle \delta B/B \rangle$. The Bohm diffusion limit of quasi-isotropic diffusion, presumably corresponding to $\langle \delta B/B \rangle \sim 1$, is realized when $\eta \sim 1$. As will become apparent, η is a parameter that critically controls the injection efficiency of low energy particles, and the upstream diffusion scale of accelerated ions. The simulation outputs particle fluxes and phase space distributions at any location upstream or downstream of the shock, and in any reference frame including that of the Ulysses spacecraft. This capability renders it ideal for comparison with observational data.

Figure 1 displays downstream distributions for thermal, pick-up and accelerated protons from the Monte Carlo simulation and the SWICS and HI-SCALE measurements (see Fig. 1 of [16]) taken in the frame of the spacecraft on the downstream side of the Day 292, 1991 shock. The solar wind and pick-up proton parameters are fairly tightly specified, so that the model has one largely free parameter, the ratio of the particle mean free path to its gyroradius, $\eta = \lambda/r_g$. The efficiency of acceleration of thermal ions in oblique shocks, i.e. the normalization of the non-thermal power-law, is sensitive [13, 15] to the choice of η, so this parameter was adjusted to obtain a reasonable "fit" to the data. Here, the accelerated pick-up ion phase space density is about a factor of 30 greater than that of the solar wind ions, denoted in the figure by the "SW only" histogram.

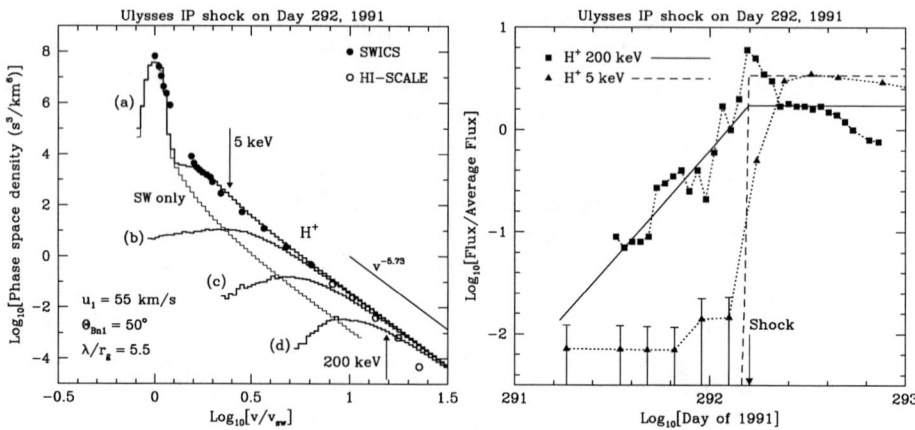

FIGURE 1. *Left panel:* Comparison between phase space velocity distribution functions for data collected by the Ulysses mission for the shock on day 292 of 1991, and Monte Carlo model results. The data are for H^+ (filled circles for SWICS data; open circles for HI-SCALE points) solar wind and pickup ions, and are taken from Gloeckler et al. [16]. The heavyweight histograms are the corresponding Monte Carlo models of acceleration of protons for $u_1 = 55$ km/sec, using the optimal choice of plasma shock parameters from [16] and sources indicated in the text (see also [19]). These four spectra correspond to (a) downstream, and successively increasing times upstream of the shock encounter, i.e. (b) 14 minutes, (c) 69 minutes and (d) 278 minutes upstream. The lighter weight histogram marked "SW only" was for a run where pick-up ions were omitted. The velocity axis is the ratio of the ion speed v, as measured in the spacecraft frame, to the solar wind speed. The model assumed $\eta = \lambda/r_g = 5.5$ and a shock of compression ratio $r = 2.1$, corresponding to diffusive acceleration power-laws of index -5.73, is indicated.
Right panel: The flux variations of accelerated pick-up ion populations as a function of time near the shock. The data for 5 keV and 200 keV pick-up H^+ are depicted by filled triangles and squares, respectively, and are taken from [16]. The Monte Carlo model generated fluxes at different distances normal to the shock, and were converted to spacecraft times by incorporating solar wind convection. The 5 keV and 200 keV pick-up H^+ traces are displayed as dashed and solid curves, respectively, and exhibit an exponential decline upstream of the shock that is characteristic of diffusive shock acceleration.

The downstream fit in the left hand panel of Fig. 1 models the accelerated protons well, for $\eta = 5.5 \pm 1.5$, a value that is slightly higher than those inferred in the fits of [15] for shocks at around 2–3 AU, yet consistent with a moderate level of field turbulence. The uncertainty in the inferred value of η is due mostly to the observational uncertainty in the shock obliquity θ_{Bn1}. The non-thermal proton distribution is composed virtually entirely of accelerated pick-up ions: the accelerated thermal H^+ ions are injected much less efficiently in the simulation than in the observations. The efficiency of acceleration of thermal ions could be increased via several means: (i) by lowering the shock obliquity angle θ_{Bn1}, for which there is a large observational uncertainty; (ii) by decreasing η, corresponding to increased turbulence, without altering the pick-up ion acceleration efficiencies substantially, and (iii) increasing the temperature of the thermal ions somewhat, though this would reduce the compression ratio and accordingly steepen the non-thermal continuum. Note that the distribution of accelerated He^+ pickup ions reported by [16] for this shock can be modeled by the *same* scattering parameter $\eta = 5.5$. This enticing property is addressed by Summerlin & Baring [19], where it is observed that the He^{++} distribution requires lower η for a fit of comparable quality.

An instructive diagnostic on the acceleration model is to probe the spatial scale of diffusion. This is routinely performed with the Monte Carlo simulation by placing flux measurement planes upstream of the shock at different distances, as well as downstream. Results are illustrated in the left hand panel of Fig. 1 via the display of upstream distributions of high energy particles at different times, i.e. distances from the shock. The Figure exhibits the characteristic "peel-off" effect where superthermal ions become depleted at successively high energies the further the detection plane is upstream of the shock; this signature was first identified by Lee [24]. Gloeckler et al. [16] discussed an energy-dependent rise in fluxes of non-thermal particles *prior* to the shock crossing. This was cited as indicating the existence of a pre-acceleration mechanism. Fluxes for two different H^+ ion energies, 5 keV and 200 keV, were obtained from spectra like those in the left hand panel in the Fig. 1, and are displayed in the right hand panel of the Figure, together with corresponding data from Fig. 3 of [16] for identical energy windows. Note that the Ulysses data normalization was established by averaging over 3 days of data, whereas the model normalization was adjusted to match observed fluxes around 1/2 day downstream of the shock.

It is clear that the spatial scale of the exponential decline of ions upstream of the shock is more or less identical to that of the model, for our choice of $\eta = 5.5$. High energy particles with a mean free path $\lambda \propto r_g$ establish an exponential dilution in space/time due to random scattering of the particles as they leak upstream against a convective flow. For the 200 keV ions with their relatively long mean free paths, the simulation results are clearly well correlated with the data prior to the shock, modulo plasma fluctuations, and in particular the overshoot just downstream. On the other hand, for the lower energy 5keV ions, the exponential decay has a very short time scale, around a factor of 40 smaller than for the 200 keV ions, realizing background levels upstream until very close to the shock. So, although the simulation results are consistent with the observed results, it is impossible to draw more definitive conclusions without an improvement in data time resolution, or a focus on ions of intermediate energy, say around 50 keV. Note that while this comparison is suggestive, it does not conclusively prove that diffusion is the dominant operating mechanism in this system. Yet alternative explanations must generate exponential declines that are consistent with convective loss scales of the order of a few gyroradii, with the physical mechanism responsible for transport upstream being also a direct cause of injection into the acceleration process.

CONCLUSIONS

This paper has compared the phase space distributions for protons from the Monte Carlo simulation of diffusive shock acceleration with those observed by the Ulysses instruments SWICS and HI-SCALE in the Day 292, 1991 shock. There is a good deal of consistency between theory and experiment for the energetic protons above speeds around 600 km/sec, an agreement that is extended to include He^+ pick-up ion spectra in [19]. At these speeds, the injection of pick-up protons dominates that of solar wind protons. The normalization of the energetic proton power-law establishes $\eta = 5.5$, where $\lambda = \eta r_g$ is the diffusive mean free path. This provides substantial cross-field diffusion ($\kappa_\perp/\kappa_\parallel \approx 0.03$), the prerequisite for efficient injection and acceleration in this diffusive model, when highly oblique shocks are being simulated.

The upstream spatial scales of the acceleration were also probed, with the flux increases of energetic protons seen upstream of the shock being well-modeled by the expected upstream "leakage" associated with diffusive acceleration. The value of $\eta = 5.5$ inferred from the spectral fit scales the upstream diffusive lengthscale, and the accompanying exponential decline in predicted flux is commensurate with the Ulysses data presented in Gloeckler et al. [16]. Hence, the observed upstream flux precursor is not clear evidence of a pre-acceleration mechanism, as claimed by [16], though it is quite possible that some pre-acceleration mechanism may be acting. The flux fluctuations in time clearly indicate the contribution of a non-diffusive process in the plasma shock, effects that are not incorporated in the simulation. Yet, the fact that the diffusive model works so well in coupling the spectral and spatial properties suggests that diffusion is an integral part of the acceleration process at this shock.

REFERENCES

1. Scholer, M., Ipavich, F. M., Gloeckler, G., & Hovestadt, D. 1980, J. Geophys. Res. 85, 4,602.
2. Gosling, J. T., Thomsen, M. F., Bame, S. J., & Russell, C. T. 1989, J. Geophys. Res. 94, 3555.
3. Sarris, E. T., & Van Allen, J. A. 1974, J. Geophys. Res. 79, 4,157.
4. Decker, R. B., Pesses, M. E., & Krimigis, S. M. 1981, J. Geophys. Res. 86, 8819.
5. Tan, L. C., Mason, G. M., Gloeckler, G., & Ipavich, F. M. 1988, J. Geophys. Res. 93, 7,225.
6. Quest, K. B. 1988, J. Geophys. Res. 93, 9,649.
7. Winske, D., Omidi, N., Quest, K. B. & Thomas, V. A. 1990, J. Geophys. Res. 95, 18,821.
8. Giacalone, J., Burgess, D., & Schwartz, S. J. 1992, in *Study of the Solar-Terrestrial System*, (ESA Special Publication, Noordwijk) p. 65.
9. Kucharek, H. & Scholer, M. 1995, J. Geophys. Res. 100, 1,745.
10. Kang, H., & Jones, T. W. 1995, ApJ 447, 944.
11. Ellison, D. C., Jones, F. C. & Eichler, D. 1981, J. Geophys. - Zeitschrift fuer Geophysik, 50, 110.
12. Jones, F. C. & Ellison, D. C. 1991, Space Sci. Rev. 58, 259.
13. Ellison, D. C., Baring, M. G. & Jones, F. C. 1995, ApJ 453, 873.
14. Ellison, D. C., Möbius, E., & Paschmann, G. 1990, ApJ 352, 376.
15. Baring, M. G., Ogilvie, K. W., Ellison, D., & Forsyth, R. 1997, ApJ 476, 889.
16. Gloeckler, G., Geiss, J., Roelof, E. C., et al. 1994, J. Geophys. Res. 99, 17,637.
17. Balogh, A., et al. 1995, Space Sci. Rev. 72, 171.
18. Hoang, S., et al. 1995, Adv. Space Res. 15 (8/9), 371.
19. Summerlin, E. J. & Baring, M. G. 2005, Adv. Space Res. in press. [astro-ph/0505569]
20. Ellison, D. C., Jones, F. C. & Baring, M. G. 1999, ApJ 512, 403.
21. Vasyliunas, V. M., & Siscoe, G. L. 1976, J. Geophys. Res. 81, 1247.
22. le Roux, J. A., Potgieter, M. S., & Ptuskin, V. S. 1996, J. Geophys. Res. 101, 4,791.
23. Forman, M. A., Jokipii, J. R. & Owens, A. J. 1974, ApJ 192, 535.
24. Lee, M. A. 1982, J. Geophys. Res. 87, 5063.

The Importance of Field-Line Meandering in Particle Acceleration at Shocks

J. Giacalone

Dept. of Planetary Sciences, University of Arizona, Tucson AZ

Abstract. We discuss the physics of particle acceleration by shocks emphasizing the importance of the large-scale fluctuating magnetic field. In particular, we discuss the implications for shock acceleration of large cross-field transport due to particles moving nearly along meandering magnetic field lines. It is found that this effect significantly aids in the acceleration of low-energy particles at perpendicular shocks and that there is no injection problem. New results from test particle and self-consistent "hybrid" simulations are presented that support this.

BACKGROUND

The acceleration of charged particles by collisionless shock waves is known to be important in a wide variety of astrophysical plasmas. The basic physics of the mechanism was described in the initial works of Krymksy [1], Axford et al. [2], Blandford & Ostriker [3], and Bell [4]. These authors demonstrated the importance of particle diffusion in the converging flows across the shock. Later, Jokipii [5, 6] extended these calculations to arbitrary angles between the magnetic field and shock normal. He applied the theory of diffusive shock acceleration (DSA) and showed that the acceleration rate is considerably higher for quasi-perpendicular shocks. For this reason, particle acceleration at perpendicular shocks is important to study further. One aspect of this problem that has received considerable attention is the well-known injection problem. It was pointed out by Jokipii [6, 7] that it is difficult to extend DSA theory to low particle energies, especially for quasi-perpendicular shocks. For this case, it is essential that the particles move across the mean magnetic field in order to be accelerated. Unfortunately, cross-field transport is not well understood. However, there has been recent progress made in this area and this is what motivates this brief paper.

In general, the motion of particles across an irregular magnetic field is composed of two parts: (1) the actual displacement of the particle guiding center off of any given magnetic line of force, and (2), motion along these lines of force which themselves meander on larger scales and move normal to the mean field. For particles with small gyroradii compared to the turbulence coherence scale, this last form of cross-field transport dominates. This is important to our understanding of how low-energy particles are accelerated by shocks.

APPLICATION TO THE INJECTION PROBLEM

The injection problem is the difficulty in describing the acceleration of particles near the thermal part of the distribution incident on a shock. The standard theory of DSA does not directly address this. However, by examining the limits of the theory, it is found that the fluctuations in the magnetic field over scales that are much larger than the low-energy particle gyroradii can lead to significant cross-field transport and lead to more efficient acceleration at low energies.

DSA theory is valid if the distribution is quasi-isotropic. One can readily derive a general expression for the diffusive streaming anisotropy [8]; and by requiring this to be small, one can derive an expression for the "injection velocity," v_{inj}. For the case of a perpendicular shock, the most general form v_{inj} is given by[9]

$$v_{inj} = 3U_1 \left[1 + \left(\frac{\kappa_A}{\kappa_\perp} \right)^2 \right]^{1/2} \tag{1}$$

where κ_A is the antisymmetric component of the diffusion tensor. Note that by assuming $\kappa_A = vr_g/3 \gg \kappa_\perp$, this is exactly the same limit as that derived by Jokipii [7].

In general, the ratio $\kappa_A/\kappa_\perp$, need not be restricted to values larger than unity. For the case of hard-sphere scattering, this ratio is simply the ratio of the scattering mean-free path to the particle gyroradius which is, reasonably, larger than unity. However, as was shown by Giacalone and Jokipii [8], hard-sphere scattering is *not* a good approximation for cross-field diffusion. Figure 1 shows a plot of the ratio of $\kappa_A/\kappa_\perp$ as a function of energy determined from a study of particles moving in large-scale magnetic fluctuations [8]. The lower (dashed) curve and open symbols are from [8], whereas the top curve uses the simulated $\kappa_\perp$ (from [8]) and assumes that $\kappa_A = vr_g/3$ which is reasonable for the turbulence spectra considered. Clearly at low energies, the ratio becomes less than unity. Thus, at low-energies, Equation 1 shows that the injection velocity at a perpendicular shock approaches $3U_1$, which is the same as that obtained for a parallel shock [9].

Thus, we conclude that the effect of field-line random walk is to sufficiently increase the perpendicular diffusion coefficient of low-energy particles so that the distribution, averaged over scales large compared to the coherence scale of the fluctuations, is isotropic. Thus, the theory predicts that there should not be an injection problem at nearly perpendicular shocks.

NUMERICAL CALCULATIONS

For particles with speeds on the order of the plasma convection speed, non-diffusive effects may be important. Moreover, low-energy particles may escape downstream by convection over a timescale that is shorter than the characteristic timescale for diffusion. Thus, to better address the physics of acceleration at low energies, we now discuss the results of recent numerical calculations. Note that the results presented here are new, but are related to our recent paper [10] where more details can be found, particularly the description of the models.

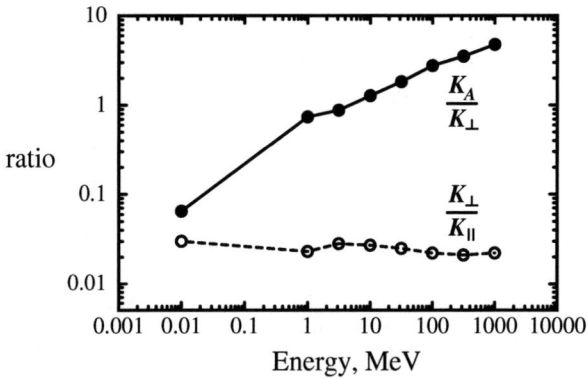

FIGURE 1. Ratios of the components of the diffusion tensor as a function of energy assuming a magnetic field, and coherence length that are typical of the heliospheric magnetic field at 1 AU. The dashed curve is adapted from Giacalone and Jokipii [8] and includes one extra point at the low-energy end. The solid curve assumes that $\kappa_A = v r_g / 3$, which is valid for the turbulence spectra considered in their study.

Test-particle approximation

In these calculations, the particles are assumed to not influence the shock flow or fields. Their motion is governed by the Lorentz force using electric and magnetic fields that are pre-specified. The magnetic field consists of a mean, which is perpendicular to the shock-normal direction, and a superimposed fluctuating component. Both components satisfy Maxwell's equations. The fluctuating component is derived from an assumed power spectrum which is Kolmogorov-like ($\propto k^{-5/3}$, where k is the wavenumber) for high wavenumbers and flat below a characteristic wavenumber that defines the coherence scale. The integrated power in the random component is typically taken to be equal to that in the mean component, with most of the power residing in scales larger than the coherence scale. Particles are released with an energy of 3 times the plasma-ram energy (in the local fluid frame), at the shock, and followed until they either escape downstream by convection, or reach an arbitrary energy cutoff.

Figure 2 shows the energy spectra downstream of the shock obtained from three different simulations of the acceleration of test-particles by a perpendicular shock. Here, the particle mean-free path, $\lambda_\|$, is approximately the same for all runs. $\lambda_\|$ is determined from the quasi-linear theory for a given form of the turbulence power spectrum which is used to synthesize the magnetic field for the test-particle runs. L_c is the coherence scale of the fluctuations and is varied. r_c is the upstream convection gyroradius (upstream plasma flow speed divided by the proton cyclotron frequency).

Figure 2 shows that the large-scale field has a profound effect on the resulting flux at high energies. This is because the cross-field diffusion at low particle energies depends on the coherence scale and is larger for larger-scale fluctuations. Thus, there is increased cross-field diffusion which leads to an increased number of accelerated particles. See

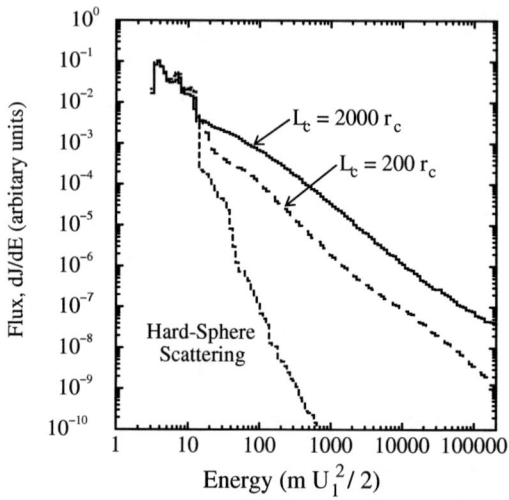

FIGURE 2. Differential intensity as a function of energy obtained from three simulations of test particles accelerated by a perpendicular shock. The top (solid) curve and the lower (hard-sphere scattering) curve are from Giacalone [10]. The middle (dashed) curve is a new simulation for the case of a smaller correlation scale, as indicated.

Giacalone [10] for more results and discussion.

Hybrid simulations

The test-particle simulations presented above consider initial particle energies that are somewhat higher than the peak in the thermal distribution that is incident on the shock. Thus, the issue of the acceleration of thermal particles to high energies remains to be addressed. Thus, we now discuss the results from recent hybrid simulations. This method is fully self-consistent, and describes the non-linear feedback of the accelerated particles on the shock fields. Moreover, the particle velocities are determined from a Maxwellian distribution; therefore, by definition, all particles start out in the thermal population.

In these calculations, we perform a two-dimensional hybrid simulation in which plasma flows in the positive x direction, and is reflected by a rigid wall. A shock forms that propagates back into the flow, in the $-x$ direction. We superimpose on the background field (which is in the z direction) a set of large-scale fluctuations determined from a Kolmogorov spectrum in a manner similar to the test-particle simulations discussed above. The total power is 1/2 of that in the mean field. These fluctuations are also injected at the upstream boundary and carried along with the injected plasma. The resulting wave field is a combination of that injected at the boundary, that which was

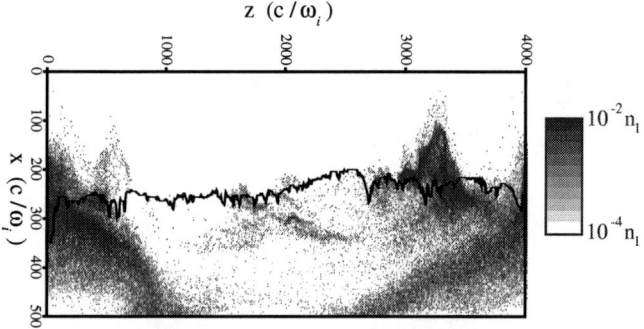

FIGURE 3. Density of energetic particles obtained from a large-scale two-dimensional hybrid simulation that includes field-line random walk on a scale much larger than the convection gyroradius. The gray-scale coding is shown at the upper right where n_1 is the total plasma density upstream of the shock. Only particles with energies greater than $10E_p$ (where E_p is the upstream ram energy of the plasma) are shown.

imposed at the start of the simulation, and that which evolves self-consistently. We note that in this two-dimensional model, the particles are unphysically tied to the magnetic field lines on which each particle starts[11]. However, particles can still move normal to the mean magnetic field by following along meandering magnetic field lines. This turns out to be very important.

Figure 3 shows results from a massive-scale two-dimensional hybrid simulation. As thermal particles encounter the shock a fraction can be accelerated to high energies. This happens non-uniformly along the shock front as indicated in Figure 3 which shows the density of particles with energies above 10 times the plasma ram energy. The solid line is the shock front itself. It has been known for some time that a fraction of the plasma incident on a shock is specularly reflected by the shock, and these ions then gyrate downstream. This is responsible for the energy dissipation required for high-Mach shocks. These ions are presumably the source of the high-energy particles shown in Figure 3, but further study is required to confirm this. Nevertheless, this figure shows that perpendicular shocks can accelerate particles from thermal energies to much higher energies. The acceleration efficiency is difficult to determine because the downstream distribution is spatially nonuniform, however, a reasonable estimate is that it is similar to that obtained ffor parallel shock with a similar Mach number. It was shown by Giacalone et al. [12] that high-Mach parallel shocks convert about 10-20% of the incident plasma ram energy into downstream enthalpy flux in the energetic particles. This has been shown to be consistent with spacecraft observations of particles accelerated by interplanetary shocks [13].

SUMMARY

We have shown that the large-scale fluctuating magnetic field is important with regards to the acceleration of low-energy particles incident on a collisionless shock. The basic idea is that the large-scale field enhances the particle motion normal to the shock and allows for particles to remain near the shock and be accelerated. This has been demonstrated from a simple analytic argument based on the isotropy of the particle distribution, and non-diffusive numerical simulations. Moreover, recent hybrid simulations also indicate that thermal particles can be accelerated to high energies as well.

ACKNOWLEDGMENTS

This work was supported in part by NASA under grants NAG5-7793 NAG5-12919, and NNG04GA79G, by NSF under grants ATM0327773 and ATM0447354.

REFERENCES

1. Krymsky, G. F., *Dokl. Akad. Nauk, SSSR*, **234**, 1306 (1977).
2. Axford, W., Leer, E., and Skadron, G., *Proc. 15th ICRC*, **11**, 132 (1977).
3. Blandford, R. D., and Ostriker, J. P., *Astrophys. J.*, **221**, L29 (1978).
4. Bell, A. R., *Mon. Not. R. Atron. Soc.*, **182**, 1476 (1978).
5. Jokipii, J. R., *Astrophys. J.*, **255**, 716 (1982).
6. Jokipii, J. R., *Astrophys. J.*, **313**, 842 (1987).
7. Jokipii, J. R., *AIP Conf. Proc.*, **264**, 137 (1991).
8. Giacalone, J., and Jokipii, J. R., *Astrophys. J.*, **520**, 204 (1999).
9. Giacalone, J., *Planet. & Space Sci*, **51**, 659 (2003).
10. Giacalone, J., *Astrophys. J.*, **624**, 765 (2005).
11. Jokipii, J. R., Kota, J., and Giacalone, J., *Astrophys. J.*, **20**, 1759 (1993).
12. Giacalone, J., Burgess, D., Schwartz, S. J., and Ellison, D. C., *J. Geophys. Res.*, **102**, 19789 (1997).
13. Ellison, D. C., and Eichler, D., *Astrophys. J.*, **286**, 691 (1984).

Energetic Particles Accelerated by Shocks in the Heliosphere: What is the Source Material?

G. M. Mason[*†], M. I. Desai[*], J. E. Mazur[¶], and J. R. Dwyer[§]

[*]Department of Physics, Univ. of Maryland, College Park MD 20742 USA
[†]I.P.S.T., Univ. of Maryland, College Park, MD 20742 USA
[¶]The Aerospace Corp., El Segundo, CA 90245
[§]Dept. Physics and Space Sci., Florida Inst. Tech., Melbourne, FL 32901

Abstract. Theoretical studies of particle acceleration by shocks have successfully modeled many observed features of the particle energy spectra and temporal variations; however, spectral slopes do not agree with *in situ* observations, and intensities show huge variations not explained by the models. Recent observations have shown that trace ions in the solar wind, such as ^{3}He and He$^+$, are present in the accelerated particles with enhanced abundances. The average heavy ion composition is also found to differ significantly from the solar wind. It is observed that these features correlate significantly with ambient suprathermal material abundances, giving evidence that the suprathermals are the seed population that is actually energized. This raises important new questions such as why the suprathermals are favored over the much more abundant solar wind. Since the suprathermal ion population has many more contributors, and much larger variability than the solar wind, this population needs much more detailed study to allow a closure of theory and observations in interplanetary space.

Keywords: acceleration of particles - interplanetary medium - shock waves
PACS: 96.50.Pw, 96.50.Fm

INTRODUCTION

The acceleration of particles by collisionless shocks has been widely investigated. Analytical models in one dimension have found steady-state solutions in which particles are accelerated in the compression region near the shock, and show decreasing intensities upstream, and constant intensities downstream that are qualitatively consistent with observations [e.g., 1]. However, the predicted spectral forms for energetic particles above a few 10s of keV/n do not fit the data satisfactorily [2, 3]. Although the fastest shocks tend to produce the largest energetic particle intensities, nevertheless for a given shock speed there is roughly a factor of 10,000 variation in intensities of MeV particles at 1 AU [4]. More recently, numerical simulations have incorporated more realistic shock geometries and effects of propagation to the observer, and have begun exploring notional events [e.g., 5].

Because the shocks presumably accelerate the particles in interplanetary (IP) space, the seed population must be drawn from the reservoir of particles in the IP medium. Figure 1 [adapted from 6] shows fluences of major heavy ions at 1 AU. In addition to the solar wind peak at low energies, there is a suprathermal tail which is continuously

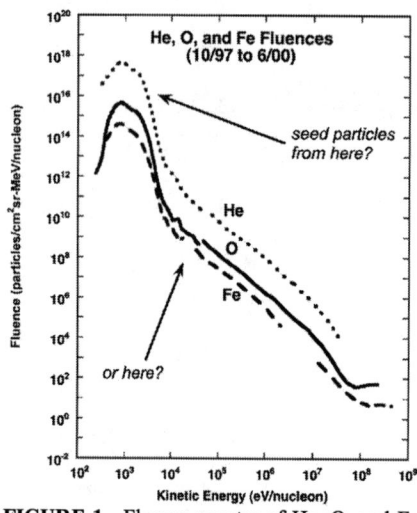

FIGURE 1. Fluence spectra of He, O, and Fe in the interplanetary medium near 1 AU

present [7], extending out to cosmic ray energies. Do the seed particles come from the more numerous solar wind particles, or from the suprathermal tail? Both sources have been suggested [e.g., 1, 8-11], but since the ion composition is roughly similar over the whole energy range it was not possible to tell the difference.

OBSERVATIONS

The Advanced Composition Explorer (ACE) mission [12] was designed to explore such questions by measuring energetic particle ion compositions with sufficient accuracy to enable distinguishing particle populations with small differences. The approach described here uses the fact that certain ions such as ^{3}He and He$^+$ are extremely rare in the solar wind, but are much more numerous in the suprathermal ions. Since no shock acceleration mechanism has been proposed that could lead to large enrichments of these ions, their presence with enhanced abundance in the energetic particles is evidence that the accelerated particles came from the suprathermal ion pool.

Particles Associated with Shocks Near 1 AU

In principle, understanding the energetic particle population associated with shock passages near 1 AU should be relatively simple since the shock properties can be measured directly, the acceleration is presumably nearby so transport effects are minimized, and the seed population should also be nearby. Desai et al. [13] surveyed 48 interplanetary shocks between Oct. 1997 and Nov. 2000 using the ULEIS instrument on the ACE spacecraft [14]. During 23 of the events, only upper limits of ^{3}He were obtained; however, in the remaining 25 events there were substantial enhancements of ^{3}He, as shown in Table 1. Figure 2 shows mass histograms of 0.5-1.0 MeV/nuc He ions in the 4 most enhanced events, where the ^{3}He was hundreds of times more abundant than in the solar wind. Where was the ^{3}He coming from? It is remnant material from previous impulsive solar particle events, which occur often during solar active periods [15-17]. Figure 3 shows that the frequency of occurrence of shocks enriched in ^{3}He correlates reasonably well with the fraction of the time that suprathermal ^{3}He is present at 1 AU [18].

Table 1

^{3}He/^{4}He ratio range	Number of Events
0.001<R<0.01	12
0.01<R<0.1	11
R>0.1	2

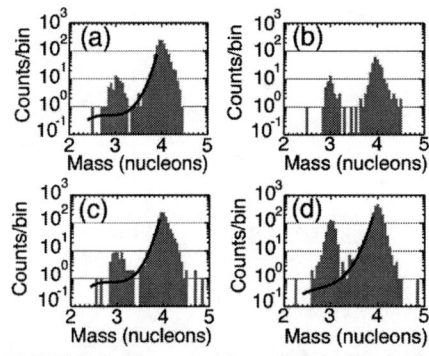
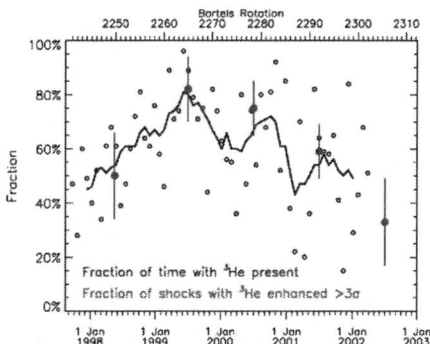

FIGURE 2. He mass histograms in IP shocks with very large enrichments of ^{3}He. Events are (a) 1998 Jan. 28 (b) 1999 Sept. 15, (c) 2000 Jan. 22, and (d) 2000 Apr. 24

FIGURE 3. Fraction of the time ^{3}He is present at 1 AU (open circles and solid curve) and frequency of occurrence of shocks showing enrichment of ^{3}He (larger points with error bars).

The presence of the ^{3}He in the shock accelerated material shows that suprathermals are being accelerated by the shock if it passes through regions of IP space, where by chance prior solar activity has left a population of impulsive solar flare material.

The other unique tracer ion is He^+, which is extremely rare in the bulk solar wind but which is present at 1 AU as a pickup ion from interstellar neutrals that penetrated into the inner solar system and then became ionized [19, 20]. Figure 4 shows spectra of H^+, He^+, and He^{++} measured in a corotating interaction region (CIR) on April 7, 1995 using an instrument on the Wind spacecraft [21]. At speeds around the solar wind peak, the He^+ is orders of magnitude less abundant than He^{++}, while above ~1.8-2 times the solar wind speed, it is 16-17% of the abundance of He^{++}, an enormous enrichment that extends out to many times the solar wind speed. Other instruments have observed similar enrichments at 1 AU [22, 23]. At greater radial distances where the pickup ions are more numerous relative to the solar wind, the ratio of He^+ to He^{++} in CIRs greatly exceeds 1 [24]. Since the pickup ion enhancements begin just about ~twice the solar wind speed, this appears to be the injection energy threshold. This is at least qualitatively consistent with several recent theoretical investigations that have shown that pickup ions can be preferentially injected at shocks [25-28].

Common Ion Abundances

Although the presence of tracer elements such as ^{3}He and He^+ establish the importance of suprathermals for understanding the presence of rare elements in the accelerated particle population, the question remains whether the more common ions never-

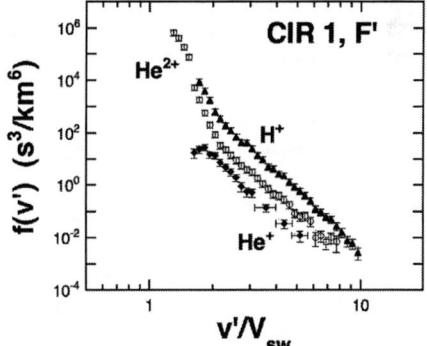

FIGURE 4. Energetic H^+, He^+, and He^{++} plotted as a function of speed relative to the solar wind. Notice that the large abundance of He^+ at higher speeds falls off at speeds near the solar wind peak.

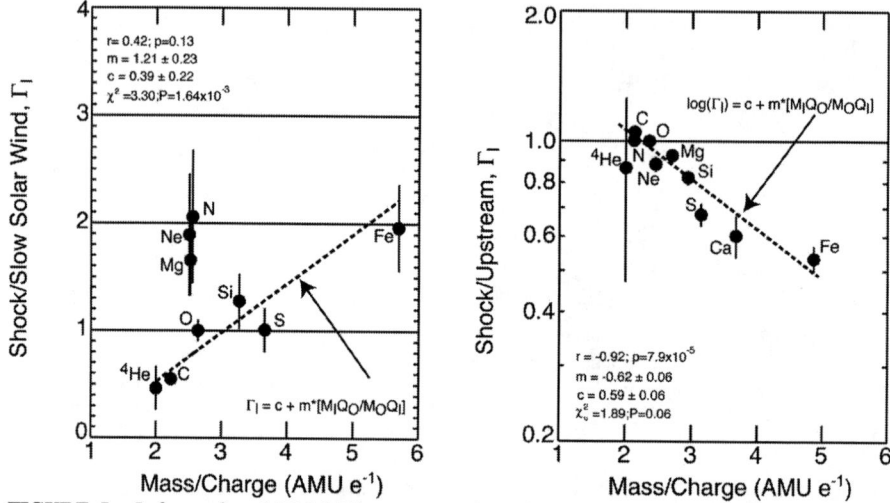

FIGURE 5. *Left panel:* ~0.75 MeV/nuc average interplanetary shock composition compared to slow solar wind vs. particle mass to charge ratio; *Right panel:* average interplanetary shock composition compared to the ambient composition preceding the shock. Abundances are normalized to O.

theless reflect the bulk solar wind composition. Desai et al. [29] investigated this by surveying the 0.1-10 MeV/nuc energetic particle abundances at 72 IP shocks. The left panel of Fig. 5 shows a comparison of the IP shock abundances compared to slow solar wind and plotted vs. Mass/Charge (M/Q) using typical ionization states from other studies. The correlation is not monotonic, and in particular there are odd variations for elements such as N, O, Ne, and Mg which all have virtually the same M/Q ratio. However, Desai et al. found that if the IP shock abundances were compared with the average ambient abundances preceding the shocks (Fig. 5 right panel) the odd variations were removed, and the general trend showed a decrease in efficiency for accelerating particles of high M/Q.

Desai et al. [3] also investigated this same set of shocks on a case by case basis to see if the average properties remained visible. It was found that the spectra of heavy elements such as O did not correlate well with the value calculated using simple acceleration theory and the shock compression ratio M, as can be seen in Fig. 6. A similar lack of correlation had been seen previously for protons [2]. Desai et al. also found that the characteristic roll-over energy of the

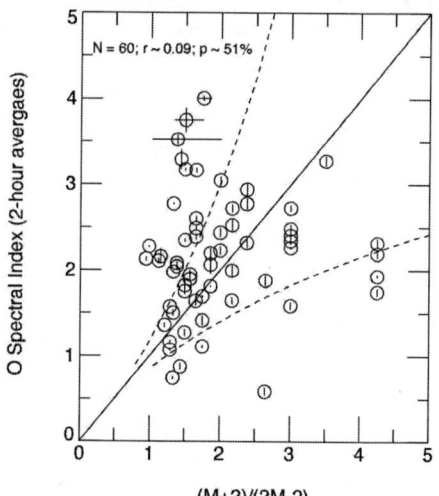

FIGURE 6. O spectra measured within 2 hours of shock passage vs. predictions of simple model.

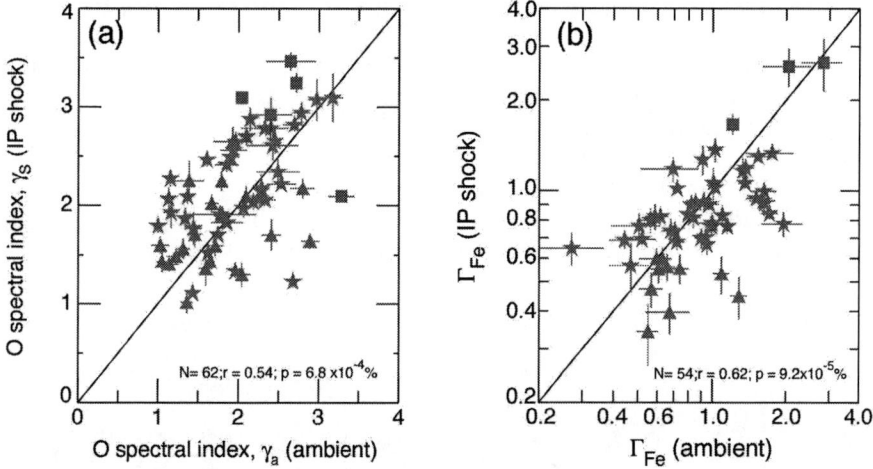

FIGURE 7. *Left panel:* Spectral index of O at individual IP shocks vs. the spectral index of O in the ambient population of the same shocks; *Right panel:* roll-over of the Fe/O ratio at IP shocks vs. the ambient population (roll-over is Fe/O at 0.62 MeV/n divided by Fe/O at 0.22 MeV/n).

shock spectra was uncorrelated with shock parameters such as shock normal angle or shock speed. On the other hand, the spectral index of O and the rollover of the Fe/O ratio both showed highly significant correlations with the same quantities measured in the ambient population preceding each shock, as shown in Fig. 7. The correlations shown in the Fig. are extremely unlikely from a random population (p = 7×10^{-4} % and 9×10^{-5} %, respectively), and we emphasize that the simple theory predicts that these quantities would be uncorrelated.

If these spectral features are seen in the ambient material, then it would be expected that the abundance ratios would correlate as well. Figure 8 shows this correlation for 62 shocks [29]. The point marked #29 is an obvious outlier that came from the complex series of events in July 2000. If this point is removed, the remaining correlation is again highly significant, having a very low probability (2×10^{-3} %) of being from a random population. Again, the simple acceleration theory would predict that the variables in Fig. 8 are uncorrelated.

We conclude that the composition and spectral features of heavy ions in IP shocks do not agree well with the simple model, but rather are significantly correlated with the ambient population.

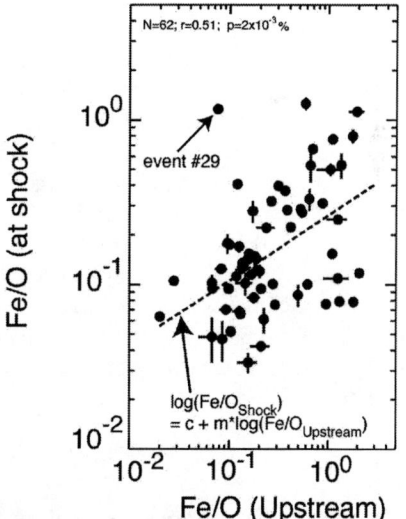

FIGURE 8. Fe/O ratio at individual IP shocks vs. Fe/O ratio in the ambient population preceding the same shock.

Particles Associated with CME-related Solar Particle Events

Large solar energetic particle (SEP) events are generally associated with Coronal Mass Ejection (CME) driven shocks. Investigating the seed population in this case is more difficult than for the traveling IP shocks for several reasons: the shock properties can't be measured; the acceleration is far away and transport effects need to be considered; and the seed population is far away. However, since these events also produce the largest intensity enhancements at 1 AU and are critical for space weather applications, it is important to understand the processes involved.

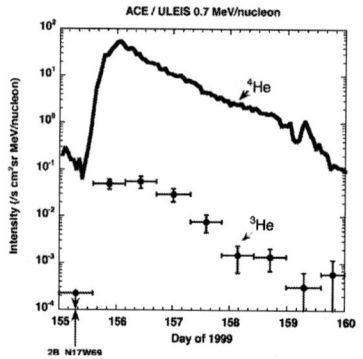 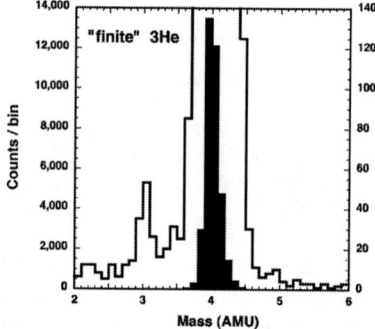

FIGURE 9. Time-intensity profiles of 0.7 MeV/n ^{3}He and ^{4}He in the June 4, 1999 event.

FIGURE 10. Mass histogram of 0.5-2.0 MeV/nuc He showing entire peak (left scale), and expanded scale (right) which shows ^{3}He.

Figure 9 shows time-intensity profiles for ^{4}He and ^{3}He in the June 4, 1999 large SEP event [15]. The shape of the profiles of the two species is very similar, indicating that they have the same acceleration and transport history. The ^{3}He in this event is enriched by a factor of 16±3 over solar wind values, an unexpected result since this same event also shows large Fe enrichment. Since Fe has a large M/Q ratio, and ^{3}He a small M/Q ratio it is not expected that an acceleration mechanism would simultaneously enrich both. The event in Fig. 9 was selected from a list of NOAA/SEC events that produced 10 MeV proton intensity enhancements, and most of the events on that list are CME related. A substantial fraction of events of this type have been found to have ^{3}He enrichments [15, 16, 30]. This is also attributed to the same remnant population of ^{3}He from prior impulsive SEP events [15]. Figure 10 shows the low energy He mass histogram from several such events; the ^{3}He is clearly resolved from ^{4}He and the background.

Common Ion Abundances

Surveys of SEP ion abundances from the ACE mission are still underway, but studies at higher energies also are showing that the large SEP abundances do not correlate well with solar wind abundances [31]. So at this stage the situation is similar to that in the IP shocks: remnant impulsive SEP ^{3}He material is found to be reaccelerated in these events, and the common ion abundances show unsystematic differences from the solar wind.

DISCUSSION

The heavy ion composition, abundances, and spectra at interplanetary shocks exhibit the following properties:
- the presence of tracer elements ^{3}He and He$^+$ that are virtually absent in the solar wind,
- average abundances that show unsystematic differences from the solar wind,
- spectral forms that correlate significantly with the ambient population, and not with simple shock acceleration theory predictions.

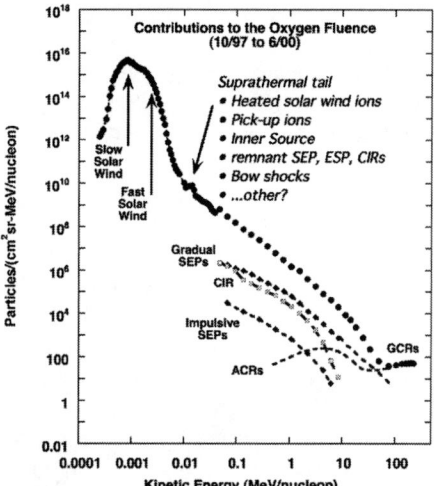

FIGURE 11. Long term oxygen fluence at 1 AU showing different contributors to the ion population.

CME associated SEP events also show the presence of ^{3}He and have common ion abundances with significant differences from the solar wind.

Taken together, the weight of the evidence is that the suprathermal population is the main seed reservoir for He and heavy ions. Numerical simulations show that pickup ions are easily injected at shocks due to their velocity distribution which is a shell in velocity space [e.g. 25, 32]. Other suprathermals, e.g. remnant SEPs, CIRs, impulsive flare ions, also have broad distributions in velocity space since they have had time to become isotropized. So the suprathermal ions would have similar advantages (compared to bulk solar wind) in acceleration at IP shocks.

Figure 11 shows the O fluence at 1 AU, and identifies the sources of ions at different energies [6]. The suprathermal tail has many contributors, and they in turn have significant spatial and temporal variations. Figure 12 shows that the suprathermal intensities show much larger time variations than the solar wind. This large variation may play a critical role in determining the peak intensities in shocks which show an extremely large range for shocks of

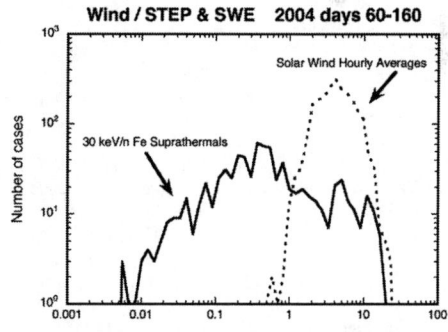

FIGURE 12. Hourly average suprathermal (solid line) and solar wind (dotted line) values for a 100 day period in 2004. While the solar wind shows a factor of 10 variation, the 30 keV/nuc suprathermals shown a factor of 1000 variation.

the same speed.

Fully understanding the role of suprathermals will be essential in realistically modeling interplanetary shocks, exploring the actual mechanisms of injection and acceleration, and producing results that can be compared in detail with observations. This would allow a better closure of theory and observations and would give the prospect of more accurate predictions of space weather events. It could also give important insights into astrophysical shocks and the particles they energize.

ACKNOWLEDGMENTS

We thank the authors and the following organizations for permission to publish copywrited material: American Astronomical Society/University of Chicago (Figs. 2,5,6,7,8,9,10); American Institute of Physics (Figs. 1, 11); American Geophysical Union (Fig. 4). Figure 12 used Solar Wind data from the Wind/SWE experiment, kindly provided by K. Ogilvie and the SWE team. This work was supported by NASA under Caltech grant 44A1055749 at the University of Maryland.

REFERENCES

1. Forman, M.A., et al., in *Geophysical Monograph No. 34*, R.G. Stone, et al., Editors. 1985, *American Geophysical Union*. p. 91-114.
2. van Nes, P., et al., J. Geophys. Res., 1984. 89: p. 2122.
3. Desai, M.I., et al., Ap.J., 2004. 611: p. 1156-1174.
4. Kahler, S., J. Geophys. Res., 2001. 106: p. 20947-20956.
5. Li, G., et al., J. Geophys. Res., 2003. 108 (doi: 10.1029/2002JA009666): p. SSH 10-1 - 10-22.
6. Mewaldt, R.A., et al., A.I.P. CP #598, ed. R.F. Wimmer-Schweingruber. 2001, A.I.P.: NY. 165.
7. Gloeckler, G., et al., in A.I.P. CP #528, R.A. Mewaldt, et al., Editors. 2000, AIP: NY. p. 221-228.
8. Dröge, W., et al., Astrophys. J. (Letters), 1992. 387: p. L97-L100.
9. Gosling, J.T., et al., J. Geophys. Res., 1981. 86: p. 547-554.
10. Tsurutani, B.T., et al., J. Geophys. Res., 1985. 90: p. 1-11.
11. Smart, D.F., et al., Proc. 21st ICRC, 1990. 7: p. 324-327.
12. Stone, E.C., et al., Space Sci. Rev., 1998. 86: p. 1-22.
13. Desai, M.I., et al., Astrophys. J. (Lett.), 2001. 553: p. L89-L92.
14. Mason, G.M., et al., Space Sci. Rev., 1998. 86: p. 409-448.
15. Mason, G.M., et al., Astrophys. J. (Letters), 1999. 525: p. L133-L136.
16. Wiedenbeck, M.E., et al., A.I.P. CP #528, ed. R.A. Mewaldt, et al. 2000, AIP: Melville, NY. 131-134.
17. Wiedenbeck, M.E., et al. in A.I.P. Conf. Proc. #679. 2003. New York: A.I.P.
18. Stone, E.C., et al., personal communication. 2003.
19. Möbius, E., et al., Nature, 1985. 318: p. 426.
20. Gloeckler, G., et al., Science, 1993. 261: p. 70-73.
21. Chotoo, K., et al., J. Geophys. Res., 2000. 105: p. 23107-23122.
22. Hilchenbach, M., et al., in Solar Wind Nine, S.R. Habbal, et al., Editors. 1999, A.I.P.: NY. p. 605-608.
23. Kucharek, H., et al., J. Geophys. Res., 2003. 108(A10): p. LIS 15-1.
24. Gloeckler, G., et al., J. Geophys. Res., 1994. 99: p. 17637-17643.
25. Giacalone, J., et al., J. Geophys. Res., 1994. 99: p. 19351-19358.
26. Scholer, M., et al., J. Geophys. Res., 1980. 85: p. 1743.
27. Ellison, D.C., et al., Astrophys. J., 1999. 512: p. 403-416.
28. Scholer, M., et al., A.I.P. CP #528, ed. R.A. Mewaldt, et al. 2000, AIP: Melville, NY. 250-257.
29. Desai, M.I., et al., Astrophys. J., 2003. 588: p. 1149-1162.
30. Mason, G.M., et al., Astrophys. J., 2002. 574: p. 1039-1058.
31. Mewaldt, R.A., et al., Adv. Space Res., 2002. 30: p. 79-84.
32. Scholer, M., et al., Geophys. Res. Let., 1999. 26: p. 29-32.

Solar Energetic Particle Spectral Breaks

R. A. Mewaldt[a], C. M. S. Cohen[a], G. M. Mason[b], A. W. Labrador[a],
M. L. Looper[c], D. E. Haggerty[d], C. G. Maclennan[e], A. C. Cummings[a],
M. I. Desai[b], R. A. Leske[a], G. Li[f], J. E. Mazur[c], E. C. Stone[a],
and M. E. Wiedenbeck[g]

[a] *California Institute of Technology, Pasadena CA 91125*
[b] *University of Maryland, College Park, MD 20742*
[c] *The Aerospace Corporation, Los Angeles, CA 90245*
[d] *Johns Hopkins University, Applied Physics Laboratory, Laurel MD 207232*
[e] *Bell Laboratories, Lucent Technologies, Murray Hill, NJ 07974*
[f] *IGPP, University of California, Riverside, CA 92521*
[g] *Jet Propulsion Laboratory, California Institute of Technology, Pasadena, CA 91007*

Abstract. The five large solar particle events during October-November 2003 presented an opportunity to test shock acceleration models with in-situ observations. We use solar particle spectra of H to Fe ions, measured by instruments on ACE, SAMPEX, and GOES-11, to investigate the Q/M-dependence of spectral breaks in the 28 October 2003 event. We find that the break energies scale as $(Q/M)^b$ with $b \leq 1.56$ to 1.75, somewhat less than predicted. We also conclude that SEP spectra >100 MeV/nucleon are best fit by a double power-law shape.

INTRODUCTION

The Halloween 2003 period provided the opportunity to study the spectra of five large solar energetic particle (SEP) events within a 10-day period, each associated with an X-class flare and very fast (>1500 km/sec) CME. An overview of this period is shown in Figure 1, which includes oxygen intensities in several energy intervals. In the two largest events (events 2 and 3), particles were still being accelerated to ~10 MeV/nuc when the shock reached 1 AU. Two studies of these five events [1, 2] have shown that elements from H to Fe have rather prominent spectral breaks in the energy range from ~3 to ~30 MeV/nuc, similar to those in earlier studies by Tylka et al. [3, 4]. Li, Zank and Rice [5; hereinafter LZR] have recently extended the model of Zank, Rice and Wu [6] to heavier ions. In their model the break energy is proportional to the square of the charge-to-mass ratio (Q/M), where Q is the charge state and M is the ion's mass number. In this paper we combine measurements from instruments on ACE, SAMPEX and GOES-11 to obtain SEP energy spectra for the five Halloween events and the Jan. 20, 2005 event. We test the Q/M dependence predicted by LZR with observations of spectral breaks for nine elements following the strong shock from the Oct 28, 2003 SEP event. We also show that a double power-law shape provides excellent fits to the high-energy (>100 MeV/nuc) spectra in these events.

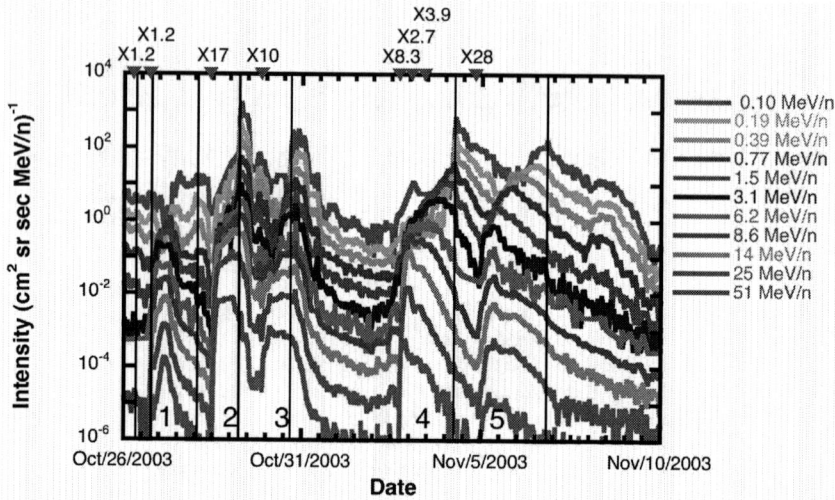

Figure 1: Time history of energetic oxygen nuclei measured by the SIS and ULEIS instruments on ACE (from Cohen et al. [1]). Five large SEP events are evident (numbered 1 to 5). Interplanetary shocks are indicated by vertical solid lines and X-class flares are shown along the top.

OBSERVATIONS

The shock from the 28 October 2003 event reached Earth within 19 hours followed by driver gas traveling at ~2000 km/sec [7]. Upon arrival at 1 AU this shock was still accelerating oxygen to ~10 MeV/nuc (see Figure 1) and protons to ~30 MeV [2]. Figure 2 shows energy spectra measured during the 6 hours following the arrival of the shock. Note that all species show a relatively sharp break at energies ranging from a few MeV/nuc to ~30 MeV/nuc. The location of the breaks depends on species – lighter species have higher break energies than heavier species like Fe (see also Tylka et al. [3]). These spectra provide an excellent opportunity to test the prediction of LZR [5] that the location of the breaks should scale as $(Q/M)^2$.

The spectra in the left panel of Figure 2 were fit by a two-step procedure. In the first step the points above and below the apparent break were individually fit with power-laws, giving average power-law indices of –1.43 below the break and –4.1 above the break. In the second step power-laws with these fixed, average slopes were individually scaled to fit the points above and below the break, and the intersection of the two power-laws was defined as the break energy. The break energies ranged from 31 MeV for protons to 4.1 MeV/nuc for Fe.

Ellison and Ramaty [8] proposed that shock-accelerated SEPs would have spectra of the form:

$$dJ/dE = KE^{-\alpha}\exp(-E/E_o). \qquad (1)$$

Here J is the intensity, K, E_o, and α are constants, and we have treated E as energy/nuc. This spectrum, which is a power-law at low energies with an exponential roll-over at high energies, has been used extensively by Tylka et al. [3, 4] to fit SEP spectra, and we use this form to obtain a second estimate of the break energy. The average value of α was 1.3. Using α = 1.3 the spectra were then refit to determine E_o for each species, which ranged from 31 MeV for protons to 3.5 MeV/nuc for Fe (see Figure 3).

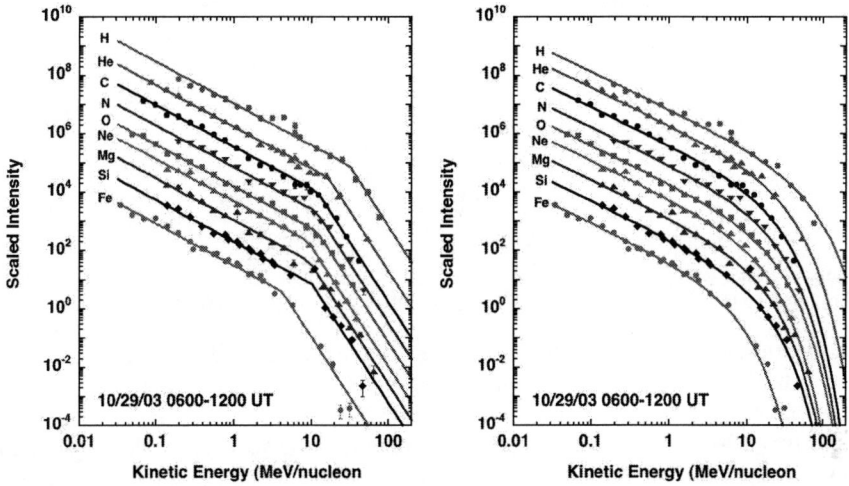

FIGURE 2. (Left) Spectra from the period following the shock on 29 October are fit with two power-laws using a fixed slope of −1.43 at low energy and −4.1 at high energy. The intersection of these fits determines the break energy. (Right) The same data are fit with the Ellison-Ramaty spectral form using a fixed value of −1.3 for the power-law index. Data are from the SIS, ULEIS and EPAM instruments on ACE, and the EPS sensor on GOES-11 (see Mewaldt et al. [2] for additional details). Each element has been multiplied by a scale factor to separate the spectra.

In order to investigate the Q/M-dependence of the breaks we use charge-state measurements from the same event by Labrador et al. [9] obtained using the geomagnetic method with the MAST instrument on SAMPEX [10]. The energy intervals range from 15 to 60 MeV/nuc for O to 27 to 90 MeV/nuc for Fe. The two estimates of the break energies are plotted against Q/M in Figure 3.

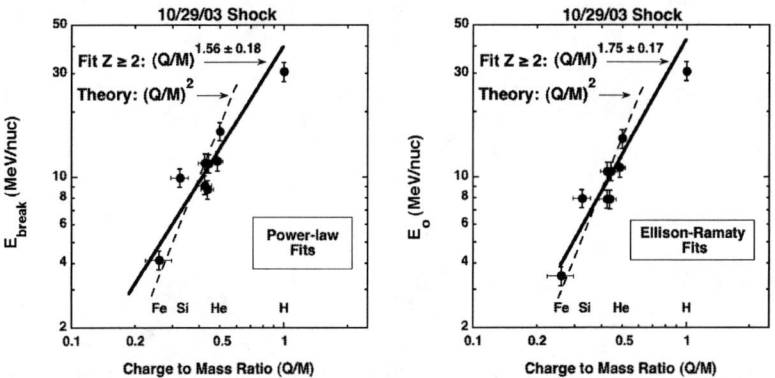

FIGURE 3. Break energies from Figure 2 are plotted versus Q/M. Fits to the Z≤2 data give somewhat weaker dependence on Q/M than is expected from the theory of LZR [6].

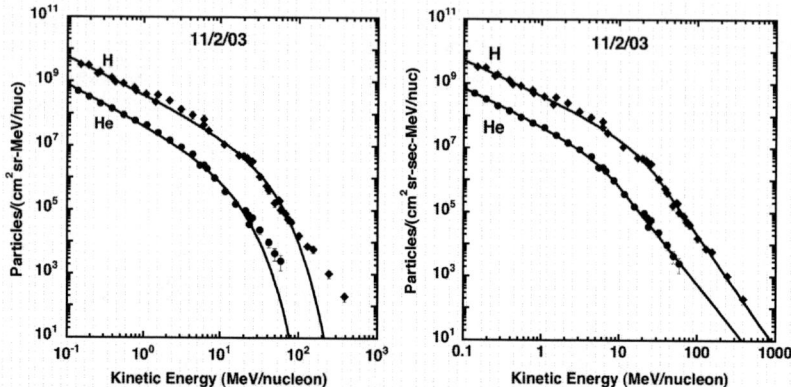

FIGURE 4. H and He Fluence spectra for the Nov. 2, 2003 event are better fit with the double power-law form of Band et al. [11; Right] than with the Ellison-Ramaty spectral form (Left).

The fluence spectra for the five Halloween events (integrated over the events) also have spectral breaks [2, 1]. For these spectra the Ellison-Ramaty form is adequate up to ~100 MeV/nuc, but fails to fit the spectra at higher energies [2]. Figure 4 illustrates that spectra >100 MeV/nuc are better fit by a double power-law formula from Band et al. [11], given by:

$$dJ/dE = CE^{-\alpha_a} \exp(-E/E_o) \text{ for } E \leq (\alpha_b - \alpha_a)E_o;$$

$$dJ/dE = CE^{-\alpha_b}\{[(\alpha_b - \alpha_a)E_o]^{(\alpha_b - \alpha_a)} \exp(\alpha_a - \alpha_b)\} \text{ for } E \leq (\alpha_b - \alpha_a)E_o. \quad (2)$$

Here α_a is the low-energy power-law slope and α_b is the high-energy power-law slope. Spectra from the five Halloween events are shown in Figure 5, along with the Jan. 20, 2005 event, where the breaks are at lower energy, and the spectra above the breaks are much harder. In this event the H and He spectra above the breaks extend as power-laws with an index $\leq$ -2.2 for about two decades in energy or more.

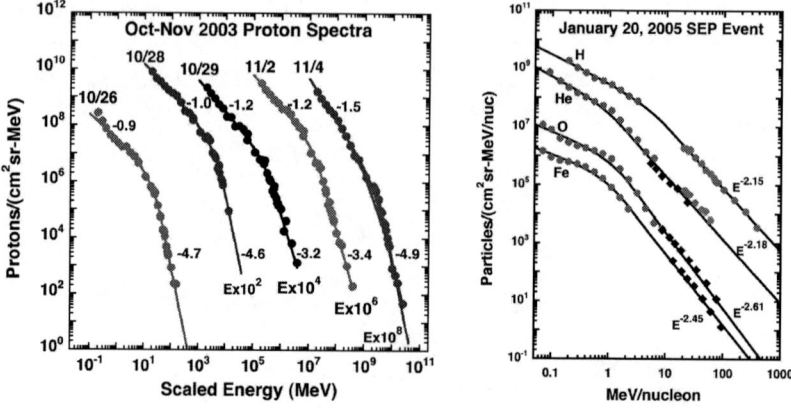

FIGURE 5. (Left) Fluence spectra for the Oct.-Nov. 2003 events fit with the Band et al. shape [11]. Spectral indices above and below the break are indicated. (Right) The 20 Jan. 2005 event had a double power-law shape with a very hard spectrum. Data are from ACE, SAMPEX and GOES-11 [2].

DISCUSSION

In the model of LZR [5] streaming protons escaping upstream from the shock generate enhanced turbulence in the form of Alfven waves that extends over ~2 decades in wave number (k; see Lee [12] and Ng, Reames, and Tylka [13], hereinafter NRT]). The proton-amplified Alfven waves play a key role in scattering particles and keeping them near the shock where they can be efficiently accelerated. According to LZR, the break in the power-law spectrum for a given species occurs at the maximum achievable momentum/nuc (p_{max}) for which there is efficient acceleration, which corresponds to the momentum/nuc that resonates with the minimum k-value (k_{min}) for which there is enhanced turbulence. Thus, for species i,

$$k_{min} = (Q_i/M_i)(eB/c\, p_{max}), \quad (3)$$

where B is the magnetic field strength, e the electron charge, and c the speed of light. Converting from momentum/nuc to energy/nuc (E) the break-energy of species i ($E_{max,i}$) is related to that of protons by

$$E_{max,i} = (Q_i/M_i)^2 E_{max,p}. \quad (4)$$

In the LZR model the sudden decrease in the turbulence level at k values below k_{min} leads to a sudden increase in the diffusion coefficient, thus allowing higher-energy particles to freely escape upstream from the shock. There are similar, though less sharp, breaks in the wave spectra calculated by NRT. Recently, evidence has been reported for proton-amplified Alfven waves at 1 AU in large SEP events [14, 15].

The location of the breaks for He to Fe (Figure 3) are reasonably consistent with the power-law behavior predicted by LZR, although the slope that we find (1.56 to 1.75) is somewhat less than predicted. Note that the proton break energy is low compared with the fits to the heavy ions. Li and Zank (personal communication) have pointed out that while heavy ions should act like test particles, protons may not, because protons produce the waves that govern this process [5, 13].

Cohen et al. [1] analyzed the location of spectral breaks for $8 \leq Z \leq 26$ ions in the fluence spectra of the five Halloween events. Using average SEP charge states measured at lower energy they find that the breaks scale like $(Q/M)^b$ with b = 0.9 to 1.5. Mewaldt et al. [2] found that breaks in the H and He fluence spectra suggest b = 0.7 to 1.3, similar to Tylka et al. [3]. It is possible that the expected Q/M dependence in fluence spectra is washed out somewhat because the maximum energy (E_{max}) is greater closer to the Sun than at 1 AU. Fluence spectra from 1 AU inevitably include particles accelerated close to the Sun as well as those accelerated locally [5].

There are also likely to be longitudinal differences in the break energies that may depend on shock geometry [4]. In fluence spectra the magnetic connection point to the shock evolves with time, and this evolution depends on the event location. These considerations, not accounted for in most theoretical models, complicate comparisons such as these. Presumably, contributions from particles accelerated closer to the Sun or at different longitudes are minimized in the Oct. 29, 2003 shock period, where local acceleration dominates, and the Q/M-behavior comes closest to the model predictions.

Cohen et al. [1] suggested that the location of spectral breaks in fluence spectra is governed by diffusion processes, such as escape upstream from the shock. In this case, spectral breaks for different species should occur at the same value of the diffusion coefficient [16]. Cohen et al. [1] found that the fluence spectra of $8 \leq Z \leq 26$ ions in the Halloween events shared a common spectral shape (see also [2]). Assuming a diffusion coefficient that scales as $(M/Q)^\alpha$, they found $\alpha = 0.8$ to 2.7. Following Droege [17], Cohen et al. related the value of α to the spectrum of interplanetary turbulence (taken to be a power-law in wave number, k^{-q}). They concluded that SEP spectral breaks in these events were organized by wave spectra in the range from $k^{-1.2}$ to $k^{0.7}$ (see also [2]), significantly flatter than a $k^{-5/3}$ Kolmogorov spectrum, consistent with a source of turbulence near the shock where the ions were accelerated. It is interesting that calculations by NRT [13] show that proton-amplified Alfven waves in SEP events can produce broad features in the wave spectra with $q \leq 0$, and that similar wave spectra have been reported in SEP events at 1 AU [14, 15].

Finally, we point out that the nature of spectra beyond the breaks has space-weather implications. The Band et al. form fits the Halloween spectra better than the Ellison-Ramaty form [2], and a power-law above the break is clearly required for the Jan. 20, 2005 event. Extrapolating these spectral forms results in significantly different estimates of SEP radiation doses, especially behind several g/cm^2 of shielding. It is therefore important to continue efforts to understand and eventually forecast not only the location of spectral breaks, but also the nature of the spectra beyond the breaks.

ACKNOWLEDGMENTS

This work was supported by NASA under grants NNG04GB55G, NNG04088G, and NAG5-12929. We appreciate the availability of GOES-11 data from NOAA's Space Environment Center at http://www.sec.noaa.gov/Data/index.html.

REFERENCES

1. Cohen, C. M. S., et al., to be published in JGR, (2005).
2. Mewaldt, R. A., et al., submitted to JGR, (2005).
3. Tylka, A. J., et al., Ap. J., **558**, L59-L63, (2001).
4. Tylka, A. J., et al., to be published in Ap. J., (2005).
5. Li, G., Zank, G. P., and Rice, W. K. M., to be published in JGR, (2005).
6. Zank, G. P., Rice, W. K. M., and Wu, C. C., JGR, **105**, 25079-25095 (2000).
7. Skoug, R. M., et al., JGR **109**, A9, DOI: 10.1029/2004JA010494 (2004).
8. Ellison, D. C., and Ramaty, R., Ap. J., **298**, 400-408 (1985).
9. Labrador, A. W., et al., to be submitted to the 29th Internat. Cosmic Ray Conf., Pune, India, (2005).
10. Leske et al., Ap. J., **452**, L149-L152 (1995).
11. Band, D., et al., Ap. J., **413**, 281-292, (1993).
12. Lee, M. A., JGR, **88**, 6109-6119, (1983).
13. Ng, C. K., Reames, D. V. and Tylka, A. J., **591**, 461-485, (2003).
14. Bamert, K., et al., Ap. J., **601**, L99-L102 (2004).
15. Kallenbach, R., et al., this volume (2005).
16. Cummings, A. C., Stone, E. C., and Webber, W. R., Ap. J. **287**, L99-L103 (1984).
17. Droege, W., Ap. J. Suppl. **90**, 567-576 (1994).

Upstream turbulence and the particle spectrum at CME-driven Shocks

Gang Li*, Q. Hu* and G. P. Zank*

Institute of Geophysics and Planetary Physics, University of California Riverside, Riverside CA 92521, USA

Abstract. Particle spectra at a CME-driven shock often exhibit a power law to certain energies, then roll over exponentially beyond. However, there are cases where a spectrum evolves to another power law above a certain energy (e.g. the Oct. 29th, 2003 event). Here we introduce an effective "loss term" into the particle transport equation and study the consequent particle spectra behavior at a CME-driven shock. The loss term represents the effect of particle leaking out from a finite shock and is related to the turbulence power at and near the shock. We show that the shape of particle spectra are tightly related to the form of upstream turbulence. Under certain circumstances, broken power-law spectrum can be obtained. The physical meaning of the "loss term" and its relationship to the upstream turbulence is discussed.

INTRODUCTION

It is now widely accepted that large gradual Solar Energetic Particle (SEP) events are due to Coronal Mass Ejections (CMEs). As coronal masses are ejected, a shock is formed and driven in front of the ejected matter and particles are accelerated at the shock front as they cross the shock back and forth multiple times. Such a process, known as diffusive shock acceleration ([1], [2], [3], and [4]) or first-order Fermi acceleration is proposed in late 1970's. In this mechanism, scattering centers assure that particles scatter back and forth between the upstream and downstream medium, and upon completing every traversal, a (non-relativistic) particle will gain momentum proportional to $m\Delta v$, where m is the particle mass and $\Delta v = |v_2 - v_1|$ is the background velocity difference between the upstream and downstream.

For this mechanism to be applicable to a CME-driven shock, however, some form of turbulence in the upstream and downstream regions must be present to provide the necessary particle scattering. For a parallel propagating CME-driven shock, the upstream and downstream turbulence is generally assumed to be Alfvén waves [5] that are driven by protons streaming away from the shock front. Since the turbulence is generated by the energetic particles themselves, the wave intensity is coupled to the anisotropic part of the particle distribution function f. Thus, a proper understanding of particle acceleration at a CME-driven shock requires a self-consistent model in which the accelerated particle spectrum and the stimulated wave intensity can be solved simultaneously.

Early observations of upstream turbulence at interplanetary shocks can be found, for example, in [6]. Using ISEE3 data, [6] studied the spectrum of low energy particles and the energy density of the magnetic field. Their analysis showed that there are simultaneous increases of energetic particle intensities and wave power density when an interplanetary shock approaches the spacecraft. Furthermore, the increases occur at particle momentum p and wave number k satisfying $p \cdot k \approx \Omega$ where $\Omega = |q|B/m$ is the particle gyro-frequency. More recent observations that aim to understand the relationship between particle spectra and wave power in large SEP events can be found in [7]. There, the wave power density is plotted as a function of distance in front of the shock. The study of[7] showed that the upstream wave intensities agree qualitatively well with the theory proposed in [5] and [8].

In this paper, we study the particle transport equation at a moving CME-driven shock under the assumption that the upstream turbulence power is decreasing with distance and drops to the background level within a finite distance. To represent such an effect, we add a "loss term" to the transport equation. We derive the steady state particle spectrum and discuss its physical meaning. We show that under certain conditions, particle spectra can be "broken power" laws.

TRANSPORT EQUATION WITH A LOSS TERM

The transport equation that governs the motion of particles, such as cosmic rays in the interstellar medium and energetic particles in the solar system, is [9],

$$\frac{\partial f}{\partial t} + \underline{u} \cdot \nabla f - \nabla \cdot (\underline{\kappa} \cdot \nabla f) - (\frac{1}{3}\nabla \cdot \underline{u})p\frac{\partial f}{\partial p} = 0. \tag{1}$$

In (1), f is the particle distribution function, u the background fluid speed, t the time, and p the particle momentum. Terms in equation (1) describe convection ($u\frac{\partial f}{\partial x}$), diffusion ($\frac{\partial}{\partial x}(\kappa\frac{\partial f}{\partial x})$) and particle energy change ($\frac{1}{3}\frac{du}{dx}p\frac{\partial f}{\partial p}$ for a non-zero $\frac{du}{dx}$) respectively (For charged particles moving within a background magnetic field B, a "drift" term $V_d = (pv/3q)\nabla \times (\vec{B}/B^2)$ needs to be added to the convection term).

When a shock is present, one can solve the transport equation (1) with the matching conditions that both the distribution function f and the current $S = -(p/3)u\partial f/\partial p - \kappa \cdot \nabla f$ are continuous at the shock front.

Consider the simplest case of a planer shock as shown in the left panel of figure 1. For the 1-D case, the general steady state solution of f is [10],

$$f = A(p) + B(p)exp\int_0^x d\zeta \frac{u}{\kappa}. \tag{2}$$

If the diffusion coefficient κ is independent of x, applying the matching conditions will yield the well-known power law spectrum,

$$A(p) = B(p) \sim p^{-3s/(s-1)}, \tag{3}$$

where $s = u_1/u_2$ is the compression ratio.

This simple and elegant theory, when proposed in the 1970's, is believed to be responsible for accelerating cosmic rays up to 10^{14-15} eV at a supernova blast shock wave. On a much smaller scale (both spatially and energetically), the same mechanism is recognized to be responsible for gradual solar energetic particles (SEPs) events [11].

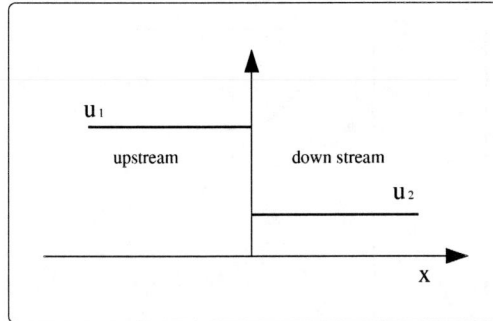 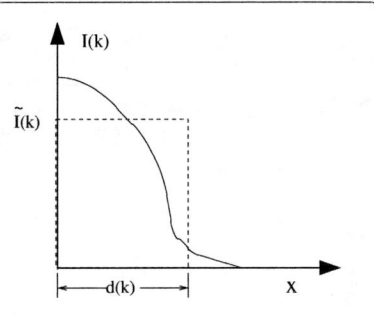

FIGURE 1. Left: Cartoon showing the upstream and downstream of an ideal planar shock. Right: Cartoon showing the upstream turbulence as a function of distance from the shock.

Unlike the supernova case, where the relationship between high energy cosmic rays and the supernova blast shock waves can only be inferred by, for example, synchrotron observations [12], solar system shocks and energetic particles can be directly observed simultaneously, thanks to spacecraft such as ACE, WIND, etc. Furthermore, magnetometers onboard these spacecraft also provide us unambiguous knowledge of the turbulence. Thus, studying CME-driven shocks and SEP events provides us an unique approach of examining the underlying diffusive shock acceleration as well as the relationship between energetic particles and the stimulated magnetic turbulence.

In the simple picture that yields the equation (2), the upstream and downstream medium are assumed to be homogeneous. The diffusion coefficient κ is assumed to be independent of x, and particles can travel very far upstream and downstream and still return to the shock. These assumptions are of course not true for a propagating CME-driven shock. When a CME-driven shock propagates out from the Sun, the scattering centers (which are Alfvén waves stimulated by the energetic particles themselves) are only "populated" close to the shock. The wave power will decrease as distance from the shock increases. Thus, those particles that move to some distance upstream or downstream of the shock can escape from the shock. To represent the escape effect, we add a loss term f/τ to the transport equation, which now becomes,

$$\frac{\partial f}{\partial t} + u\frac{\partial f}{\partial x} - \frac{\partial}{\partial x}(\kappa\frac{\partial f}{\partial x}) - \frac{1}{3}\frac{du}{dx}p\frac{\partial f}{\partial p} + \frac{f}{\tau} = \frac{f_\infty}{\tau_\infty}. \quad (4)$$

This equation was considered by [13]. However, [13] considered physical reactions such as ionization, coulomb and nuclear collisions, which change the identity of a particle and thus remove it from the shock front. This is different from the escape effect we consider here, which is due to the finite extent of the turbulence region near the shock. Furthermore, in [13], the initial particle spectrum is assumed to be a δ function, which, as we will see, does not apply to the case of a propagating CME-driven shock.

The escape time scale τ introduced in equation (4) depends on the escape length scale d, which is energy dependent. In principle, particles can escape from both downstream and upstream, so

$$1/\tau = 1/\tau_{up} + 1/\tau_{dn}, \quad (5)$$

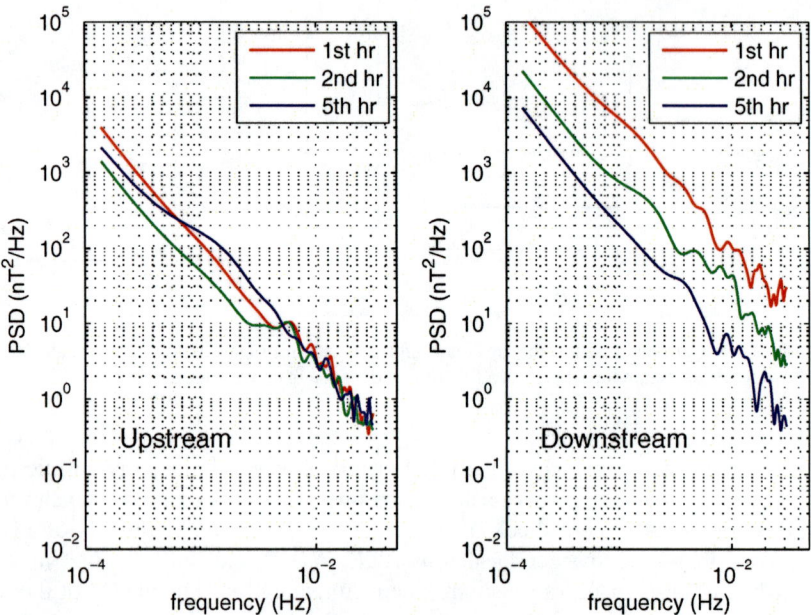

FIGURE 2. Upstream and down stream wave intensities for the Oct. 29, 2003 event. It is clear that the downstream wave power intensity is much stronger than that upstream. This observation agrees qualitatively with theoretical work by [14]. For a thorough description of the October 29, 2003 event, see *Mewaldt et al.* in this volume.

where τ_{up} and τ_{dn} represent the escape time scale for upstream and downstream respectively. However, the turbulence downstream are much stronger than that upstream of the shock (see [14] and [7]), usually by a factor ~ 10. Thus we can ignore $1/\tau_{dn}$ in equation (5) and consider only the escaping effect of particles from the upstream. Analyses of upstream and downstream wave intensities of large SEP events, as shown in figure 2, for the October 29 2003 event, verify this 10-fold increase.

The right panel of figure 1 is a cartoon for the wave intensity $I(k)$ as a function of distance upstream of the shock. The solid line represents the upstream turbulence, which decreases with distance and reaches the background value beyond some distance $d(k)$. One can define the distance $d(k)$, for example, through, $I(k,d(k)) = 1/10 I(k,0)$. Using $d(k)$, one can then define another quantity $\tilde{I}(k)$, such that the area of the rectangular $\tilde{I}(k) \times d(k)$ equals that underneath the curve $I(k)$. Clearly, $d(k)$ represents the effective length scale of the upstream turbulence and $\tilde{I}(k)$ represents the effective strength of the wave intensity at wave number k.

We can estimate the escape time scale τ and compare it with the acceleration time scale τ_{acc}. Consider a particle at the shock front with energy T and momentum p, which resonates with wave number k according to $p \times k \approx \Omega$. The mean free path of the particle $\lambda(p)$ is proportional to $1/\tilde{I}(k)$ in quasi-linear theory. The escape time scale τ will be the time for particles to reach $d(k)$ from the shock front. If $\lambda(p)/d(k) \gg 1$, particles will undergo a random walk before reaching $d(k)$. Thus $\tau \sim (d(k)/\lambda(p))^2 \lambda(p)/v$. On the other hand, if $\lambda(p)/d(k) \sim 1$, particles will experience "free-streaming" and we have $\tau \sim d(k)/v$. Compared with the "free-streaming" case, the "random walk" case has an extra factor of $d(k)/\lambda(p)$. Intuitively, we expect the ratio $\lambda(p)/d(k)$ for low energy particles to be smaller than that of high energy particles. Thus, $1/\tau$ is larger for high energy particles and the loss term becomes more important for high energy particles.

The relative size of the loss time scale τ and the acceleration time scale τ_{acc} is also important. For a particle with an initial momentum p_0, the time to reach a momentum q is [15],

$$\tau_{acc} = 3s/(s-1) \int_{p_0}^{q} \kappa(p)/u_1^2 \Delta p/p. \tag{6}$$

Thus the "loss term" can only manifest itself when $\tau \leq \tau_{acc}$. By equating τ_{acc} with τ, one can find the momentum $\tilde{p}$, above which the loss effect is important. At a propagating CME-driven shock, the momentum $\tilde{p}$ is time dependent. This is because as the shock propagates and weakens, the acceleration time scale t_{acc} increases with time [16]. Thus, particles can be accelerated to higher energies when the shock is closer to the Sun. Consequently, the particle spectrum will reach a higher momentum p_h at an early time t_1 rather than at a late time t_2. This spectrum near p_h at t_2 will then be subject to the "loss effect" should the loss time scale $\tau(p_h, t_2)$ became comparable and/or smaller than $\tau_{acc}(p_h, t_2)$. The following sections study the behavior of the spectrum near p_h.

STEADY STATE SOLUTION

Consider the steady state solution of equation (4). The solution should yield equation (3) if $\tau \to \infty$. Furthermore, since we ignore escape downstream, we expect that the downstream solution is still given by $f = A(p)$. In the upstream region, we expect that escape will lead to a smaller f. Based on these considerations, we assume the solution of equation (4) has the following form,

$$f = \begin{cases} A(p) & \text{downstream} \\ B(p)exp[(1+\delta)x\frac{u}{\kappa}] & \text{upstream} \end{cases}. \tag{7}$$

On applying the same matching condition, one finds

$$A(p) = B(p) \sim p^{-(1+\delta)[3s/(s-1)]}; \tag{8}$$

$$\delta = \frac{-1 + \sqrt{1 + 4\alpha\kappa/u_1^2}}{2}, \tag{9}$$

where $\alpha = 1/\tau$. Equations (8) and (9) are not surprising. Indeed, $4\alpha\kappa/u_1^2$ is the ratio of the acceleration time scale τ_{acc} to the loss time scale τ. If $\tau_{acc} \ll \tau$, $\delta \to 0$ and the classical result is retained. Physically, this corresponds to "particles not seeing the boundary because their mean free path is too small". Note that the spectrum becomes softer due to the presence of the "loss term". This is of course because higher energy particles will escape more easily.

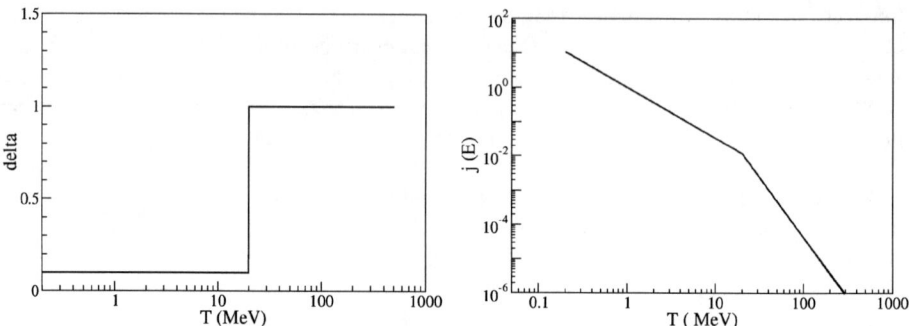

FIGURE 3. A step-like δ (left) will lead to a break in the particle spectrum (right) through a loss term "f/τ".

A special case: broken power law spectra

We now consider a special case where δ is a step function of energy,

$$\delta = \begin{cases} 0.1 & T < 20 \text{ MeV} \\ 1.0 & T > 20 \text{ MeV} \end{cases}. \quad (10)$$

Since the value of δ depends on the ratio of τ_{acc}/τ, equation (10) corresponds to the case where the "escaping effect" sets in suddenly, instead of gradually, above a characteristic energy. With equation (10), the solution (8) now becomes a "broken" power law. The current $j(E)$ at the shock and δ are plotted in figure 3. Below 20 MeV, $\delta = 0.1$ and the "loss" term in (7) can be ignored; above 20 MeV, $\delta = 1.0$ and the "loss" term becomes important and the spectrum is bent over and becomes softer. Note that we have assumed the initial spectrum is also a power law with a spectral index $3s/(s-1)$, which corresponds to $\delta = 0$. We do not expect the choice of initial condition will change our conclusion, especially when the initial spectral index is harder than $(1+\delta)[3s/(s-1)]$ ($\delta = 1$) above T 20 MeV, which is normally satisfied at a propagating CME-driven shock.

Spectra with a "broken" power law do exist. *Mewaldt et al.*, in this volume, reported several large SEP events occurred in late October, 2003, which show "broken" power law features. If one believes that the break is due to the loss effect, then one can deduce the value of δ from the difference of the spectral indices around the breaking energy. Furthermore, from the breaking energy, one can gain insight into the upstream wave intensities. Thus, studying the spectrum at the shock provides us another way of investigating upstream turbulence of shocks. Of course, the exact energy dependence of δ must depend on the characteristics of the upstream turbulence. A step-like δ is probably rare and a slowly changing δ may be more common. In the latter case, we expect a sharp break will be smeared out and an exponential roll-over at high energy will result.

DISCUSSION AND CONCLUSIONS

We have studied the transport equation in the context of a propagating CME-driven shock with a loss term. The loss term is due to the fact that the upstream turbulence, which is stimulated by the accelerated streaming energetic particles themselves, decreases with distance from the shock front. Consequently, particles that move to some distance ahead of the shock can escape. The loss time scale τ is energy dependent and is determined by the upstream turbulence. For a steady state case, the solution of the transport equation with a loss term is still a power law, with an index given by equations (8) and (9). If the acceleration time scale τ_{acc} (see equation 6) is smaller than the escape time scale, the classic solution is retained. Under certain conditions, the spectrum can be represented by a "broken" power law. The location of the break, and the difference of the spectral indices above and below the break can be used to infer the characteristics of the upstream turbulence and compare to observations, which has only recently begun (see [7]).

ACKNOWLEDGMENTS

This work is partially support by a NASA grant NAG5-10932 and an NSF grant ATM-0296113. One of the authors, GL, would like to acknowledge useful discussions with Yan Wang.

REFERENCES

1. W. I. Axford, E. Leer and G. Skadron, Proc. 15th Int. Cosmic Ray Conf. (Plovdiv), 11, 132, 1977.
2. A. R. Bell, MNRAS, 182, 147& 443, 1978.
3. G. F. Krymsky, *Sov. Phys. Dokl. 23, 327*, 1977.
4. Blandford, R.D., and J.P. Ostriker, *ApJ.*, 221, 269-280, 1978.
5. Lee, M. A., *J. Geophys. Res.*, 88, p. 6109, 1983.
6. Sanderson, T. R., Reinhard, R., van Nes, P., and Wenzel, K.-P. *J. Geophys. Res.*, 90, p19, 1985.
7. Bamert, K., Kallenbach, R., Ness, N. F., Smith, C. W., Terasawa, T., Hilchenbach, M., Wimmer-Schweingruber, R. F., Klecker, B., *ApJ.*, 601, p. L99, 2004.
8. Gordon, B. E.; Lee, A. M.; MÃűbius, E.; Trattner, K. J., *J. Geophys. Res.*, 104, 28,263, 1999.
9. Parker, E. N., Rev. Geophys. Space Phys., 13, 9 1965.
10. J.G. Kirk, D.B. Melrose, E.R. Priest, *Plasma astrophysics*, Edited by A.O. Benz and T.J.-L. Courvoisier, Springer-Verlag, 1994.
11. Kahler, S.W., *Ann. Rev. Astron. Astrophys.*, 30, 113-141, 1992.
12. Ratkiewicz, R.; Axford, W. I.; McKenzie, J. F. *Astron. & Astrophys.*, 291, p. 935-942, 1994.
13. Voelk, H. J.; Morfil, G. E.; Forman, M. A. *ApJ* 249, p. 161-175, 1981.
14. McKenzie, J. F., and K. O. Westphal, *Phys. Fluids, 11*, 2350, 1968.
15. Drury, L.O'C., *Rep. Prog. Phys., 46*, 973-1027, 1983.
16. Zank, G.P., W.K.M. Rice, and Wu, C.C., *J. Geophys. Res. (Space)*, 105, 25079-25095, 2000.

Generation of Turbulence at Shocks

Martin A. Lee

Space Science Center, Morse Hall, University of New Hampshire, Durham, NH 03824 USA

Abstract. A review of upstream waves at collisionless shocks is presented including their instability due to the cyclotron-resonant interaction with the energetic shock-accelerated protons, and their observed sites throughout the heliosphere. A new calculation of the power spectrum, $I(k, z)$, of these waves is described which reveals that $I(k \ll k_0(z)) \propto |k|^{\beta-6}$ and $I(k \gg k_0(z)) \propto k^{-2}$, where $\beta = 3X/(X-1)$, X is the shock compression ratio, and $k_0(z)$ is a decreasing function of increasing z. This form of $I(k, z)$ has important consequences for ion fractionation at shocks and the form of the high-energy cutoff.

INTRODUCTION

The title of this talk was suggested by the conference organizers. Although I decided not to change the title, turbulent processes do not appear to play a crucial role in the formation of the magnetic fluctuation intensity upstream and downstream of collisionless shocks in the heliosphere. Linear and quasilinear theory of the wave/particle interactions responsible for the fluctuations appear able to describe the observed features of all but the strongest shocks.

This talk starts with a brief review of upstream waves including their instability, their resonant interaction with energetic ions, and their observed sites throughout the heliosphere. Then I shall outline a new calculation of the upstream intensity spectrum, which I completed on the airplane flight to Palm Springs.

The enhanced magnetic fluctuations are a crucial feature of collisionless shocks. They are excited from the background waves by the streaming of the energetic protons accelerated at the shock by the process of diffusive shock acceleration. The unstable waves propagate in the upstream plasma frame away from the shock but are swept into the shock ramp or subshock where they attain their maximum upstream intensity. The waves are then transmitted to the downstream flow where they decay slowly by a partial reacceleration of the energetic ions by stochastic acceleration and by doing work on the expanding flow of the solar wind. The upstream waves and energetic ions provide the channel of dissipation at a collisionless shock which operates on the largest spatial scale.

We have seen many examples of upstream waves at this Conference. Upstream of Earth's bow shock we find the beautiful sinusoidal waves associated with the "intermediate" ion distributions. In one example shown by Hoppe et al. [1] we observe a spacecraft-frame period of ~24 s. We also find the more complex wave structures associated with the "diffuse" ions, which exhibit "shocklet" structure with leading whistler wave trains. The normally high-Mach-number bow shock is one case where nonlinear processes modify the upstream waves close to the shock.

Upstream waves are also prominent at the largest quasi-parallel interplanetary traveling shocks. The famous event of 11, 12 November 1978 described by Kennel et al. [2] already featured large-amplitude waves 1.5 hr upstream of the shock ramp. The intensity spectrum near the shock showed a broad rounded enhancement in the frequency range $10^{-2} - 10^{-1}$ Hz [3]. Waves are less prominent at quasi-perpendicular shocks. They are observed at other planetary bow shocks and the shocks bounding corotating interaction regions in the solar wind.

The instability mechanism is easy to visualize. We consider transverse hydromagnetic waves propagating parallel to **B** with the Alfvén velocity $V_A \hat{e}_z$, and an initially isotropic distribution of energetic protons in a frame with velocity $V_p \hat{e}_z$. The protons are now allowed to scatter elastically in the wave frame in which there is no electric field. If $V_p < V_A$ it is clear that, as the protons scatter to reduce their anisotropy in the wave frame, on average they gain energy in the plasma frame. Since the massive bulk plasma experiences negligible energy change in this frame, the waves damp to compensate the energetic ion energy gain. In contrast, if $V_p > V_A$ the scattering protons on average lose energy in the plasma frame and the waves grow. Since at a stationary shock the upstream ions are approximately isotropic in the shock frame and the inflowing plasma is superalfvénic, the latter case ($V_p > V_A$) is appropriate and the transverse hydromagnetic waves are unstable.

The wave-ion interaction is resonant. Parker [4] showed that a particle, which traverses a small-amplitude nearly monochromatic variation of **B** = **B**(z) such that **B**(z) = $B_0 \hat{e}_z$ as $z \to \pm\infty$, experiences a change in pitch-angle whose sign depends on the phase of the particle orbit relative to the phase of the variation. The change is only substantial if $k v'_\parallel \cong \Omega$, where k is the dominant wavenumber of the fluctuating field, Ω is the particle gyrofrequency, and $v'_\parallel$ is the particle velocity component parallel to $\mathbf{B}_0$ in the frame of the fluctuation. This is the cyclotron resonance condition satisfied by the upstream waves and ions for parallel wave propagation.

At Earth's bow shock the spacecraft-frame frequency ω_S is dominated by the Doppler shift so that $\omega_S \cong \mathbf{k} \cdot \mathbf{V}_{SW} \cong k V_{SW}$. Substituting the resonance condition and noting that the lowest energy protons, which dominate the wave spectrum, satisfy $v'_\parallel \cong 2 V_{SW}$, we obtain $\omega_S \cong \Omega_p/2$. With $\Omega_p(1\text{AU}) \cong 0.5$ s^{-1} we obtain the wave period $T_S \cong 24$ s, in excellent agreement with the characteristic period observed upstream of Earth's bow shock and quoted earlier in this talk!

A beautiful example of the resonant nature of the wave/proton interaction is shown by Russell et al. [5], which shows a plot of the dominant ω_S versus solar wind field strength for all the planetary foreshocks where waves had been observed before 1990. Since V_{SW} is approximately constant we expect $\omega_S \cong \Omega_p/2 \propto |\mathbf{B}_{SW}|$, a proportionality which is indeed observed over two orders of magnitude from Mercury's to Uranus's bow shock!

The upstream waves play a crucial role in the process of diffusive shock acceleration: (i) They dramatically enhance the rate of acceleration at quasi-parallel shocks. The solar wind wave spectrum would lead to a very sluggish process. (ii) They modify the effective shock compression ratio as sensed by the energetic ions. (iii) They control the escape of ions upstream of quasi-parallel shocks. (iv) In a related role they are responsible for the form of the high-energy cutoff as a function of energy. (v) They control the lengthscale of the decrease of ion intensity with distance from the shock, and the resulting fractionation between species with different A/Q. (vi) They vary with ψ, the angle between the shock normal and the upstream **B**, and contribute to the marked ψ-dependence of shock structure. For all these reasons it is essential to predict the magnitude and form of the wave intensity correctly.

FORM OF THE WAVE INTENSITY

The equation controlling the wave intensity, $I(k, z)$, is easily derived, where we assume wave propagation parallel to **B** and we take $\langle |\delta \mathbf{B}|^2 \rangle = \int_{-\infty}^{\infty} dk I(k)$. I consider CME-driven shocks only for which we need only consider upstream waves propagating away from the shock in the solar wind frame, which is the dominant propagation direction of the solar wind wave spectrum. According to quasilinear theory the pitch-angle diffusion coefficient for a particle species of mass $m_i = m_p A_i$ and charge $q_i = e_p Q_i$ is

$$D_{\mu\mu} = \pi Q_i^2 e_p^2 (2 A_i^2 m_p^2 c^2 |\mu| v)^{-1} I(\Omega_i v^{-1} \mu^{-1}), \tag{1}$$

where v is ion speed, μ is the cosine of ion pitch angle, c is the speed of light, and $\Omega_i = Q_i e_p B_0 / (A_i m_p c)$ is the ion cyclotron frequency. For strong scattering, for which the gyrotropic ion distribution function may be written $f_i(v, \mu) = f_{0i}(v) + g_i(v, \mu)$ with $|g| \ll f_0$ and $0.5 \int_{-1}^{1} d\mu f(v,\mu) = f_0(v)$, we have

$$\partial g_i / \partial \mu = -(v/2) D_{\mu\mu}^{-1} \, \partial f_{0i} / \partial s, \tag{2}$$

where s is arclength along **B**. The spatial diffusion coefficient parallel to **B** is then given by [6]

$$K_{\|i}(v) = (v^2/8) \int_{-1}^{1} d\mu (1 - \mu^2) D_{\mu\mu}^{-1} . \tag{3}$$

From the kinetic dispersion relation for the transverse hydromagnetic waves we obtain the wave growth rate for wavenumber k [7]

$$\gamma = 2\pi^3 |k|^{-1} (V_A/c^2)(e_p^2/m_p) \int_0^{\infty} dv v^2 \int_{-1}^{1} d\mu (1-\mu^2) \delta(\mu - \Omega_p k^{-1} v^{-1}) \partial f_p / \partial \mu, \tag{4}$$

where we assume that protons dominate the wave growth rate, and the δ-function enforces cyclotron resonance. The wave kinetic equation is given by

$$-(V - V_A \cos\psi) \partial I / \partial z = 2\gamma I, \tag{5}$$

where z measures distance upstream of the shock along the shock normal, and $-V$ is the z-component of the upstream plasma flow in the shock frame. Substituting equation (2) into (4), and (4) into (5), equation (5) may be integrated with respect to z to yield

$$I = I^0(k) + 4\pi^2 k^{-2} V_A |\Omega_p| m_p \cos\psi (V - V_A \cos\psi)^{-1} .$$

$$\cdot \int_{|\Omega_p/k|}^{\infty} dv v^3 (1 - \Omega_p^2 k^{-2} v^{-2}) f_{0p}(v,z). \tag{6}$$

Noting that the sign of k corresponds to polarization (right or left circularly polarized), the wave enhancement is unpolarized if $I^0(k) = I^0(-k)$.

According to the theory of diffusive shock acceleration

$$f_0(v,z) = A(v/v_0)^{-\beta} \exp\{-8Vv^{-2} \int_0^z dz' [\cos^2\psi \int_{-1}^1 d\mu(1-\mu^2)D_{\mu\mu}^{-1}]^{-1}\}, \tag{7}$$

where v_0 is the injection speed, $\beta = 3X/(X-1)$, X is the wave-frame compression ratio of the shock, and we neglect diffusion perpendicular to $\mathbf{B}_0$. Since $D_{\mu\mu}$ depends on I, $f_0(v, z)$ depends on I, and equation (6) is an integral equation for $I(k, z)$. Actually the nasty right-hand side involves three integrals: the v-integral sums over all protons which contribute to wave growth at wavenumber k; the z'-integral determines the proton distribution at z; and the μ-integration determines $K_{\|}$. At first sight it appears to be intractable. However, one special case is easy to derive. When $z = 0$, a straightforward integration yields

$$I(k,0) = I^0(k) + 8\pi^2 k^{-2}(V_A/V) |\Omega_p| m_p \cos\psi A v_{0p}^\beta (\beta-4)^{-1}(\beta-2)^{-1} |k/\Omega_p|^{\beta-4}$$
$$\equiv I^0(k) + C|k|^{\beta-6}. \tag{8}$$

The solution of equations (6) and (7) has been approximated in different ways: Zank et al. [8] took equation (8) with $I^0 \cong 0$ to represent $I(k, z)$ out to a value of z corresponding to 4 e-folding scalelengths. Lee [9] and Gordon et al. [7] assumed that the growth rate integral in equation (6) is controlled by $f_{0p}(|\Omega_p/k|,z)$, and that extreme resonance broadening dictates that $K_{\|i}(v) \propto [I(|\Omega_i/k|,z)]^{-1}$. These assumptions effectively eliminate the v- and μ-integrations and facilitate an analytical solution. Ng et al. [10] considers time-dependent wave excitation without treating the acceleration self-consistently. The difficulty with these ad hoc approximations is that $f_i(v, z)$ is very sensitive to the form of $I(k, z)$, which appears in an exponential function.

In spite of its complexity, equation (6) exhibits a simple dependence on k as $k \to \infty$ [7]. Since the argument of the exponential function in equation (7) becomes large and negative as $v \to 0$, we anticipate that the v-integration in equation (6) is independent of k as $k \to \infty$. Thus $I - I^0(k) \propto k^{-2}$ as $k \to \infty$. This symmetry arises from the rapid decay of the low-energy distribution with distance upstream of the shock so that the assumption of Lee [9] that the growth rate is controlled by $f_{0p}(|\Omega_p/k|,z)$ is not valid at large k. Thus, neglecting $I^0(k)$, we anticipate

$$I[k \ll k_0(z)] = C|k|^{\beta-6}$$
$$I[k \gg k_0(z)] \propto k^{-2}. \tag{9}$$

Because $\beta > 4$, equation (9) exhibits a hump at $k \cong k_0(z)$. Since this limiting behavior at small and large k is largely independent of $D_{\mu\mu}$ in equation (7), we anticipate that we may insert a trial intensity, $I_t[k, k_0(z)]$, based on equation (9) into equation (7) and calculate an improved value of the wave intensity using equation (6). As a preliminary consistency requirement both sides of equation (6) must undergo the transition between the two asymptotic forms at the same $k = k_0(z)$. This requirement yields

$$[k_0(z)]^{-(\beta-3)} = V\pi C |\Omega_p|^{-1} B_0^{-2} (\beta-4)(\beta-2)(\beta-3)^{-1} Jz, \qquad (10)$$

where J is a specified definite integral of order 1. Thus $k_0(z)$ decreases as z increases and the k^{-2} portion of the spectrum reduces I over a larger range of k with increasing z. The improved value of $I(k, z)$ reveals the next term in the asymptotic expansions, whose lead terms for large and small $|k|$ are given in equation (9).

It is clear from equations (1) and (7) that $I \propto k^{-2}$ yields a form of $f_{0i}(v, z)$ which is independent of A_i/Q_i. Thus, where the wave intensity has that form no subsequent ion fractionation is possible. The fractionation occurs in a sheath adjacent to the shock where $I \propto |k|^{\beta-6}$, which below the high-energy cutoff enhances ions with larger A/Q. Of course for weaker shocks, and/or at higher energies, and/or far from the shock, the background spectrum, $I^0(k) \propto |k|^{-5/3}$, controls the power spectrum, in which case larger A/Q ions are favored weakly. The derived form of $I(k \gg k_0) \propto k^{-2}$ differs markedly from that obtained by Lee [11], $I \propto |k|^{-3}$, upstream of the immediate vicinity of the shock. The form $I \propto |k|^{-3}$ results in large fractionation in favor of those ions with small A/Q in contrast with observed trends (e.g. [12]).

CONCLUDING REMARKS

As mentioned in the Introduction, the proton-excited upstream wave enhancement at quasi-parallel shocks plays a crucial role in ion acceleration and shock structure. Energetic ion composition in the foreshock depends sensitively on the form of $I(k, z)$, which also controls ion escape [11] and the form of the high-energy ion cutoff. Therefore, from a theoretical point of view it is essential to calculate $I(k, z)$ accurately. I have shown in the previous section that $I(k, z)$ exhibits a double power-law spectrum with a transition wavenumber $k_0(z) \propto z^{-(\beta-3)^{-1}}$. In future work we shall explore the consequences of this form of the wave intensity at planar stationary quasi-parallel shocks.

I have focused here on quasi-parallel shocks. Quasi-perpendicular shocks are much less well understood. They have higher injection-energy thresholds and require energetic or suprathermal seed particles for diffusive shock acceleration to operate [13, 14]. With the exception of "shock surfing" acceleration which requires a large electrostatic field in the shock ramp and incident particles occupying a particular small volume of phase space [15, 16], fluctuations are necessary for large energy gains at quasi-perpendicular shocks (as at all shocks). Ambient fluctuations in the solar wind plasma may suffice. However, fluctuations at quasi-perpendicular shocks may also be driven by proton gradients normal to $\mathbf{B}_0$ and be quite different from the waves described above for quasi-parallel shocks. Most current progress on acceleration at quasi-perpendicular shocks results from numerical simulations [e.g. 17, 18].

ACKNOWLEDGMENTS

The author wishes to thank Gang Li, Gary Zank, and their colleagues at IGPP-UCR for the expert organization of the Conference as well as their kind hospitality in Palm Springs. This work is supported, in part, by NSF Grant ATM-0091527.

REFERENCES

1. M.M. Hoppe, C.T. Russell, L.A. Frank, T.E. Eastman, and E.W. Greenstadt, *J. Geophys. Res.* **86**, 4471-4492 (1981).
2. C.F. Kennel, F.W. Coroniti, F.L. Scarf, W.A. Livesey, C.T. Russell, E.J. Smith, K.-P. Wenzel, and M. Scholer, *J. Geophys. Res.* **91**, 11917-11928 (1986).
3. B.T. Tsurutani, E.J. Smith, and D.E. Jones, *J. Geophys. Res.* **88**, 5645-5656 (1983).
4. E.N. Parker, *J. Geophys. Res.* **69**, 1755-1758 (1964).
5. C.T. Russell, R.P. Lepping, and C.W. Smith, *J. Geophys. Res.* **95**, 2273-2279 (1990).
6. J.R. Jokipii, *Rev. Geophys.* **9**, 27-87 (1971).
7. B.E. Gordon, M.A. Lee, E. Möbius, and K.J. Trattner, *J.Geophys.Res.* **104**, 28263-28277 (1999).
8. G.P. Zank, W.K.M. Rice, and C.C. Wu, *J. Geophys. Res.* **105**, 25079-25095 (2000).
9. M.A. Lee, *J. Geophys. Res.* **88**, 6109-6119 (1983).
10. C.K. Ng, D.V. Reames, and A.J. Tylka, *Astrophys. J.* **591**, 461-485 (2003).
11. M.A. Lee, *Astrophys. J. Suppl.* **158**, 38-67 (2005).
12. A.J. Tylka, D.V. Reames, and C.K. Ng, *Geophys Res. Lett.* **26**, 2141-2144 (1999).
13. Jokipii, J.R., *Astrophys. J.* **313**, 842 (1987).
14. Webb, G.M., G.P. Zank, C.M. Ko, D.J. Donohue, *Astrophys. J.* **453**, 178 (1995).
15. Zank, G.P, H.L. Pauls, I.H. Cairns, and G.M. Webb, *J. Geophys. Res.* **101**, 457-478 (1996).
16. Lee, M.A., V.D. Shapiro, and R.Z. Sagdeev, *J. Geophys. Res.* **101**, 4777-4789 (1996).
17. Giacalone, J., and D.C. Ellison, *J. Geophys. Res.* **105**, 12541-12556 (2000).
18. Giacalone, J., J.R. Jokipii, and J. Kota, *J. Geophys. Res.* **99**, 19351-19358 (1994).

Relationship of solar flare accelerated particles to solar energetic particles (SEPs) observed in the interplanetary medium

R. P. Lin

Physics Department & Space Sciences Laboratory, University of California, Berkeley, CA 94720-7450, USA

Abstract. Observations of hard X-ray (HXR)/gamma-ray continuum and gamma-ray lines produced by energetic electrons and ions, respectively, colliding with the solar atmosphere, have shown that large solar flares can accelerate ions up to many GeV and electrons up to hundreds of MeV. Solar energetic particles (SEPs) are observed by spacecraft near 1 AU and by ground-based instrumentation to extend up to similar energies, but these appear to be accelerated by shocks associated with fast Coronal Mass Ejections (CMEs). The Ramaty High Energy Solar Spectroscopic Imager (RHESSI) mission provides high-resolution spectroscopy and imaging of flare HXRs and gamma-rays. Here we review RHESSI observations for large solar flares and SEP events. The 23 July gamma-ray line flare was associated with a fast, wide CME but no SEPs were observed, while the 21 April 2002 flare had no detectable gamma-ray line emission but a fast CME and strong SEP event were observed. The October-November 2003 series of large flares and associated fast CMEs produced both gamma-ray line emission and strong SEP events. The spectra of flare-accelerated protons, inferred from the gamma-ray line emission observed by RHESSI, is found to be essentially identical to the spectra of the SEPs observed near 1 AU for the well-connected 2 November and 20 January events.

Keywords: RHESSI, SEPs, gamma-rays, flares, CMEs, electrons
PACS: 96.50Fm, 96.50Pw, 96.60Rd, 96.60Wh, 96.60Vg, 96.60Tf

INTRODUCTION

The Sun is the most energetic particle accelerator in the solar system. In large solar flares, nuclear gamma-ray lines and pion decay emission have been detected that are produced by energetic ions with $\sim$10–100 MeV and GeV energies, respectively, in nuclear collisions with the solar atmosphere [8]. HXR/gamma-ray continuum emissions, produced by bremsstrahlung collisions of energetic electrons with the atmosphere, have been observed up to $>\sim$100 MeV. Ions and electrons up to about the same energies have been directly detected by in situ space observations (and ground-based observations for the most energetic ions) in the interplanetary medium (IPM) near 1 AU in SEP events. The observations indicate that the energetic ions and electrons that produce the HXR and gamma-ray emissions in solar flares are accelerated by a different process than the SEPs observed near 1 AU—those appear to be accelerated by fast coronal mass ejections (CMEs) and the shocks they drive, or possibly by high coronal acceleration processes. These extremely energetic particle acceleration phenomena are associated with enormous transient releases of energy, $>\sim 10^{31-32}$ ergs, by the Sun. For large solar flares the energy in accelerated particles can be a significant fraction, $\sim$10–50%, of the total energy released [7]. For the fast CMEs, however, most of the energy released is

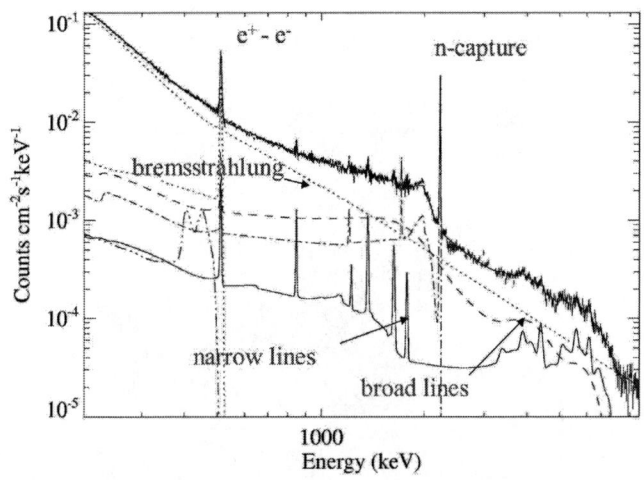

FIGURE 1. Gamma-ray count rate spectrum observed by RHESSI in the 28 October X17 solar flare [12].

TABLE 1. Gamma-ray line – SEP Comparison

Date of Event	Oct 28, 2003	Nov 2, 2003	Jan 20, 2005
	Energetic Proton Power Law Exponents		
Gamma-ray line ratio (Energy range)	Gamma-ray/SEP	Gamma-ray/SEP	Gamma-ray/SEP
Ne/C^+O (~2–20 MeV)	−2.0 to −3.2/1.33	−1.6 to −3.2/1.68	
e^+/C^+O (~10–50 MeV)	−2.2 to −3.3/1.95	−2.3 to −3.3/2.77	
n-capture/C^+O (~10–100 MeV)	−2.8 to −3.8/2.47	−2.8 to −3.8/2.97	
e^+/n-capture (~30–>100 MeV)			~2.2/2.15

contained in the $\sim 10^{15}$–10^{16} grams of coronal material ejected at high (>~1000 km/s) speed, although of order ~10% appears to go into SEPs. [3]

Here we review RHESSI observations of large gamma-ray line flares, and of flare/CME events associated with large SEP events. We compare the SEP proton measurements from spacecraft near 1 AU with RHESSI gamma-ray line observations that provide information on the energetic proton populations at the Sun for the October 28 and November 2, 2003 large flares.

ENERGETIC IONS AT THE SUN AND SEPS AT 1 AU

RHESSI has detected gamma-ray line emission from eleven solar flares from February 2002 through January 2005, with SEP events detected near 1 AU for at least seven of them: October 28, 29, and November 2, 2003; November 10, 2004; and January 15, 17, and 20, 2005. As an example, Figure 1 shows the 0.3–10 MeV gamma-ray spectrum from the 28 October 2003 X17 flare.

A major SEP event occurred following the first X-class flare (GOES X1.5) observed

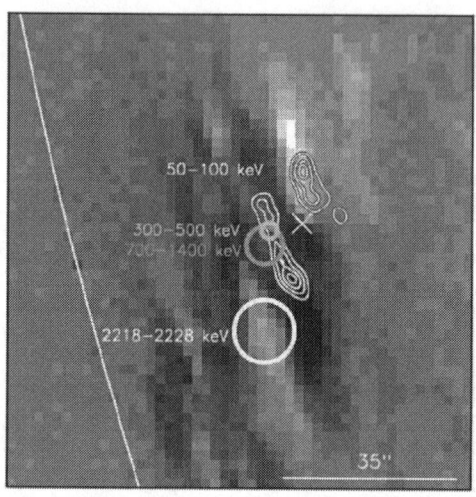

FIGURE 2. Locations of the gamma-ray sources. The thick circles represent the 1 σ errors for the 300–500 keV (*light gray*), 700–1400 keV (*dark gray*), and 2218–2228 keV (*white*) maps made with identical parameters. The 35″ FWHM angular resolution is shown in the lower right. The field of view is 96″×96″, centered 880″ east and 240″ south of Sun center with a white arc indicating the solar limb. The white contours show the high-resolution 50–100 keV map with 3″ resolution. The cross shows the centroid of the 50–100 keV emission made with the same lower resolution as the gamma-ray maps. The background image is a SOHO/MDI magnetogram acquired at 00:12 UT, 15 minutes prior to the flare. [5]

by RHESSI, on April 21, 2002, beginning at ∼0040UT. This was a long duration flare located at S14W84 that was observed by RHESSI to produce significant >10 keV emission for at least 12 hours [4]. A fast (∼2500 km/s) CME was detected whose initiation was closely related in time to the flare. The flare impulsive hard X-ray emission appears to trigger a wave seen in TRACE moving outward from the flare that starts the CME eruption. No gamma-ray line emission was detected above background, however, indicating that no significant acceleration of energetic (>∼10 MeV) ions occurred in the flare.

Solar gamma-ray lines were detected by RHESSI [8] for the first time from the GOES class X4.8 flare on 23 July 2002, located at S13 E72 close to the east limb of the Sun. RHESSI was able to make the first ever image of a solar gamma-ray line (Fig. 2) and obtain the first high resolution spectrum of gamma-ray lines. Energetic neutrons, produced by collisions of accelerated ions with the ambient solar atmosphere, are thermalized by elastic collisions with hydrogen atoms in the dense photosphere. They are then captured by hydrogen to form deuterium in an excited state, which decays by the emission of a gamma-ray at 2.223 MeV deuterium line. This is typically the strongest and narrowest line in solar flares. The 2.223 MeV line image showed for the first time where the flare-accelerated ions are located. The source was small (<∼30 arcsec) and located in the vicinity of the optical flare. Surprisingly, the gamma-ray line source was separated from the HXR sources (produced by energetic electrons) by ∼15,000 km. The HXR imaging and spectroscopy of this flare indicate that magnetic reconnection plays a major role in the energy release and electron acceleration process(es) for this flare (and very likely other flares).

This flare was accompanied by a very fast (∼2180 km/s) and wide CME traveling outward to the east. Since the flare was near the east limb, it was not surprising that no SEPs were detected by the near-Earth ACE and Wind spacecraft. The Electron Reflectometer instrument on the Mars Global Surveyor (MGS) spacecraft, located on the opposite side of the solar system very close to the nominal Parker spiral field from the flare region, however, did not detect any penetrating SEPs (>∼20 MeV), even though they were detected from two previous events (July 16 and 19) in the same active region. Since gradual SEP events are generally observed over a wide longitude range, especially the magnetically well-connected longitudes, this indicates that even a very fast and wide CME may not always accelerate SEPs.

The series of flares that occurred in late October – early November 2003 were the most intense observed since GOES soft X-ray measurements began about thirty years ago. RHESSI detected gamma-ray line emission from the 28 October (X17, S16E08), 29 October (X11, S15W02), and 2 November (X8, S15W56) flares, but missed the 4 November flare (S15W90) that saturated the GOES soft x-ray detectors (estimated to be X28). These flares were accompanied by fast CMEs, and intense SEP events were detected by spacecraft near 1 AU [4]. As the shock waves driven by these fast CMEs passed by 1 AU, increases in the SEP fluxes were observed up to energies of tens of MeV, but as in previous observations near 1 AU, no increases were detected at higher energies, however, even though these shock waves were traveling at speeds up to >∼2600 km. Thus, something is different close to the Sun to allow acceleration to energies of 10s to 100s of MeV and higher.

For the 28 October and 2 November SEP events, the spectrum of the energetic protons observed near 1 AU (using ACE, GOES 10, and SAMPEX spacecraft to provide full energy coverage), were integrated over the entire event to obtain the fluences [4]. The observed proton spectra are double power-law with exponents near -1 below ∼10 MeV breaking downward to much steeper power law, exponents of -3.5 to -4.5 at energies above ∼30–100 MeV (Table 1).

Since the cross-sections for the production of the different gamma-ray lines have different energy dependencies, ratios of the line intensities can provide information on the spectrum of the high energy particles that produce them. For example, the threshold for protons to excite the 6.13 MeV Oxygen line is ∼10 MeV, while it is ∼30 MeV for the 2.223 MeV neutron-capture line, and only ∼3 MeV for the 1.634 MeV Neon line. The power-law exponents derived from these gamma-ray line ratios for the 28 October and 2 November flares are shown in Table 1.

For 2 November, where the flare is magnetically well-connected (W56), the flare-accelerated protons have a spectrum that is the same as that of the SEPs at 1 AU, within the limits of the measurements; while the two spectra differ significantly for the 28 October event, where the flare is near central meridian (E08) and not well connected.

At least three of the recent January 2005 gamma-ray line flares also had associated SEP events observed by near-Earth spacecraft. The 20 January 2005 SEP event produced the largest sea-level neutron monitor increase since the February 1956 event [1]. Preliminary SEP measurements from spacecraft [10] show an extremely hard proton spectrum, with a power law exponent of -2.15 extending from ∼10 MeV to >500 MeV, for this well-connected event (flare located at N14 W61). The preliminary gamma-ray line measurements from RHESSI for this flare also indicate an extremely hard spectrum for the flare-accelerated energetic protons, with a spectral exponent of -2.3, essentially identical to that of the SEPs. Thus, for the two well-connected flare-SEP events, 2 November 2003 and 20 January 2005, the spectrum of the energetic protons in the flare is identical, within the measurement uncertainties, to the spectrum of the energetic protons measured near 1 AU.

DISCUSSION

From these comparisons, it is clear that SEP acceleration (by a fast CME) can occur without energetic ions being accelerated in the associated flare (April 21, 2002). On the other hand, there can be flare acceleration of energetic ions and a fast CME without SEP acceleration (July 23, 2002). Finally, in the two gamma-ray line flares observed by RHESSI that are magnetically well-connected (November 2, 2003 and January 20, 2005), the spectrum of the energetic protons producing the gamma-ray lines is essentially the same as the spectrum of the SEP protons observed at 1 AU. These two events had quite different spectral slopes, so this agreement is unlikely to be a chance coincidence. It suggests that these two populations of energetic protons may have the same source. Given our current paradigm that the gamma-ray producing protons are accelerated by a different process (flares) from the SEP protons (fast CMEs), this is very surprising. Previously, Cliver [2] suggested that SEP events might be "mixed" or hybrid, with flare-accelerated particles dominating in the region where the flare was well-connected, and CME-accelerated particles outside this region. More detailed analysis of the gamma-ray spectrum, lines and continuum, for these flares will provide information on the composition of the energetic heavy ions (since they produce broad lines), which can then be compared to the observed composition of the SEPs. Furthermore, comparison of the timing of the flare emissions and SEP arrival times, and imaging observations of both the flare and CME near the Sun, as well as imaging of the gamma-ray line emissions, are needed to illuminate these acceleration phenomena.

ACKNOWLEDGMENTS

I'm pleased to acknowledge useful discussions with A. Y. Shih and S. Krucker at Berkeley; R. Mewaldt at Caltech; and G. Share and R. Murphy at NRL. This research was supported in part by NASA contract NAS5-98033 and grant NAG FDNAG5-11804.

REFERENCES

1. Bieber et al., 2005, Spring 2005 AGU meeting abstract.
2. Cliver, E.W., Solar Flare Photons and Energetic Particles in Space, AIP Conf. Proc., 528(1), 21–31, 2000.
3. Emslie, A.G., H. Kucharek, B.R. Dennis, N. Gopalswamy, G.D. Holman, G.H. Share, A. Vourlidas, T.G. Forbes, P.T. Gallagher, G.M. Mason, T.R. Metcalf, R.A. Mewaldt, R.J. Murphy, R.A. Schwartz,

and T.H. Zurbuchen, Energy partition in two solar flare/CME events, J. Geophys. Res., 109, A10104, 2004.
4. Gallagher, P., B. Dennis, S. Krucker, R. Schwartz, and K. Tolbert, RHESSI and TRACE observations of the 21 APRIL 2002 X1.5 Flare, Solar Phys., 210, 341–356, 2002.
5. Hurford, G.J., R.A. Schwartz, S. Krucker, R.P. Lin, D.M. Smith, and N. Vilmer, First gamma-ray images of a solar flare, Astrophys. J., 595, L77–L80, 2003.
6. Li, G., G.P. Zank and W.K.M. Rice, Energetic particle acceleration and transport at coronal mass ejection-driven shocks, J. Geophys. Res., 108(A2), 1082: SSH 10-1 to 10-22, 2003.
7. Lin, R.P. and H.S. Hudson, Non-thermal processes in large solar flares, Solar Phys., 50, 153–178, 1976.
8. Lin, R.P., S. Krucker, G.J. Hurford, et al., RHESSI observations of particle acceleration and energy release in an intense solar gamma-ray line flare, Astrophys. J. Lett., 595, L69–76, 2003.
9. Mewaldt et al., submitted to J. Geophys. Res., 2005.
10. Mewaldt et al., private communication, 2005.
11. Ramaty, R., and R.J. Murphy, Nuclear processes and accelerated particles in solar flares, Space Sci. Rev., 45, 213, 1987.
12. Smith, D.M., A.Y. Shih, R.J. Murphy, et al., RHESSI Spectroscopy of Nuclear De-excitation Lines in X-class Flares, American Astronomical Society 204th Meeting, Denver, Colorado, 30 May – 3 June, paper 2.01, 2004.

Pickup Ions Upstream and Downstream of Shocks

G. Gloeckler[*,†], L. A. Fisk[†] and L. J. Lanzerotti[**]

[*]*Department of Physics and IPST, University of Maryland, College Park, MD 20742, USA*
[†]*Department of Atmospheric, Oceanic and Space Sciences, University of Michigan, Ann Arbor, MI 481092, USA*
[**]*Center for Solar-Terrestrial Research, Department of Physics, New Jersey Institute of Technology, Newark, NJ 071022, USA*

Abstract. The evolution of differential energy spectra and distribution functions of H^+, He^+ and He^{++} across shock boundaries is examined to gain insight into the acceleration of solar wind and pickup ions by various types of shocks. Data from the SWICS and HISCALE instruments on Ulysses are combined in order to span the wide energy range from ~0.6 keV/e to almost 5 MeV. We study two different types of shock crossings, the reverse shock of a Corotating Interaction Region, and the Jovian bow shock. Upstream ion velocity distributions are found to have hard suprathermal tails extending to high energies. Downstream ion distributions have higher particle densities in the tails, especially at energies corresponding to a few times the solar wind speed. At velocities greater than about ten times the solar wind speed the density increase from upstream to downstream is relatively small in the case of Jupiter's bow shock. The presence of pre-existing suprathermal tails provides particles with sufficiently high speeds to be readily injected into the shock acceleration mechanism.

Keywords: Particle acceleration; Shock waves; Solar wind interactions with planets.
PACS: 96.50.Pw; 96.50.Fm; 96.50.Ek

INTRODUCTION

It is well known that astrophysical shocks accelerate particles. Relatively little is known, however, about the details of the injection and acceleration processes. What is the input spectrum of ions delivered to the shock and how is this spectrum modified by the shock? How does this depend on the nature of the shock? We begin to investigate these questions by examining in some detail particle acceleration and injection by two shocks of distinctly different origin, the reverse shock of a Corotating Interaction Region (CIR) and the Jovian bow shock, using measurements of velocity distributions of protons, He^{++} and He^+ (pickup He) from solar wind energies to about 5 MeV.

Particle intensity increases downstream of shocks and between the forward-reverse shocks of Corotating Interaction Regions are common in the interplanetary medium as is illustrated in Fig. 1. In this figure we also demonstrates the remarkable fact that in regions far removed from shocks the intensities of suprathermal (~5 to ~100 keV) and even ~MeV ions are relatively large. While the origin of these ~5 to ~1000 keV ions that are not clearly associated with shocks remains unknown, these suprathermal particles are likely candidates for the material that is being accelerated by shocks.

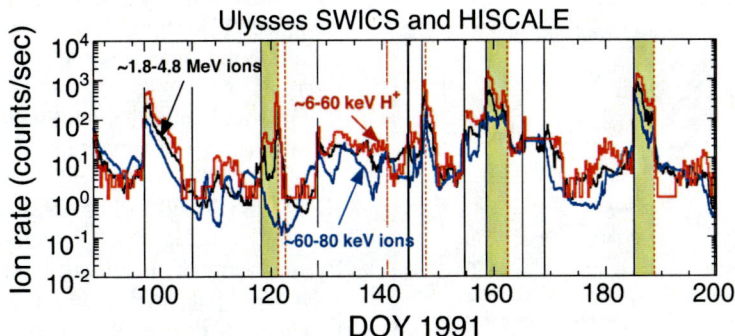

FIGURE 1. Spin-averaged counting rates versus time of 6-60 keV protons (red) from SWICS, and 60-80 keV (blue) and 1.8-4.8 MeV (black) ions from HISCALE. Forward and reverse shocks are marked by solid and dotted vertical lines respectively. Yellow shaded regions indicate Corotating Interaction Regions (CIRs) bounded by forward-reverse shock pairs. While the intensity in CIRs and downstream of shocks is usually far above average, there are numerous intensity increases, complicated spectral features and strong anisotropies (not shown) in regions far removed from observable shocks.

REVERSE SHOCK OF 11 OCTOBER 1992

The quasi-perpendicular reverse shock recorded [1] at 09:16:24 on day 285, 1992 was the stronger of the forward-reverse pair of the CIR with $B_{down}/B_{up} = 2.46\pm0.25$, $M_s=5.21$ and $\theta_{BN} = 68\pm11°$. Fig. 2 shows that at the reverse shock the differential intensity of pickup H^+ (1.3 to 2.1 times the solar wind speed) increases sharply by about 100. For 56 - 78 keV ions the overall increase is as large but begins ~15 hours before the shock. For >0.5 MeV ions the increase starts about three days upstream of the shock.

For about 1.5 days behind the shock the intensities remain nearly constant at the lowest energies and decrease slightly with distance from the shock at higher energies. The time profiles of the ion differential intensities measured in several energy intervals between 56 keV and 4.8 MeV show many increases accompanied by complicated spectral and anisotropy features, especially in the upstream region of the shock. In our present analysis we ignore these small-scale features when

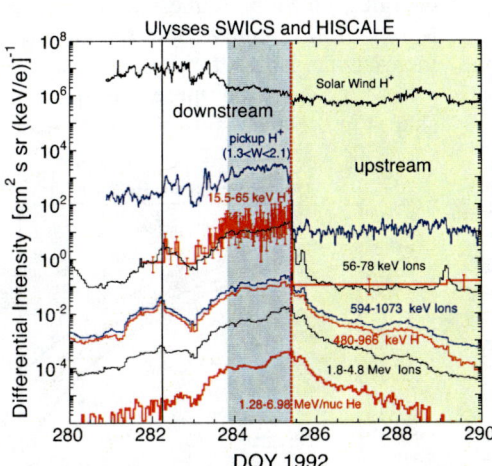

FIGURE 2. Differential intensity (from top to bottom) of solar wind H^+, pickup H^+ and 15.5-65 keV protons from SWICS, 56-78 keV, 594-1073 keV and 1.8-4.8 MeV ions from HISCALE, and 480-966 keV H and 1.28-6.98 MeV/nuc He from HISCALE during a 10 day time period around the day 282 – 285, 1992 CIR.

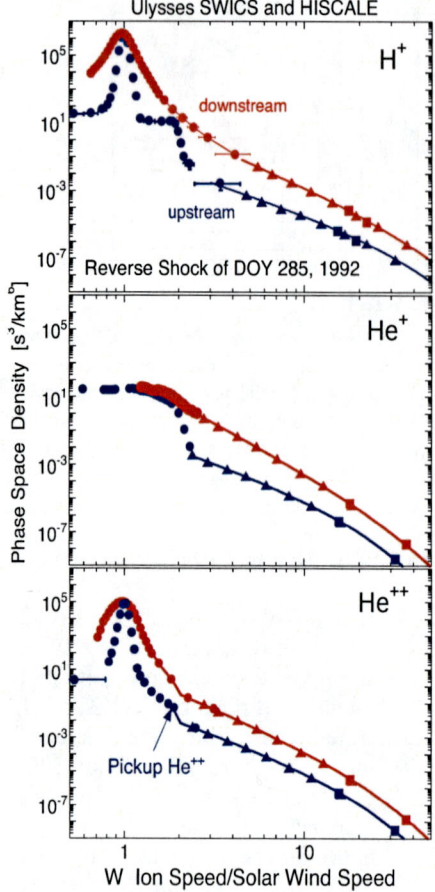

computing velocity or differential intensity spectra by using spin-averaged data accumulated over relatively long time intervals.

Time and spin-averaged velocity distributions in the upstream (DOY 285 10:00 to 290 10:00) and downstream (DOY 283 20:00 to 285 09:00) regions are shown in Fig. 3. The average spectra upstream of the shock have narrow solar wind peaks (H^+ and He^{++} spectra) and pickup ion features (primarily H^+ and He^+). All have strong suprathermal tails. In the downstream region both the solar wind and pickup ions are strongly heated and the pickup ion suprathermal tails harden and dominate.

The tails are well represented by $f(W) = f_o W^{-\gamma} \exp(-W/W_o)$ with $\gamma = 3.3, 3.5$ and 3.5 upstream and $\gamma = 4.4, 5.5$ and 4.3 downstream for H^+, He^+ and He^{++} respectively. Based on simple shock theory a value of 5.2 would be expected for γ. Values of W_o range from about 6 to 14. Large increases in downstream phase space density are observed at all energies, up to the highest energies (~5 MeV) of this study, and the density increases are largest for pickup ions.

In Fig. 4 we compare the spectral shapes in the upstream and downstream

FIGURE 3. Phase space density of H^+, He^+ and He^{++} (top to bottom) upstream (blue) and downstream (red) of the reverse shock of day 285, 1992. Filled circles are from SWICS and filled triangles and squares from HISCALE measurements. To separate the HISCALE ion data (assumed to be the sum of only H^+, He^+ and He^{++}) we matched the lowest energy HISCALE phase space densities to the highest energy SWICS values for H^+, He^+ and He^{++}, and the high energy HISCALE ion density to the HISCALE H and He phase space density values. For the upstream He^+ and He^{++} spectra we assumed the same spectral shape because the sharp drop in the SWICS He^+ spectrum made the match between of the SWICS and HISCALE He^+ spectrum ambiguous.

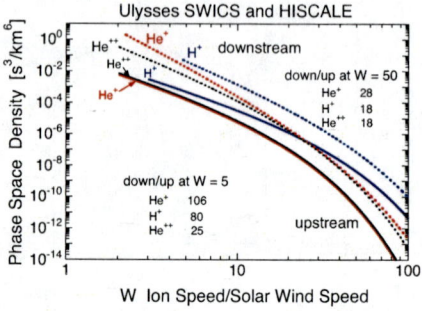

FIGURE 4. Fits to tail portions of the upstream (solid curves) and downstream (dotted curves) spectra of Fig. 3. The downstream/upstream density ratios at W of 5 and 50 for H^+, He^+ and He^{++} are shown in the figure.

regions of the shock using fits to the tail portions of the distributions of Fig. 3. The spectra roll over and are well represented by power laws below speeds W_o and exponentials above W_o. In the downstream region the density of He^+ is above that of He^{++} below $W \approx 10$ and comparable to that of He^{++} above $W \approx 10$. At $W \approx 5$ the downstream to upstream increase is largest (106) for pickup He^+ and smallest (25) for solar wind He^{++}. Protons comprising both solar wind and pickup ions increase by a factor of 80. At $W \approx 50$ the increases are smaller, ~18 for H^+ and He^{++} and somewhat larger (28) for pickup He^+. In the upstream region the separation between the proton spectrum and the spectra of He^+ and He^{++} increases with increasing speed. He^+ and He^{++} spectra roll over much more than does the proton spectrum.

JUPITER'S BOW SHOCK

The second shock we study is the Jovian bow shock observed during the inbound pass of Ulysses through Jupiter's magnetosphere. The bow shock was crossed [2] at 17:33 on day 33 of 1992. In Fig. 5 we show the time profiles of the differential intensities for day 33, 1992. The lowest energy (1.93-15.5 keV) protons increase by about 100 at the shock and then gradually decrease in the magnetosheath behind the shock. The increases near the shock are much smaller for 15.5-65 keV protons and 59 to 337 keV ions, start hours before the bow shock and peak behind the shock. Higher energy (~0.5 to ~5 MeV) ions show no intensity increases near the shock. In the magnetosheath the intensities of particles of all energies used in this study generally decrease but often have complicated structures.

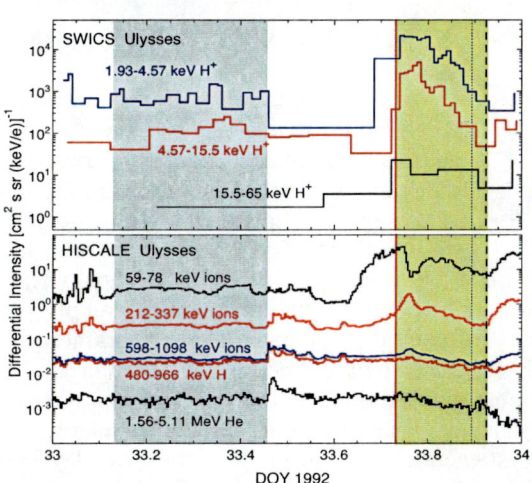

FIGURE 5. Differential intensities of H^+ from SWICS (top panel) and low energy ions as well as ~1 MeV/nuc H and He from HISCALE (bottom panel) for day 33, 1992. The averaging intervals for the SWICS data are variable because they were adjusted to include at least 5 counts. Solid and dashed vertical lines indicate bow shock and magnetopause crossings respectively. The shock was quasi-parallel with a compression ratio (B_{down}/B_{up}) of 3.8 and $\theta_{BN} = 36°$.

Velocity distributions for H^+, He^+ and He^{++} averaged over the time period indicated by the shaded regions in Fig. 5 (DOY 033 3:12-10:59 upstream and 033 17:35-22:04 downstream) are shown in Fig. 6. The velocity spectra in the upstream region again show narrow solar wind peaks for H^+ and He^{++}. The pickup H^+ and He^+ cutoffs are not as sharp as in Fig. 3. The tails are power laws to the highest energies used in this study, and the pickup He^+ tail is softer than the tails of H^+ and He^{++}. The phase space densities of H^+ and He^+ are comparable in the pickup ion range (~1.3 < W < 2.1) and, except for solar wind energies, the density of He^+ exceeds that of He^{++}.

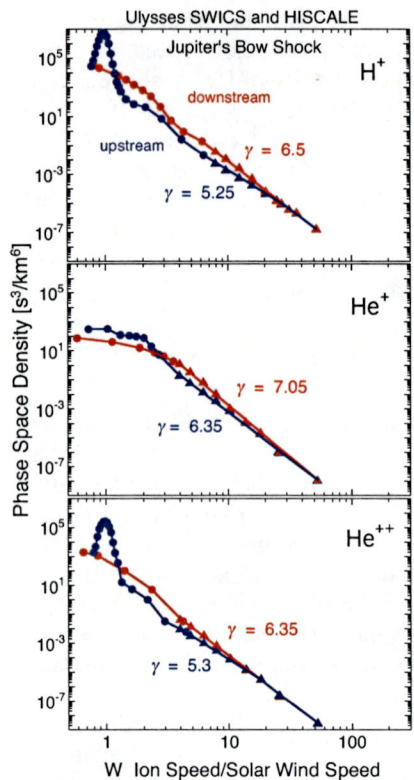

FIGURE 6. Same as Fig. 3 but for the Jovian bow shock of DOY 033. The tail portions of the He$^+$ and He^{++} spectra derived from the HISCALE ion data could be separately determined because there was sufficient overlap between the SWICS and HISCALE data.

In the magnetosheath, downstream of the bow shock, both solar wind and pickup ions are strongly heated. Because the convective speed in the sheath is low (~130 km/s) the bulk of the solar wind H$^+$ and He^{++} is now below the energy threshold of SWICS and thus not detectable. Pickup H$^+$ and He$^+$ tails dominate, and, as in the upstream region, the He$^+$ tail is softer than the H$^+$ and He^{++} tails. There is more He$^+$ than He^{++} above $W \approx 4$.

By comparing the upstream to downstream (magnetosheath) spectra it becomes clear that this quasi-parallel bow shock is a modest ion accelerator. In fact, acceleration stops at about ~500 keV for pickup ions (He^{++} and H$^+$) and at ~300 keV for solar wind He^{++}. The tails remain power laws in normalized speed W in the downstream re-gion to the highest energies of this study (~5 MeV), and the power law indices range from $\gamma \approx 6.4$ (He^{++}) to $\gamma \approx 7$ (He$^+$).

SUMMARY AND DISCUSSION

We have examined injection and acceleration of ions for two shocks: a quasi-perpendicular corotating reverse shock of a CIR with compression ratio of ~2.5 and a quasi-parallel, nearly stationary, strong (compression ratio of ~3.8) Jovian bow shock. Both shocks heat the solar wind and pickup ions. Ahead of both shocks is a plentiful supply of suprathermal ions in the tails of these upstream distributions. These suprathermal particles are presumably injected and accelerated by the shocks. The difference is in the efficiency of acceleration by, and possibly injection into these two shocks. While the day 285, 1992 corotating reverse shock is very effective in accelerating solar wind and especially pickup ions to the highest energies examined in this study, the day 33, 1992 Jovian bow shock is a weak accelerator, with acceleration confined to energies below ~300 keV for solar wind He^{++} and below ~ 500 keV for pickup ions.

The power law indices in the power law portions of the accelerated tail distributions differ from simple shock model predictions. In the case of the reverse shock the rollover shapes of the spectra are preserved from upstream to downstream of the shock. In the case of the bow shock, the spectral shapes (power laws in this case) are again preserved, but the accelerated tail portions of the spectra are softer than simple shock theories would predict.

Why is the Jovian bow shock so ineffective in accelerating ions? Perhaps it is because it is a quasi-parallel shock. Perhaps, downstream of this quasi-stationary shock the thermal speeds of solar wind and pickup ions are small, and thus only a small fraction of the heated ions are above the injection threshold.

The heliosphere's termination shock is expected to represent another type of an important astrophysical shock that is starting to be examined in-situ by the two Voyager spacecraft. It is interesting to ask how similar some local portions of the termination shock might be to Jupiter's bow shock. To address this question, we compare in Fig. 7 extrapolated solar wind and pickup proton spectra expected upstream of the termination shock at ~95 AU with measurements [3] of the proton spectra made by the LECP instrument on Voyager 1, presumably downstream of the shock (spectrum D). While the extrapolation of the solar wind and pickup proton [4] distributions to 95 AU is straightforward, that of the spectral tails is not. Depending upon whether one extrapolates the 'quiet' or 'baseline' tail spectra [5] to 95 AU, the acceleration could be significant (intensity increase of ~110) to modest (intensity increase of ~25). Since measurements of ions in the suprathermal range (few to ~40 keV) are not available from either of the Voyagers, extrapolated distributions such as shown here could be used to model the production of energetic neutral atoms (ENAs) and the acceleration of ions by the termination shock.

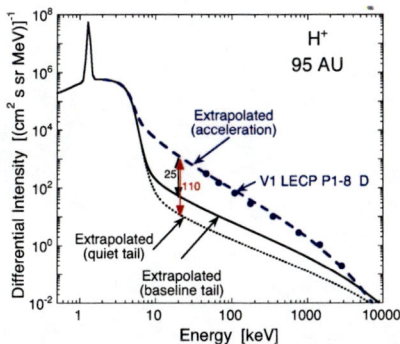

FIGURE 7. Differential intensities of protons at 95 AU. The solid curve is the extrapolated solar wind plus pickup proton and baseline tail intensity, the dotted curve extrapolates the quiet tail spectrum to 95 AU and the dashed curve is an empirical fit to the observed [3] LECP spectrum D.

ACKNOWLEDGMENTS

This work was in part supported by NASA/JPL contract 1268016, NASA/Caltech grant NAG5-6912 and NSF grant ATM0318590. The research at NJIT was supported in part by contract NAG5-11894 from NASA to The Johns Hopkins University Applied Physics Laboratory.

REFERENCES

1. A. Balogh, J. A. Gonzales-Esparza, R. J. Forsyth, M. E. Burton, B. E. Goldstein, E. J. Smith and S. J. Bame, *Space Science Reviews* **72**, 171-180 (1995).
2. A. Balogh, M. K. Dougherty, R. J. Forsyth, D. J. Southwood, E. J. Smith, B. T. Tsurutani, N. Murphy and M. E. Burton, *Science* **257**, 1515-1518 (1992).
3. R. B. Decker, S. M. Krimigis, E. C. Roelof and D. Lario, these conference proceedings (2005).
4. G. Gloeckler, J. Geiss and L. A. Fisk, "Heliospheric and Interstellar Phenomena Revealed from Observations of Pickup Ions", in *The Heliosphere near Solar Minimum: the Ulysses Perspectives*, edited by A. Balogh, E. J. Smith and R. G. Marsden, Springer-Praxis, Berlin, 2001, pp. 287-326.
5. G. Gloeckler, "Ubiquitous Suprathermal Tails on the Solar Wind and Pickup Ion Distributions" in *Solar Wind Ten: Proceedings of the Tenth International Solar Wind Conference*, edited by M. Velli, R. Bruno and F. Malara, AIP Conference Proceedings 679, American Institute of Physics, Melville, NY, 2003, pp. 583-588.

SESSION 5
SHOCKS IN THE OUTER HELIOSPHERE

Observations of Energetic Ions and Electrons in the Distant Heliosphere: 2001 - 2005.0

Frank B. McDonald[1], Edward C. Stone[2], Leonard F. Burlaga[3], Alan C. Cummings[2], Bryant C. Heikkila[3], Nand Lal[3], Norman F. Ness[4], John D. Richardson[5], and William R. Webber[6]

[1] *Institute for Physical Science and Technology, University of Maryland, College Park, MD, USA*
[2] *California Institute of Technology, Pasadena, CA, USA*
[3] *NASA / Goddard Space Flight Center, Greenbelt, MD, USA*
[4] *Bartol Research Institute, University of Delaware, Newark, DE, USA*
[5] *Center for Space Research, Massachusetts Institute of Technology, Cambridge, MA, USA*
[6] *Department of Physics and Astronomy, New Mexico State University, Las Cruces, NM, USA*

Abstract. As Voyager 1 (V1) moves closer to the heliospheric termination shock (TS), a new energetic particle population is observed: Termination Shock Particle events (TSP). Interplanetary disturbances in the form of merged interaction regions (MIRs) – identified using Voyager 2 (V2) data – have a major effect on the V1 TSP events from their onset to termination along with triggering episodic increases in higher energy ions (35 MeV H) and MeV electrons. The nature of these interactions appear to evolve as V1 moves closer to the TS.

Keywords: Heliosphere, Termination Shock, Merged Interaction Regions, Solar/Interplanetary Events, Termination Shock Particles
PACS: 96.40.Cd, 96.40.Fg, 96.40.Kk, 96.50.Bh, 96.50.Ci, 96.50.Fm

INTRODUCTION

In 2002.54 the energetic particle experiments on V1 (85AU, 34°N) observed increases in the intensity of MeV ions and electrons [1,2] that have persisted as highly variable but durable features over most of the ensuing 2.75 years as the spacecraft traveled some 10 AU further from the sun. These increases are characterized by frequent periods of streaming along the expected direction of the interplanetary magnetic field (IPB), relatively flat energy spectra, and a charge composition at low energies resembling that of anomalous cosmic rays as defined by the high O/C ratio [2,3,4]. The TS remains the most probable source of this new energetic particle population which has been termed Termination Shock Particles (TSP).

V2 (25°S) is some 19 AU closer to the sun than V1. Beginning with the Bastille day event (which arrived at V2 in 2000.9 at 62.9 AU [5,6]), there have been a series of 9 energetic particle increases that persist over some 3-4 solar rotations. These increases can generally be associated with specific periods of solar activity, are accompanied by increases in the solar wind velocity (V) and the IPB, and by moderate decreases in the galactic cosmic ray (GCR) intensity [5,6,7].

Despite a spacecraft separation of some 90 AU and 60° in heliolatitude there is a close correspondence between the time histories of the first 4 solar/interplanetary (S/IP) events at V2 and V1. After the onset of the TSP events at V1, it is observed that the S/IP transients, identified at V2, have a significant and diverse influence at V1, including large TSP decreases and impulsive increases in high energy protons and 2-15 MeV electrons. In this note we explore the properties of the V2 S/IP events and their effect on the V1 TSPs.

OBSERVATIONS

This study makes use of the data from the Voyager Cosmic Ray Subsystem (CRS) experiment (E.C. Stone, P.I.) the Magnetic Field (MAG) experiment (N. Ness, P.I.) and the Plasma Subsystem (PLS) experiment (J. Richardson, P.I.).

Of the 9 V2 S/IP increases (Fig. 1), 7 are associated with increases in V and IPB and with decreases in the >70 MeV rate. It is these four-fold indicators of S/IP intensity, V and IPB increases and >70 MeV GCR decreases that allows the identification of events 5 and 7. Event 6 inversely mirrors the >70 MeV rate but is different from the other 7 in that it is contained between 2 large MIRs as defined by the increases in V and IPB [8]. The V1/V2 event 2 is well defined but there are no associated IP changes. There are also several increases in V or IPB at V2 that are not associated with changes in the other data sets.

Because of the higher rigidity and velocity of the GCR's, the >70 MeV rate decreases (Fig. 1d) reflect the properties of the IP disturbances over a greater volume of space that extends beyond the spacecraft. The onset of the GCR decreases and the time of minimum intensity are taken as a measure of the width of the disturbances, MIRs, that are controlling the low-energy S/IP ions. These MIR boundaries are shown as shaded regions in the figures.

Events 1-4 are observed at V1 after a delay of ~ 0.19 years. For these events the initial time-history is very similar to that of V2 except the peak intensity is smaller, corresponding to an average radial intensity gradient of ~ -7% / AU. However events 1 and 3 at V1 have a unique feature not seen at V2 - a sharp peaked increase some 64 days after the event onset. For event 2 there are several peaked increases followed by two additional short increases between events 2 and 3.

TSP 1 and 2 start immediately after the predicted passage of MIRs associated with events 4 and 7. Event 6, with its flanking IPB and V peaks, is associated with the termination of TSP 1 [8]. The MIR of event 8 produces a rapid decrease in the 2.5 MeV ion intensity of TSP 2, briefly reducing it to the level of the V2 S/IP event (Fig. 1). Burlaga et al. [9] have shown that the formation of MIR 8 is dominated by the strong, corotating interaction regions observed in the inner solar system. Event 9 at V2 is the product of the Halloween series of large solar events that had occurred in the

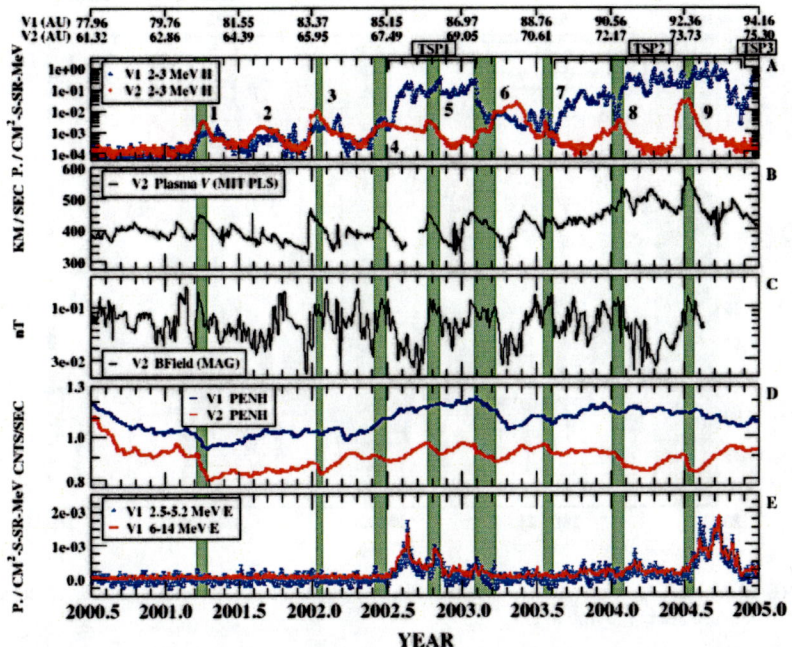

FIGURE 1. A) V1 and V2 2.5 MeV ion intensity; **B)** V2 solar wind speed, V; **C)** V2 IPB magnitude; **D)** V1 and V2 >70 MeV rate; **E)** V1 2.5 - 5.2 and 6-14 MeV electrons (Note linear intensity scale for electrons). All of the V2 data in this figure has been time shifted to the location of V1 using a V of 465 km/s for the period 2000.1 to 2003.6 and 560 km/s for the remainder of the period. The shaded regions are defined by the onset and minimum times of the V2 >70 MeV rate. The CRS and PLS data are 5 day moving averages. The MAG data are daily averages.

Sun's southern hemisphere some 6 months earlier, resulting in the largest cycle 23 S/IP ion increase yet observed at V2 [10,11].

The background corrected electron intensities from the HET (2.5-5.2 MeV) and TET (6-14 MeV) telescopes are shown in Fig. 1e. There are electron enhancements above the background level associated with the passage of MIRs 5 and 6 and following events 4, 8 and 9. The peak intensities of these two energy intervals are in reasonable agreement, indicating the spectral slope remains essentially constant.

One of the characteristics of the TSP events reported by both the CRS and LECP teams is the frequent observation of strong streaming of energetic ions along the direction of the IPB field [2,3,4] generally in the nominal direction away from the Sun. The look directions of the bi-directional HET telescope are approximately aligned along the nominal B direction, providing a means of detecting the streaming of 16-56 MeV/ions [10]. The HET J_A and J_B intensities (Fig. 2a) show episodic increases in 35 MeV H with the largest following the predicted passage of MIRs 8 and 9. Smaller increases were noted for events 5 and 6. There are periods when J_B is much larger than J_A indicating particle streaming along the field in the direction away from the Sun

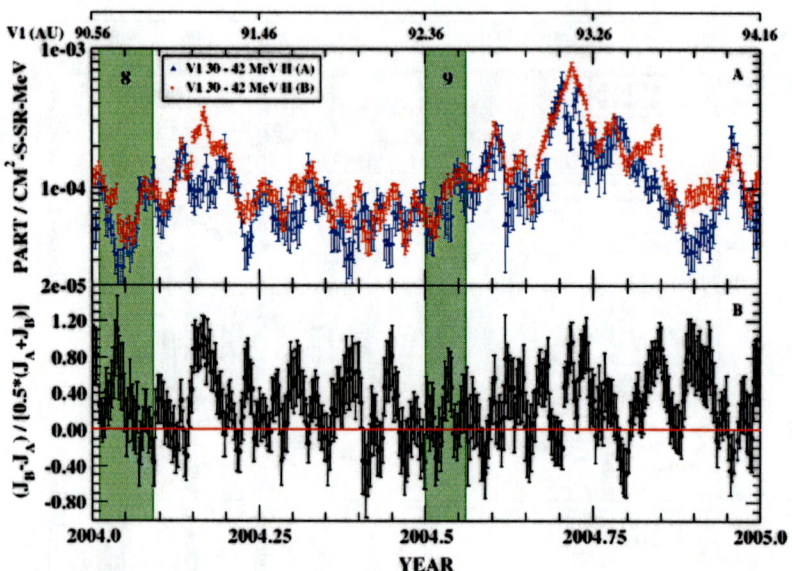

FIGURE 2. A) HET J_A, J_B 30-42 MeV Protons; **B)** The magnitude of the HET J_A and J_B anisotropy. All of the data are 5 day moving averages.

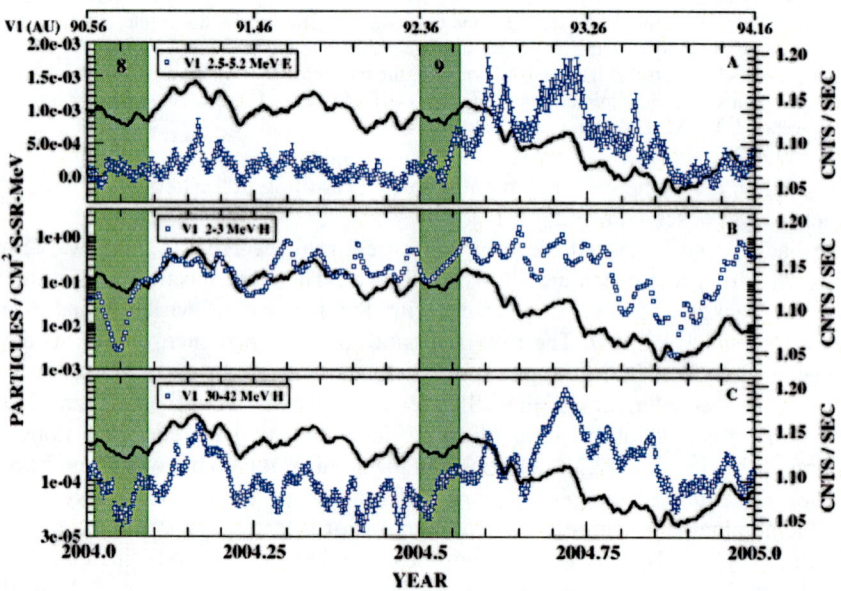

FIGURE 3. A) 2.5-5.2 MeV Electrons; **B)** 2.5 MeV ion intensity; **C)** HET J_B 30-42 MeV Protons; The >70 MeV rate is shown in each panel as a solid black line. All of the data are 5 day moving averages.

consistent with that reported for the lower energy ions. The magnitude of this anisotropy is shown in Fig. 2b.

There is a strong correspondence between intensity changes in 35 MeV H and MeV electrons as shown in Fig. 3 for the period 2004.0 - 2005.0. Also plotted are the time-histories of 2.5 MeV H and the >70 MeV rate is shown in each panel. Short-term fluctuations are assumed to be associated with the presence of local transients in the interplanetary medium. The sharp decrease in 2.5 MeV H produced by the passage of MIR 8 is followed by an increase in all 4 components (2004.09 - 2004.23) with the temporal structure of the TSPs being similar to that of the GCR.

After the predicted time-of-arrival of the Halloween event at V1 (2004.5) there are large increases and fluctuations in the intensity of the MeV electrons and the higher energy ions. Again, the fluctuations in all 3 TSP components reflect changes in the GCR rate but now there is a broad decrease in this rate in a series of discrete steps that modulate the TSPs with their minimum intensity coinciding with the GCR minimum, marking the end of TSP 2. This episodic event shows much larger variations, over a longer time period, than the previous ones, possible due to the closer proximity of the TS. The minimum, at 2004.8, is followed by the onset of TSP 3.

DISCUSSION

As V1 moves closer to the termination shock and with the ongoing recovery toward solar minimum conditions, the lower energy particle observations are dominated by a series of 3 TSP events extending from 2002.54 to 2005.0. This new particle population most probably originates at the TS. The series of S/IP events at V2 have their origin in episodes of solar activity. Thus, V2 is in the solar domain while V1 reflects the activity close to the TS.

At V2 the S/IP events are clearly controlled by MIRs. While the onset of the particle event begins on average some 3 AU before the MIR, the time of the initial peak of the MeV ion intensity coincides with the solar wind peak speed for 7 of the 9 S/IP events. The V1 TSP intensities are generally several orders of magnitude larger than those observed for the V2 S/IP events indicating they have a different source.

At V1 these MIRs play an important role in shaping the time histories of the TSP events. TSP 1 and 2 (Fig. 1) follow immediately after the extrapolated passage of MIRs 4 and 7. The termination of TSP 1 and 2 is related to the passage of MIR 6 and 9.

An unusual feature is the enhancement of MeV electrons and higher energy ions (Fig. 1, 2, 3) associated with the passage of MIRs 5 and 6 and following events 4, 8 and 9. The electrons are probably of galactic origin that have experienced local re-acceleration at the TS [12]. If this is the case then one possible source of the episodic electron and hydrogen ion increases could be further acceleration at the TS by the passage of the MIR.

There is often a difference between the decreases in the GCR rate at V2 and V1 both in form and magnitude. Richardson *et al.* [13] have examined the relation between the dynamic solar wind pressure at V2 and the V1 TSPs. These dynamic pressure profiles generally define the broad outline of the V2 events and when convected out to V1, were observed to have a significant role in modulating the V1

TSPs. Using the >70MeV rate provides a sharper definition of the MIR and the modulation effects of MIR 5, 6, 8, and 9 are in agreement with those of Richardson *et al.* [13].

The preferential streaming of TSPs along the IPB, in a direction away from the Sun, has been interpreted by Cummings and Stone [14] and Jokipii *et al.* [15] as requiring the shape of the heliosphere to be distorted in such a way that these field lines have encountered the TS prior to reaching V1.

REFERENCES

1. McDonald, F.B., Cummings, A.C., Stone, E.C., Heikkila, B.C., Lal, N and Webber W.R., Nature 426, 48, 2003
2. Krimigis, S.M., Decker, R.B., Hill, M.E., Armstrong, T.P., Gloeckler, G., Hamilton, D.C., Lanzerotti, L.J. and Roelof, E.C., Nature, 426, 45, 2003
3. Cummings, A.C. *et al.*, Proc. 28th, Int'l Cosmic Ray Conf. (Tsukuba) 7, 3777, 2003
4. Decker, R.B., Krimigis, S.M., Roelof, E.C. and Hill, M.E., Proc. 28th Int'l Cosmic Ray Conf. (Tsukuba) 7, 3773, 2003
5. Burlaga, L.F., Ness, N.F., Richardson, J.D. and Lepping, R.P., Solar Physics, 204, 399, 2001
6. Wang, C., Richardson, J.D., Burlaga, L.F., Solar Physics, 204, 411, 2001
7. McDonald, F.B., *et al.* Physics of the Outer Heliosphere, 3rd Annual IGPP Conf, (ed by V. Florinski) 139, 2004
8. Burlaga, L.F., *et al.* Journal of Geophys Res., 30, NO 20, 2072, DOI:10.1029/2003 GLO 18 291, 2003
9. Burlaga, L.F., Ness, N.F., Richardson, J.D., McDonald, F.B., and Stone, E.C. Astrophys Journal 618, 1074, 2005
10. Richardson, J.D., Wang, C., Kasper, J.C. and Liu, Y., Geophys Res Letter 32, L03503, DOI:10.1029/2004 GLO20679, 2005
11. Burlaga, L.F., Ness, N.F., Stone, E.C., McDonald, F.B. and Richardson, J.D., Geophys Res Letter 32, L03505, DOI 10.10291/2004 GLO21480, 2005
12. Potgieter, M.S. and Ferreira, S.E.S. Journal of Geophys Res., 107, A7, 10.1029/2001 JA 009040, 2002
13. Richardson, J.D., Journal of Geophys Res., 2005, JA01156R
14. Cummings, A.C. and Stone, E.C., this volume
15. Jokipii, J.R. *et al.*, Astrophys Journal 611, L141, 2004

Search for the Heliospheric Termination Shock (TS) and Heliosheath (HS)

Norman F. Ness[1], Leonard F. Burlaga[2], Mario H. Acuña[2], Edward C. Stone[3], and Frank B. McDonald[4]

[1]*Bartol Research Institute, University of Delaware, Newark, DE 19716, USA*
[2]*NASA-Goddard Space Flight Center, Greenbelt, MD 20771, USA*
[3]*California Institute of Technology, Pasadena, CA 91109, USA*
[4]*Institute for Physical Science and Technology, University of Maryland, College Park, MD 20742, USA*

Abstract. Voyager 1 continues to measure the very distant Heliospheric Magnetic Field (HMF) beyond 95 AU at ~35 North latitude. The MAG instrument data covers more than a full 22 years solar magnetic cycle. The magnitude of the observed HMF is well described, on average, by Parker's Archimedean spiral structure if due account is made for time variations of the source field strength and solar wind velocity. The V1 magnetic field observations do not provide any evidence for a field increase associated with entry into a subsonic solar wind region, such as the heliosheath is expected to be, nor an exit from this regime. We see no evidence for crossing of the Termination Shock (TS) as has been reported at ~85 AU by the LECP instrument [1]. Merged Interaction Regions are identified by an increased HMF and associated decreases in the flux of >70 MeV/nuc cosmic rays which are then followed by a flux recovery. This CR-B relationship has been identified in V1 data and studied since 1982 when V1 was at 11 AU. The variance of HMF, a direct measure of the energy**1/2 in the HMF fluctuations, shows no significant changes associated with the alleged TS crossings in 2002-2003. Thus, the absence of any HMF increase at the entry into the heliosheath appears not to be due to the onset of meso-scale turbulence as proposed by Fisk [2]. The TS has yet to be directly observed in-situ by the V1 MAG experiment in data through 2003.

Keywords: Interplanetary Magnetic Fields, Cosmic Rays, Solar Wind Plasma, Shock Waves, Turbulence
PACS: 52.35 Ra, 52.35 Te, 96.40, 96.50 Bh, 96.50 Ci

INTRODUCTION

This paper presents a status report on the continuing search for the Termination Shock (TS) theorized to be associated with the complex interaction of the Solar Wind (SW) with the Local Interstellar Medium (LISM). This study concentrates on observations of the very weak Heliospheric Magnetic Field (HMF), generally less than 0.2 nT by the dual magnetometer experiment [3,4] on the Voyager 1 (V1) spacecraft during the 2 year period 2002-2003. At that time, V1 was located at a heliographic latitude of ~35 degrees North and moved radially from 83 to 91 AU from the Sun.

An earlier report by Burlaga et al. [5] showed that the two reported crossings of the TS by the V1 Low Energy Charged Particle experiment [1] were inconsistent with the expected changes in average field magnitude at both the alleged entry 2002 into and 2003 exit from the subsonic and sub-Alfvenic SW Heliosheath (HS). Higher energy particles on V1 were studied by McDonald et al. [6] and their interpretation did not confirm any crossings of the TS.

INSTRUMENT ACCURACY AND IN-FLIGHT CALIBRATION

The most important basis of the V1-MAG interpretation is the accuracy of the data. We briefly describe and demonstrate the method which has been in standard use for several decades and discussed in a limited treatment in the Appendix A in Burlaga et al. [7]. The tri-Axial vector fluxgate magnetometer sensors are not absolute instruments in that their calibration does not depend upon atomic constants such as used in the proton precession or alkali-vapor self-oscillating scalar instruments [8].

Periodic sensitivity calibrations of all the dual triaxial sensors on both the V1 and V2 attitude controlled spacecraft occur by the traditional technique of temporarily imposing additional external fields on each sensor axis by use of carefully calibrated electrical currents.

For determination of possible zero level shifts of each sensor axis and contaminating contributions of spacecraft generated fields, both V1 and V2 are periodically placed in a roll-calibration maneuver about the S/C z axis, which is Earth pointed. There are 10-programmed complete rolls of the S/C during a period of approximately 6 hours. These are scheduled to be done every 3 months and during those periods when the S/C are being tracked and data recorded by the JPL-Deep Space Network (DSN) tracking system.

Figure 1 presents V1 calibrated data from the primary magnetometer (outboard) for a typical 2-day interval in 1991 including a roll calibration. These data also indicate the data gaps, which occur due to a lack of JPL-DSN continuous tracking. The effective zero level calibration has been determined and folded into the final data plot (see Ness [8] for a discussion of fluxgate sensor performance characteristics on a spinning S/C). The data plot includes field magnitude (B) and 2 heliographic angles, latitude (δ) and azimuth (λ), at the top. The three field components in heliographic coordinates of the magnetic field R, T and N are presented at the bottom. It is clear that the two components T and N reflect the roll modulation by the ambient external magnetic field. It is from such data that the effective zero levels of each sensor axis, which is the combined S/C field and zero level offset, is derived by least square analyses of the roll modulation of each.

As discussed by Burlaga et al. [7] the effective zero level of the z axis of the dual triaxial sensors, which is the R component in the heliographic coordinate system, is determined by assuming that the average of the R component observations for the period of 2 solar rotations, one on either side of the roll maneuver, are zero. (Note that at distances >80 AU, the R component direction is within 1° of the S/C z axis.) That is a reasonable assumption since in any theoretical model of the extrapolated heliographic field, the typical field magnitude of the R component at >50 AU is <0.001 nT and essentially zero, below the quantization step size of the digital data system of the MAG [9].

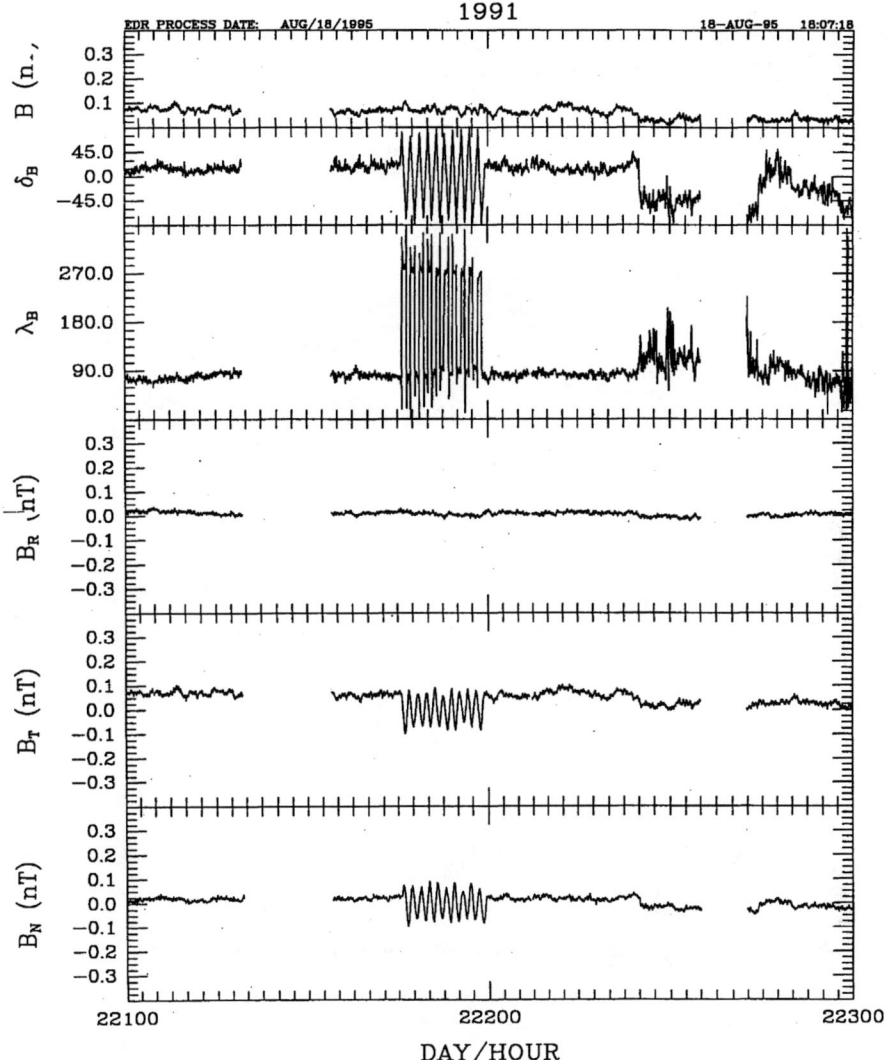

FIGURE 1. Primary triaxial fluxgate magnetometer data in 1991 during a roll calibration maneuver.

OBSERVATIONS

Figure 2 presents the daily average of the magnetic field (B) for the two years 2002-2003 along with the daily average of the 16-minute Pythagorean mean Standard Deviation (SD) or variance of the 1.92 second sampled magnetic field triaxial components. This SD quantity is a direct measure of the three-dimensional field fluctuations and is independent of coordinate system used; SD is proportional to the square root of the fluctuation energy. The lower panels in each year's plots presents the simultaneous 5 day running average of the observations of the particle flux of >70Mev/nucleon detected by the V1 Cosmic Ray System experiment.

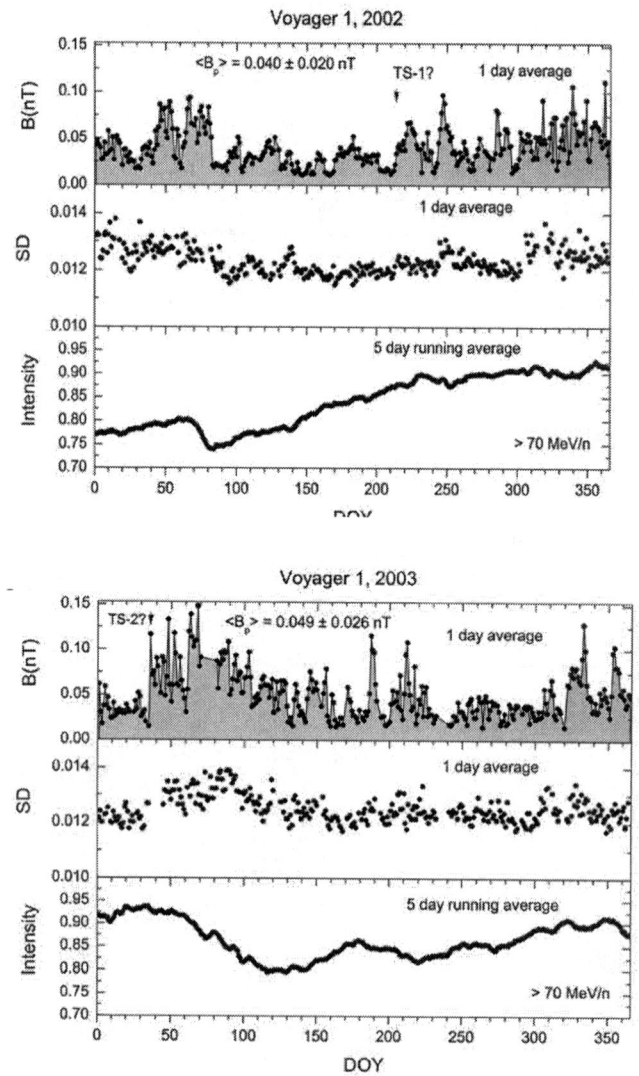

FIGURE 2. V1 magnetic field magnitude, and standard deviation, variance, in 2002 to 2003 along with cosmic ray flux of >70 Mev/nucleon particles.

A persistent long-term relationship between the HMF and cosmic ray flux has been observed since V1 was >11 AU [10, 11]. Whenever the magnetic field magnitude increases, the flux tends to decrease and when the field decreases to lower levels, the flux recovers to either a constant level or continues a recovery to a higher level. This pattern of behavior also is seen clearly in these 2 years of V1 data.

The times of the putative crossings of the TS by the LECP are shown as TS1 in 2002 and TS2 in 2003. As Burlaga et al. [5] have noted, there is no increase in field magnitude upon the claimed entry of V1 into the Heliosheath (HS) at TS1. At the claimed exit from the HS, following TS2, the field actually increases rather than decreases. Indeed, upon closer examination of the detailed data around the period following TS2, a shock might have occurred (the discontinuous data complicate identifying exactly when). That increase is followed by stronger fields along with a decreased cosmic ray flux, which is characteristic of the passage by V1 of a merged interaction region having passed by V1 [12].

ALTERNATIVE EXPLANATIONS FOR ABSENCE OF FIELD INCREASE AT TS1

Fisk [2] has recently suggested that the basic physical processes at the termination shock are altered from those typical of interplanetary propagating shocks at distances >20 AU and at planetary bow shocks due to the presence of LISM neutrals approaching the termination shock. He proposes that a turbulent HS exists in which the stronger magnetic fields are eliminated by a process of dissipation associated with macro-scale turbulence.

Fluctuating magnetic fields of all scales are observed in the HMF from seconds to minutes to hours. Frequency spectra of magnetic fluctuations in the plasmas in interplanetary space have been studied for many years. Correlated with variations in SW plasma speed and density, they indicate the presence of predominantly outwardly propagating Alfven waves. There is now no plasma data available from the PLS instrumentation on V1 to measure the SW so such a study is not possible.

Careful inspection of the HMF fluctuation data presented in Fig. 2 indicates that there is no noticeable change in the level of fluctuations associated with the alleged termination shock crossing at TS1.

The increases in field magnitude B and the fluctuation level that is evidenced in the SD data post TS2 can be understood as being associated with a propagating Merged Interaction Region (MIR) [9, 10, 11, 12]. These disturbances in the heliosphere are often preceded by a shock and the sudden spike in field magnitude seen at dcy 35 2003 is a prime candidate for that event.

This MIR interpretation is buttressed by the leveling and then decreasing flux of >70 Mev nucleons which begins near dcy 35 2003 and continues to dcy ~125 when the field magnitude decreases. The correlated field increases and flux decreases that occur thereafter are simply a manifestation of the cosmic ray/field relationship already observed and studied since V1 was at distances >11 AU.

SUMMARY

A careful study of the V1 HMF observations in 2002-2003 provides no persuasive evidence that the Termination Shock has yet been observed, as alleged by interpretation of Low Energy Charged Particle observations [1]. So the search will continue for the Heliosheath and its characteristics.

ACKNOWLEDGMENTS

We appreciate the contributions of the MAG team members at NASA-GSFC and the CRS team at University of Maryland, GSFC and Caltech in the conduct of these studies and the technical support staff at NASA-GSFC. Norman F. Ness is supported, in part, by NASA-GSFC grant NNG04GB71G to CUA.

Note added in proof: Conclusive evidence of the observation of the Termination Shock Crossing by the magnetometer experiment occurred on December 16, 2004 when V1 was at 94.0 AU. The compression ratio of the magnetic field from before to after the crossing ranges between 2 to 4 with an average best estimate of 3.0. The field increases from 0.045 to 0.136 ± .035 nT across the shock as V1 enters the heliosheath. (See Burlaga, et al [13] and associated V1 papers).

REFERENCES

1. Krimigis, S.M., Decker, R.B., Hill, M.E., Armstrong, T.P., Gloeckler, G., Hamilton, D.C., Lanzerotti, L.J., and Roelof, E.O., *Nature* **426**, 45-48, doi:10.1038/nature02068 (2003).
2. Fisk, L.A., "Mesoscale Variations in the Heliospheric Magnetic Field and their Consequences in the Outer Heliosphere" in *CP719, Physics of the Outer Heliosphere: Third International IGPP Conference*, edited by V. Florinski, N.V. Pogorelov and G.P. Zank, American Institute of Physics, 0-7354-0199-3/04, Melville, New York, 2004, pp. 365-372.
3. Ness, N. F., Behannon, K.W., Lepping, R.P., and Schatten, K.H., *J. Geophy. Res.* **76**(16), 3564-3573 (1971).
4. Behannon, K.W., Acuña, M.H., Burlaga, L.F., Lepping, R.P., Ness, N.F., and Neubauer, F.M., *Space Sci. Rev.* **21**(3), 235-257 (1977).
5. Burlaga, L.F., Ness, N.F., Stone, E.C., McDonald, F.B., Acuña, M.H., Lepping, R.P. and Connerney, J.E.P., *Geophys. Res. Lett.* **30**, doi:10.1029/2003GL018291 (2003).
6. McDonald, F.B., Stone, E.C., Cummings, A.C., Heikkila, B., Lal, N., and Webber, W.B., *Nature* **426**, 48-51 (2003).
7. Burlaga, L.F., Ness, N.F., Wang, Y.-M., and Sheeley, Jr., N.R., *J. Geophys. Res.* **107**(A11), 1410, doi:10.1029/2001JA009217 (2002).
8. Ness, N.F., *Space Science Reviews* **11**, 111-222 (1970).
9. Ness, N.F., and Burlaga, L.F., *J. Geophys. Res.* **106**(A8), 15803-15817 (2001).
10. Burlaga, L.F., Ness, N.F., *J. Geophys. Res.* **105**(A3), 5141-5148 (2000).
11. Burlaga, L.F., Ness, N.F., and McDonald, F.B., "Voyagers 1 and 2 Observe a GMIR and Associated Cosmic Ray Decreases at 61 and 82 AU" in *Proceedings of International Cosmic Ray Conference*, 2001, pp 3641-3643.
12. Burlaga, L.F., Ness, N.F., McDonald, F.B., Richardson, J.D., and Wang, C., *The Astrophysical Journal* **582**, 540-549 (2003).
13. Burlaga, L.F., Ness, N.F., Acuña, M.H., Lepping, R.P., Connerney, J.E.P., Stone, E.C., and McDonald, F.B., *Science* to appear (2005).

Characteristics of the Termination Shock: Insights from Voyager

A. C. Cummings and E. C. Stone

California Institute of Technology, Mail Code 220-47, Pasadena, CA 91125, USA

Abstract. We examine the energy spectra obtained from the cosmic ray instrument on the Voyager 1 spacecraft during 2002/215 through 2005/60. We find that the energy spectra of protons below ~20 MeV often resemble two power laws with a relatively hard index at low energies and a softer index at higher energies. The point of intersection of the two power laws is ~3 MeV. Beginning in 2005, the low-energy index is typically –1.5, corresponding to a shock strength (compression ratio) of 2.5. We attribute these characteristics to a restricted region of the solar wind termination shock that is sporadically connected to the Voyager 1 spacecraft by the interplanetary magnetic field. The absence of significant spectral variability in 2005 suggests that Voyager 1 entered a region with minimal spatial gradients of the lowest energy ions.

Keywords: Anomalous cosmic rays, solar wind termination shock, Voyager
PACS: 96.40.Cd, 96.40.De, 96.40.Kk, 96.50.Pw

INTRODUCTION

Since mid-2002 the charged particle instruments on the Voyager 1 (V1) spacecraft have been detecting very large intensities of low-energy particles. The first episode of these enhanced intensities lasted approximately six months[1,2]. The particles are believed to be coming from the termination shock, presumably from the sunward side of the shock since at the time of the first intensity enhancement in mid-2002 the magnitude of the magnetic field did not show the enhancement expected if the termination shock were crossed[3]. There is often significant beaming of the particles outward along the azimuthal magnetic field that occurs sporadically as the scattering mean free path varies along the field line connecting Voyager 1 to the particle source.

The solar wind termination shock is the likely source of anomalous cosmic rays (ACRs), which originate as interstellar neutral atoms[4] that become ionized in the heliosphere and carried to the shock where they are accelerated to cosmic ray energies[5]. The region of the shock connected to Voyager 1 is apparently not the source of ACRs, as the ACR energy spectra are strongly modulated during the periods of enhanced intensity at low energies and show little variability[2]. However, the low-energy part of the energy spectrum of the new low-energy component often appears to have the power-law shape expected from diffusive shock acceleration[6]. The power-law dependence usually extends up to ~3 MeV, where the spectrum rolls off into a steeper power law. In this paper, we examine these two properties of the energy spectra of H for data acquired from the Cosmic Ray Subsystem (CRS) instrument[7] on Voyager 1 from 2002/215 to 2005/60.

ANALYSIS TECHNIQUE

The Voyager CRS instruments on each of the Voyager 1 and 2 spacecraft have 9 telescope apertures for recording cosmic rays of various types in several energy bands. The relevant telescope bore sights are shown in Fig. 1 for day 300 of 2002. The directions change little over the period of analysis reported here. The designations A, B, C, and D refer to four essentially identical Low-Energy Telescopes (LETs), which provide the energy spectra of H from 0.5-7.8 MeV, He from 1.0-7.8 MeV/nuc, and O from 1.05-17.1 MeV/nuc. The bore sights of the High Energy Telescope (HET) apertures are also shown. The HET telescopes (H1 and H2) cover the energy ranges from 7.5-375 MeV for H, 7.5-540 MeV/nuc for He, and 11.9-125 MeV/nuc for O. Only the H2 telescope is used for H and He analysis and a combination of both H1 and H2 is used for O.

Streaming of 3.3-7.8 MeV protons along the azimuthal direction is evident in the lower panel of Fig. 2. The ratio of the average intensity in LET A and LET B to the intensity in LET C telescope is mostly >1 during the periods of enhanced intensity, indicating flow outward along the nominal azimuthal magnetic field direction. Note the intensity of ACR O is relatively steady throughout the period.

Also shown in Fig. 2 are the energy spectrum for protons observed at Voyager 1 for 2004 day 169 (92.2 AU and 34 N) and spectra of helium and oxygen for the period 2004/163-2004/175. The He spectrum shows three components: the new low-energy component, the ACR component with a peak energy at ~30 MeV/nuc, and the galactic

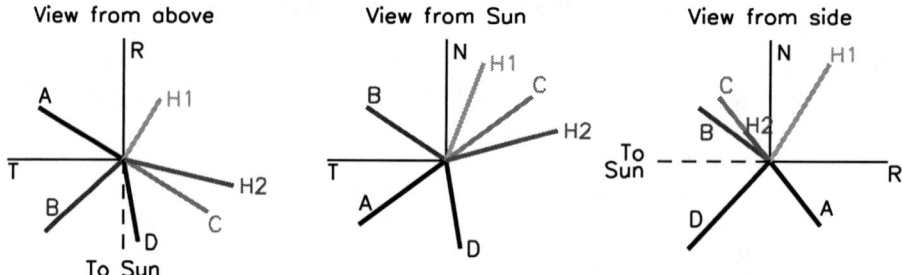

FIGURE 1. View directions of the telescope apertures of the Voyager CRS instrument for 2002 day 300 in the RTN coordinate system as described in the text. The RTN coordinate system is spacecraft centered with R pointed radially away from the Sun, T is counterclockwise in the azimuthal direction, and N completes the right-handed coordinate system.

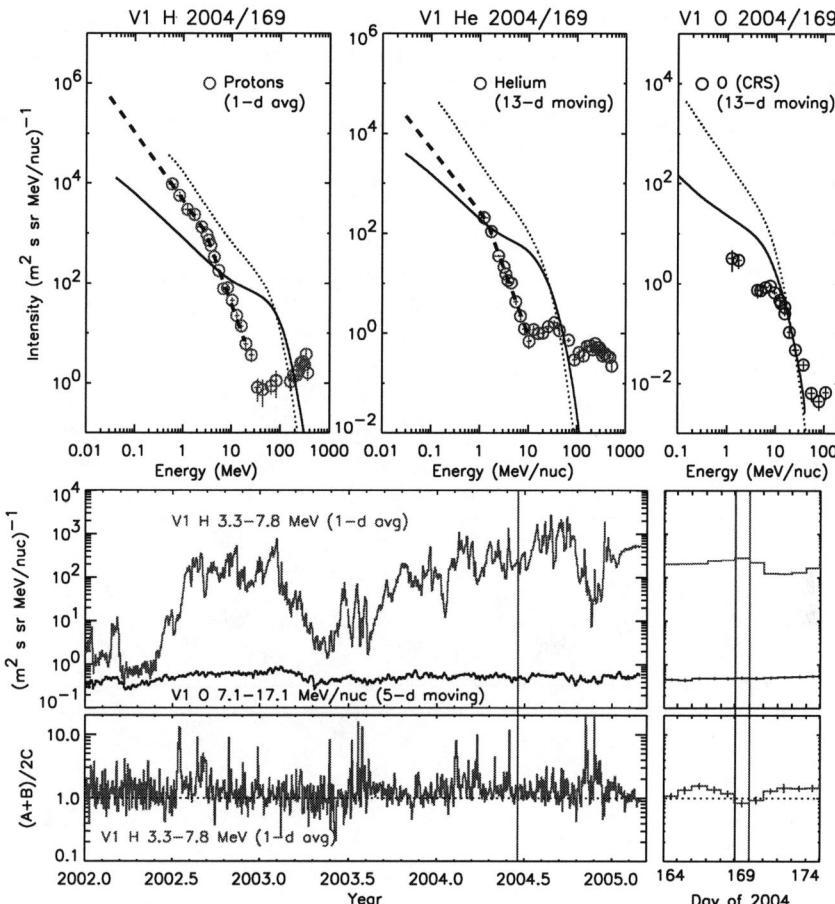

FIGURE 2. (top panels) Energy spectra of H, He, and O from V1 for the period indicated. The LET H and He points from 3.3-7.8 MeV and 3.0-7.8 MeV/nuc, respectively, are from the LET D telescope only. At lower energies for H and He and below 2 MeV/nuc for O, the data are from a single-parameter analysis of the first detector in the LET A telescope. The H points in this energy range have been corrected to the LET D direction by the ratio of the LET D to LET A intensity in the 3.3-7.8 MeV interval. The single parameter He and O points are also corrected to the LET D direction using the LET D to LET A intensities in the 3.0-7.8 MeV/nuc He interval. The dashed lines are fits to a two-power-law function described in the text. The solid and dotted lines are energy spectra expected at the termination shock as described in the text. (middle panel) Intensity of 3.3-7.8 MeV protons and 7.1-17.1 MeV/nuc ACR O vs time. (bottom panel) Azimuthal streaming index as described in the text. At the right in the bottom two panels, corresponding data for an 11-day period centered on 2004/169 are shown.

cosmic ray (GCR) component, which peaks at ~250 MeV/nuc. The dashed line in the H panel represents a fit of the low-energy data to a function consisting of two power laws, intersecting at a rolloff energy E_{o_H}. In the helium panel, the dashed line is similar except the power-law index at low energies is fixed by the fit to the H data.

Also shown for reference are spectra for ACRs expected at the termination shock for both a strong shock (r=4) and a weak shock (r=2.4)[8,9]. The differences in the spectra of the low-energy and ACR components indicate that they come from different source regions on the shock. The source of the low-energy particles is connected to V1 along the magnetic field while the source of the ACRs, which are higher energy, is from a region that is not magnetically connected to Voyager.

In Fig. 3 we show the four parameters from a least-squares fit of a two-power-law function to the H spectra from 0.5-22.3 MeV for those days from 2002/215 to 2005/60 where the estimated background in the single-parameter H data was less than 60%. The top panel shows the intensity at 0.5 MeV from the fit, together with the observed 3.3-7.8 MeV H intensity in LET D. The low-energy power-law index, a_H, shows relatively little change throughout the 2.5 year period, occupying a band principally between ~-1 to ~-1.7. The high-energy power-law index, b_H, shows more variability, ranging mainly from ~-2 to –4. The rolloff energy remains remarkably steady with a typical value of ~3 MeV. Beginning in early 2005, the energy spectra show much less daily variation than during 2002.5-2005.0, suggesting that Voyager 1 entered a region

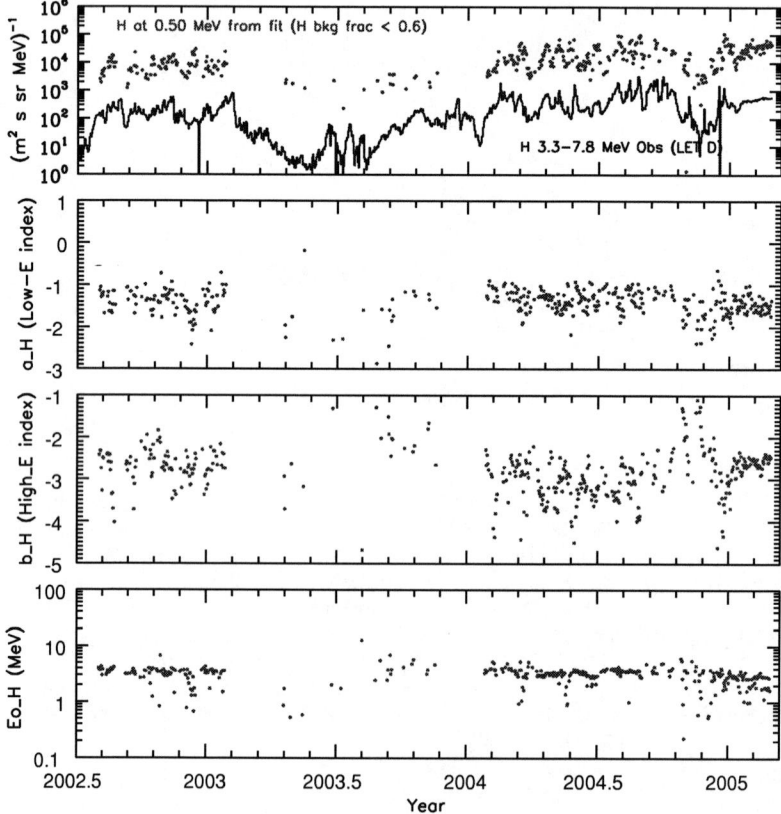

FIGURE 3. (top panel) Intensity of 3.3-7.8 MeV protons from V1 LET D and the intensity of 0.5 MeV from the fit to the two-power-law function vs time. (bottom panels) Low-energy power-law index (a_H), high-energy power-law index (b_H), and rolloff energy (Eo_H) from the model fits vs time.

with minimal spatial gradients of the low-energy ions. This new regime will be discussed in a future paper.

DISCUSSION

The energy spectra shown in Fig. 2 are but one example of many spectra during the period 2002/215-2005/60 which appear to be spectra somewhat similar to those expected from diffusive shock acceleration (see reviews by Drury[10] and Jones and Ellison[11]). The power-law dependence at low energies is directly related to the compression ratio at the shock (see, e.g., Blandford and Ostriker[6]): $a\_H = (r+2)/(2-2r)$, where a_H is the power-law index and r is the compression ratio (or shock strength). In 2005, the low-energy power-law index in Fig. 3 is typically -1.5, implying a shock with a compression ratio of 2.5. We attribute the spread in indices in prior years to modulation of the source spectrum by turbulence between the observation point and the source.

Numerical models of the spectrum of ACR H at the termination shock generally show an exponential rolloff in intensity at ~100 MeV (see, e.g., Cummings et al.[8]). Florinski and Jokipii[12] attribute the rolloff to either cooling in the supersonic solar wind, particle escape across the heliopause, or curvature of the shock seen by particles in the heliosheath. In a non-equilibrium situation, the acceleration time scale would also limit the maximum energy in the accelerated energy spectrum. The rolloff energy of the low-energy component is only ~3 MeV and shows no upward trend over the approximately 2.5 years of the observations. Hence, the part of the shock producing these particles appears to be stable over a 2.5-year period. We also do not observe an exponential drop off in intensity above this rolloff energy, but rather a transition to another power-law dependence. These aspects of the energy spectra need further theoretical study.

ACKNOWLEDGMENTS

This work was supported by NASA under contract NAS7-03001.

REFERENCES

[1] S. M. Krimigis, R. B. Decker, M. E. Hill et al., Nature **426** (6962), 45 (2003).
[2] Frank B. McDonald, Edward C. Stone, Alan C. Cummings et al., Nature **426** (6962), 48 (2003).
[3] L. F. Burlaga, N. F. Ness, E. C. Stone et al., Geophysical Research Letters **30** (20), 2072 (2003).
[4] L. A. Fisk, B. Kozlovsky, and R. Ramaty, Astrophys. J. Lett. **190**, L35 (1974).
[5] M. E. Pesses, J. R. Jokipii, and D. Eichler, Astrophys. J. Lett. **246**, L85 (1981).
[6] R. D. Blandford and J. P. Ostriker, Astrophys. J. Lett. **221**, L29 (1978).
[7] E. C. Stone, R. E. Vogt, F. B. McDonald et al., Space Sci. Rev. **21**, 355 (1977).
[8] A. C. Cummings, E. C. Stone, and C. D. Steenberg, Astrophys. J. **578**, 194 (2002).
[9] A. C. Cummings, E. C. Stone, and C. D. Steenberg, Astrophys. J. **581**, 1413 (2002).
[10] L. Drury, Rep. Prog. Phys. **46**, 973 (1983).
[11] F. C. Jones and D. C. Ellison, Space Sci. Rev. **58**, 259 (1991).
[12] V. Florinski and J. R. Jokipii, Astrophys. J. **591**, 454 (2003).

Voyager Observations of Interplanetary Shocks

John D. Richardson* and Chi Wang[†]

*Kavli Institute for Astrophysics and Space Research, Massachusetts Institute of Technology, Cambridge, MA 02139
[†]Center for Space Science and Applied Research, Chinese Academy of Sciences, P.O. Box 8701, Beijing 100080, CHINA

Abstract.
Data from Voyager 2 are used to compile a shock catalogue covering the 27 years of Voyager 2 solar wind data through the end of 2004. This catalogue is used to investigate the characteristics of shocks as a function of distance out to 75 AU. The shock occurrence frequency decreases with distance. Reverse shocks are most prevalent between 3 and 10 AU. The shock speed is solar cycle dependent with the highest speeds just after solar maxima. Shock strengths decrease with distance but do not show a strong solar cycle dependence. In the outer heliosphere shocks are often associated with merged interaction regions. As in the inner heliosphere, the relationship between the shocks and energetic particle fluxes is highly variable, with no apparent acceleration at some shocks and strong acceleration at others.

Keywords: Interplanetary shocks
PACS: 96.50.Fm, 96.50.Ci, 96.50.Pw

INTRODUCTION

The Voyager spacecraft were launched in 1977 and continue to explore the outer edge of the heliosphere. These spacecraft have sampled solar wind data over several solar cycles and at radial distances out to 95 AU. Interplanetary shocks are an important feature of the solar wind; since the solar wind is supersonic, the interaction of different speed solar wind streams results in shock formation. These shocks are generally driven either by interplanetary coronal mass ejections (ICMEs) or corotating interaction regions (CIRs) where fast coronal hole solar wind overtakes slow streamer belt solar wind (see reviews by Richter et al., 1985; Pizzo, 1985; Smith, 1985). ICME-driven shocks often form inside 1 AU whereas CIR shocks generally form at 2-3 AU. As the solar wind moves outward, the shocks interact and change in structure. This paper surveys the shocks observed by Voyager 2 out to 75 AU and describes their variation with distance and solar cycle. We also summarize the results of case studies of shocks which have been traced from the inner heliosphere to Voyager 2 and discuss the energization of particles by shocks in the outer heliosphere.

SHOCK SURVEY

Many types of shocks are present in the solar wind. Fast, slow and intermediate shocks can be both forward and reverse shocks. We searched the Voyager 2 data for discontinuities. Voyager data are only obtained when the spacecraft are tracked by the Deep

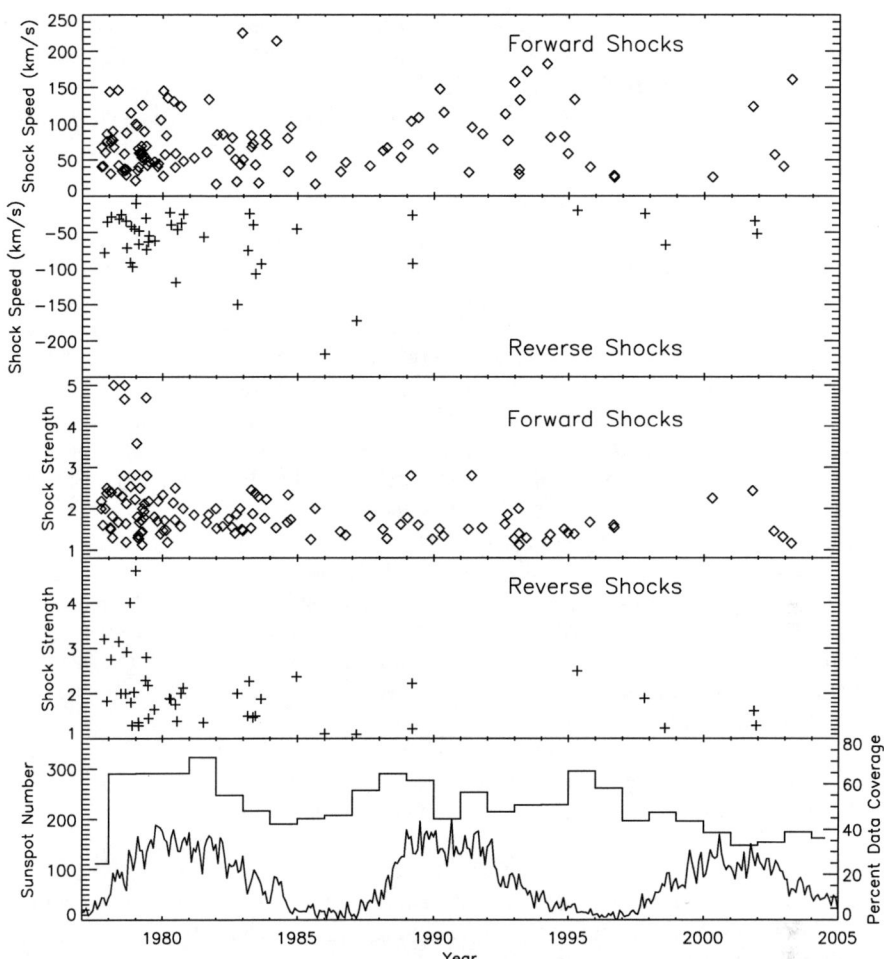

FIGURE 1. The top four panels show shock speeds and strengths observed by Voyager 2 for forward and reverse shocks. The bottom panel shows the sunspot number and the data coverage (histogram) all plotted versus time.

Space Network, 35-70% of the time. We do not include jumps in solar wind parameters which occur across data gaps since these could be gradual changes, not discontinuities. From this list of discontinuities, we select the fast forward and fast reverse shocks. Fast forward shocks are characterized by increases in the solar wind speed, density, temperature, and magnetic field magnitude. Fast reverse shocks have an increase in speed but a decrease in density, temperature, and magnetic field magnitude. Magnetic field data from Voyager 2 were used through 1989 when Voyager 2 was near 30 AU.

The shock speed is obtained from the requirement of flux conservation across the shock. We assume the shock normals are radial in the outer heliosphere. The shock strengths are the ratios of the upstream to the downstream densities. Figure 1 shows the shock speeds and strengths of the fast forward and fast reverse shocks observed by Voyager 2 as a function of time. For reference the sunspot number is also shown. The histogram at the bottom of the plot shows the percent of data coverage in each year.

The average data coverage decreases from about 70% in the inner heliosphere to about 40% after 1997. Thus the occurrence frequencies of shocks in the outer heliosphere should be roughly doubled to compare with those in earlier solar cycles. Even with this adjustment, the number of shocks clearly decreases with radial distance. From 1980-1985, 34 fast forward shocks were observed, from 1990-1995, 16 fast forward shocks, and from 2000-2005, 5 fast forward shocks. Reverse shocks are even more strongly clustered in the inner heliosphere, with very few observed outside 15 AU. The shocks form in the inner heliosphere; shocks driven by ICMEs often form well inside 1 AU. Most shocks driven by CIRs aren't observed until 2-3 AU, where forward-reverse shock pairs are formed. Both forward and reverse shocks interact with other shocks, which can weaken or strengthen the shocks and/or result in the shocks merging. Shocks also weaken with distance; both these effects reduce the shock numbers. Another possible reason for the lack of reverse shocks in outer heliosphere is that they propagate to higher latitudes and are rarely present at the Voyager heliolatitudes (Pizzo and Gosling, 1994). Forward shocks propagate equatorward and are thus more likely to be observed.

The shock speeds do not show a clear decrease with distance for either the forward or reverse shocks. The fast shocks in the outer heliosphere are often produced by the merger of many outward moving shocks (i.e., Wang et al., 2001); these mergers increase the shock speeds and strengths.

The shock occurrence frequency also varies with solar cycle, with few shocks near the solar minima. CMEs occur more frequently near and after solar maxima than during other parts of the solar cycle, driving more shocks. The tilted stream structure needed to drive CIRs is most prevalent in the declining phase of the solar cycle, giving more forward-reverse shock pairs at that time. The combination of these effects gives the result shown in Figure 1: more shocks near solar maximum and fewer near solar minimum. The shock speeds are higher near solar maxima, with the fastest forward shocks having similar speeds each solar cycle. The shock strengths do not have a strong solar cycle dependence.

SHOCK TRACKING

Several shocks have been successfully traced from the inner to the outer heliosphere. The shock preceding the Bastille day ICME (July 14, 2000) was ideal for this purpose since Earth and Voyager 2 were at nearly the same heliolongitude and thus probably observe the same ICME material. Wang et al. [2001] combined 1-D (inner heliosphere) and 2-D (outer heliosphere) models to propagate the solar wind from Earth to Voyager 2. The Wind observations of the plasma and magnetic field are used as model input. The model includes the effects of pickup ions, which slow and heat the solar wind. Since heating the wind increases the shock speed, the effect of the pickup ions on the shock arrival time is small. But the pickup ions do tend to increase the speed jump at a shock and decrease the

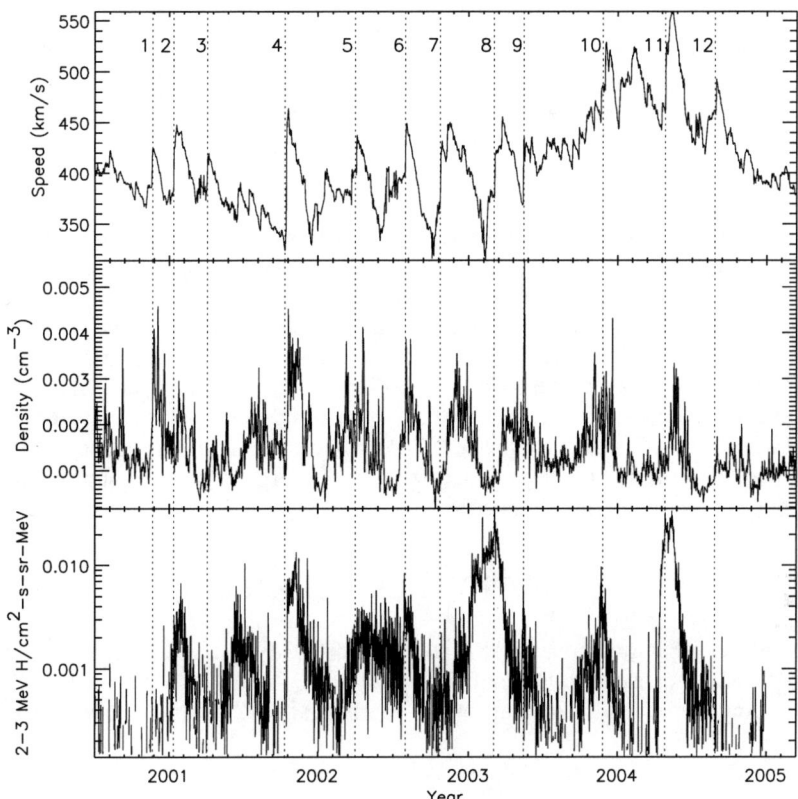

FIGURE 2. The solar wind speed and density and the CRS 2-3 MeV H intensity. The dotted vertical lines show the locations of large speed jumps, most of which are shocks.

density change [Rice et al., 1999]. The model predicted the arrival time of the Bastille shock at Voyager 2 to within a few days and did a good job with the initial speed, density, and magnetic field increases. However, the structure behind the shock is different from that predicted; in particular the density increase is very short-lived and anti-correlated with the magnetic field magnitude. The Bastille day shock was an example of a case where a large ICME quickly formed a single shock which moved outward and decayed, but still was significant at 63 AU. The shock observed at Voyager 2 in October 2001 was an example of a series of ICMEs observed at Earth from April-May 2001 which merged to form one large shock (Wang and Richardson, 2002).

The shocks which reach the outer heliosphere have a large effect on the solar wind structure. Most of the shocks which reach the outer heliosphere result from the merger of many ICME-driven shocks. These shocks lead the ICMEs, which form merged interaction regions (MIRs). These MIRs dominate the plasma structure in the outer heliosphere near solar maximum and are characterized by enhanced solar wind speed, density, dynamic pressure, and magnetic field magnitude. Richardson and Wang [2004] show that these correlated structures become stronger with distance based on results from their solar wind propagation model.

The shocks which lead these MIRs, and the MIRs themselves, can affect the energetic particle fluxes. These shocks can accelerate ions; the high magnetic field MIRs can pile up energetic particles ahead of them and exclude them from behind the MIRs. Figure 2 shows the solar wind speed, density, and the 2-3 MeV H intensity measured by the cosmic ray subsystem (CRS) on Voyager 2. The MIRs are very clear in the density data and correspond to speed increases. The dashed lines show the relationship between some of the shocks and shock-like structures and the energetic particle fluxes. In some cases the energetic particle flux shows a sharp rise simultaneous with the shock arrival (shocks 2 and 4). In some shocks (1 and 12) there is no energetic particle response to the shock passage. In some cases the energetic particles increase before the shock arrival but peak at the shock (shocks 8, 10, and 11). The reasons for these different behaviours are not clear. In this way the shocks in the outer heliosphere are similar to those in the inner heliosphere; they are often related to acceleration of energetic particles but for any given shock the amount of acceleration can vary widely.

ACKNOWLEDGMENTS

This work was supported under NASA contract 959203 from the Jet Propulsion Laboratory to the Massachusetts Institute of Technology. We thank F. McDonald and E. Stone for providing the Voyager CRS data and L. Burlaga and N. Ness for the Voyager MAG data.

REFERENCES

1. Richter, A. K., K. C. Hsieh, A. H. Luttrell, E. Marsch, and R. Schwenn, in Collisionless Shocks in the Heliosphere: Reviews of Current Research, Geophys. Monogr. Ser., vol. 35, edited by B. T. Tsurutani and R. G. Stone, pp. 3350, AGU, Washington, D. C., 1985.
2. Pizzo, V. J., in Collisionless Shocks in the Heliosphere: Reviews of Current Research, Geophys. Monogr. Ser., vol. 35, edited by B. T. Tsurutani and R. G. Stone, pp. 5168, AGU, Washington, D. C., 1985.
3. Smith, E. J., in Collisionless Shocks in the Heliosphere: Reviews of Current Research, Geophys. Monogr. Ser., vol. 35, edited by B. T. Tsurutani and R. G. Stone, pp. 69-84, AGU, Washington, D. C., 1985.
4. Pizzo, V. J., and J. T. Gosling, 3-D simulation of high-latitude interaction regions: Comparison with Ulysses results, Geophys. Res. Lett., 21, 2063, 1994.
5. Wang, C., J. D. Richardson, and K. I. Paularena, J. Geophys. Res, 106, 13,007-13,013, 2001.
6. Rice, W. K. M., and G. P. Zank, J. Geophys. Res., 104(A6), 12,563-12,576., 1999.
7. Wang, C., and J. D. Richardson, *Geophys. Res. Lett.*, **29**, noi:10.1029/2001GL014472 (2002).
8. Richardson, J. D., C. Wang, and L. F. Burlaga, *Geophys. Res. Lett.* **30**, 2207, noi:10.1029/2003GL018253 (2003).

Charged-Particle Acceleration at the Heliospheric Termination Shock

J. R. Jokipii

Department of Planetary Sciences, University of Arizona

Abstract. The paradigm of the diffusive acceleration of pickup ions by the heliospheric termination shock to form the anomalous cosmic rays (ACR) has successfully accounted for many observed features of the ACR, including electrical charge, composition, energy spectrum, time variations and spatial gradients. Recent observations from the Voyager 1 spacecraft as it approaches the heliospheric termination shock, and other data, have forced a more detailed confrontation between this theoretical paradigm and observations. This has provided new insights and constraints concerning the structure of the heliospheric termination shock. In particular, the observed three-dimensional vector anisotropy provides valuable constraints. Here it is shown that all three components of the anisotropy may be understood if the global shape of the termination shock is as expected from simulation and analogy.

INTRODUCTION

Anomalous cosmic rays (hereinafter ACR) are currently thought to be freshly-ionized interstellar neutrals [1] accelerated at the termination shock probably by the mechanism of diffusive shock acceleration [2] [3]. ACR have long been regarded as a remote probe of the termination shock. The nature of the expected variation of the ACR spectrum with position for a steady termination shock, obtained from model simulations, has been discussed by several authors [3] [4].

The location of the termination shock is expected to vary with time over a variety of time scales [5]. It is likely that the first encounter of a spacecraft with the termination shock will be at such an incursion, because the incursion can move much faster than the spacecraft. Recent papers (e.g., [6] [7]) report many-month-long increase in energetic particle fluxes observed on Voyager 1. They agree that these increases seem to be associated with the approach of Voyager 1 to the termination shock, but interpret the event differently. Krimigis, et al. suggest that the event was due to the termination shock moving inward and crossing Voyager 1 and then moving out again, ending the event. McDonald et al. argue instead that the shock was *not* crossed, but that the observed phenomena are precursors, upstream of the shock. The Voyager 1 magnetic field data of Burlaga et al. [8] argue persuasively that the shock was not crossed, and this is also the conclusion of Gurnett, et al. [9], based on radio data.

Clearly, the ACR spectrum, being accelerated at the shock, will fluctuate in time at any given location as the termination shock moves in and out, even if it does not cross the spacecraft. ACR observations can serve as a useful diagnostic of the shock proximity or crossing. A significant aspect of the energetic-particle interpretations in the papers by

McDonald et al. (ibid) and Krimigis et al. (ibid) are based on the observed streaming anisotropies.

In this paper, we address the question of the expected three-dimenstional vector streaming anisotropies of particles accelerated at the termination shock in the diffusion approximation (using the Parker transport equation) since the results are easy to interpret physically and the results for anisotropies are not well known. The approximation is correct in the limit of small anisotropy magnitudes, but gives valuable insights even if the anisotropies are significant [10]

DIFFUSIVE SHOCK ACCELERATION OF ANOMALOUS COSMIC RAYS

The mechanism of diffusive acceleration of energetic particles by the termination shock is thought to be the origin of the ACR's. The Parker transport equation [11] for the phase-space density $f(\mathbf{r}, p)$ as a function of position $\mathbf{r}$ and momentum magnitude p, may be written in terms of the diffusion tensor κ_{ij}, the fluid flow velocity $\mathbf{U}$ and any source Q as:

$$\frac{\partial f}{\partial t} = \frac{\partial}{\partial x_i}\left[\kappa_{ij}\frac{\partial f}{\partial x_j}\right] - U_i\frac{\partial f}{\partial x_i} + \frac{1}{3}\frac{\partial U_i}{\partial x_i}\frac{\partial f}{\partial \ell n(p)} + Q. \qquad (1)$$

The basic approach is to solve this equation for a model system which includes a shock, which appears as a discontinuity in the fluid velocity $\mathbf{U}$. Note that, in this form, the drift velocity $\mathbf{V_d}$ is incorporated in the antisymmetric part of κ_{ij}. Numerical solutions to (1) have been found for the steady-state ACR spectrum by a number of authors [3] [12]. The general nature of the solutions agree well with observations out to some 90 AU.

The diffusion is anisotropic, with diffusion coefficients $\kappa_\parallel$ parallel to the average magnetic field. I also consider the case where there are two different perpendicular diffusion coefficients, one ($\kappa_{\perp,1}$) in the direction normal to the magnetic field in the (r,θ) plane, and the other ($\kappa_{\perp,2}$) in the latitudinal direction (also normal to the Parker spiral magnetic field). This form was suggested first by Jokipii [13] who considered the evolution of the diffusion tensor in the radial flowing solar wind. Generally it is expected that $\kappa_\parallel \gg \kappa_{\perp,2} > \kappa_{\perp,1}$ in the outer heliosphere. This form has found observational support in Ulysses observations [14] In this case we may write for the diffusion tensor

$$\kappa_{ij} = \kappa_{\perp,1}\delta_{ij} + \left(\kappa_\parallel - \kappa_{\perp,1}\right)\frac{B_i B_j}{B^2} + \kappa_A \varepsilon_{ijk} B_k / B + \left(\kappa_{\perp,2} - \kappa_{\perp,1}\right) Q_i Q_j, \qquad (2)$$

where, if r_g is the gyroradius, $\kappa_A = r_g w/3$ is the antisymmetric part of the diffusion tensor, containing the effects of drifts and B_i is the local magnetic field vector and the vector $Q_i = \varepsilon_{ijk} V_i B_k$. Equation (1) is valid in the diffusion approximation, and should provide a reasonable indication of the expected phenomena, even though the anisotopies can be large. The basic approach is to solve this equation for a model system which includes a shock, which appears as a discontinuity in the fluid velocity $\mathbf{U}$. Note that, in this form, the drift velocity $\mathbf{V_d}$ is incorporated in the antisymmetric part of κ_{ij}.

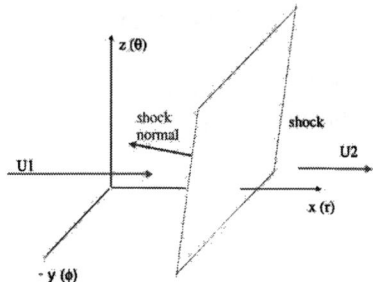

FIGURE 1. Illustration of the coordinate system used in discussing the solution. x is in the radial direction and z is in the latitudinal direction. The primed coordinate system is obtained by rotating the x axis to be normal to the shock plane, directed downstream.

The streaming flux of the particles may then be written:

$$S_i = -\kappa_{ij}\frac{\partial f}{\partial x_j} - \frac{U_i}{3}\frac{\partial f}{\partial \ln(p)} \qquad (3)$$

with the associated anisotropy

$$\delta_i = \frac{3S_i}{wf}, \qquad (4)$$

where w is the particle speed. Note that the anisotropy defined here is the direction in which the particles are flowing and is *not* the viewing direction from which they are seen. Near the termination shock, deviations from a planar shock geometry are may be neglected in a first approximation [10]. Consider then the case of diffusive shock acceleration for a one-dimensional planar shock,, where the geometry is as illustrated in Figure 1. Here the shock intersects the x axis at $x = x_{sh}$, and there are no upstream or downstream boundaries. The transport equation is readily solved analytically for this case. First, if U_1 and U_2 are the constant upstream and downstream flow speeds, respectively, and $\kappa_{1,ij}$ and $\kappa_{2,ij}$ the constant upstream and downstream diffusion coefficients, it is readily shown that the steady-state solution for the distribution function f may be written in terms of x', the direction normal to the shock, and the associated $\kappa_{x'x'}$ and $U_{1x'}$ as

$$\begin{aligned} f(x,p) &= Ap^{-\gamma} & x > x_{sh} \\ &= Ap^{-\gamma}\exp(\frac{U_{1x'}}{\kappa_{1,x'x'}}) & x \le x_{sh} \end{aligned} \qquad (5)$$

where the power-law exponent $\gamma = 3r/(r-1)$ and $r = U_1/U_2$. Note that the plane shock normal may be at an angle to the radius vector and the solar wind flow direction. $U_{1x'}$ is the component of the wind velocity normal to the shock.

ANISOTROPY UPSTREAM OF THE TERMINATION SHOCK

The full vector anisotropy associated with the solution in (5) is readily computed. Using a coordinate system as shown in Figure 1, the resulting anisotropy vector is

$$\delta_{x'} = -\frac{3U_{1x'}}{w} - \frac{U_{1x'}}{wf}\frac{\partial f}{\partial \ln p}$$
$$\delta_{z'} = \frac{3U_{1x'}}{w\kappa_{1,x'x'}}\kappa_{z'x'}$$
$$\delta_{y'} = \frac{3U_{1x'}}{w\kappa_{1,x'x'}}\kappa_{z'x'}, \qquad (6)$$

where the diffusion tensor is given by equation (2) above. This deceptively simple equation is in reality quite complicated and difficult to interpret for the general case, where the shock normal deviates significantly from the radial, or upstream flow, direction. However, this deviation is expected to be small, and we discuss that case semi-quantitatively here. The full case will be discussed elsewhere.

First, the anisotropy normal to the shock plane, $\delta_{x'}$, which is approximately the radial direction, was the topic the analysis by Jokipii and Giacalone [10], which showed that this was compatible with the Voyager 1 observations. Small deviations from the radial direction will not appreciably alter this conclusion.

The Parker spiral or average magnetic field in the far outer heliosphere is nearly transverse to the radial and latitudinal direction. The largest energetic-particle anisotropies measured by Voyager 1 were along the the magnetic field in the direction away from the Sun [6] [7]. In our coordinate system, δ_z is nearly field aligned, and is directed parallel to the shock face, in the direction in which the field line increases its distance from the shock (the -z direction in Figure 1). This consists manily of a diffusive anisotropy because the convective anisotropy is mostly in the x, or radial direction. Because the ratio $\kappa_\perp/\kappa_\parallel$ is small, this anisotropy can be much larger than that in the radial direction if the shock normal angle θ_{bn} is near 90 degrees. The analysis by Jokipii, Giacalone and Kóta [15] shows that this direction for the anisotropy will be outward from the Sun if the shape of the heliopause and termination shock is such that the shock is closest to the Sun in the direction toward the incoming interstellar plasma flow, as expected from simulations and analogies with other flexible spheroidal objects immersed in a flow. This is illustrated in the left panel figure 2.

Similarly, the anisotropy in the latitudinal direction will be governed by $\kappa_{\perp,2}$, which is related to the magnetic field line random walk, and which is expected to be larger than $\kappa_{\perp,1}$. As illustrated in the right panel of figure 2, at Voyager 1 in the northern hemisphere, the random walking magnetic field line will intersect the shock first at lower heliographic latitudes. Just as in the case of the field-aligned anisotropy, this will cause a predominantly northward anisotropy. Such an anisotropy was indeed observed on Voyager 1 [16].

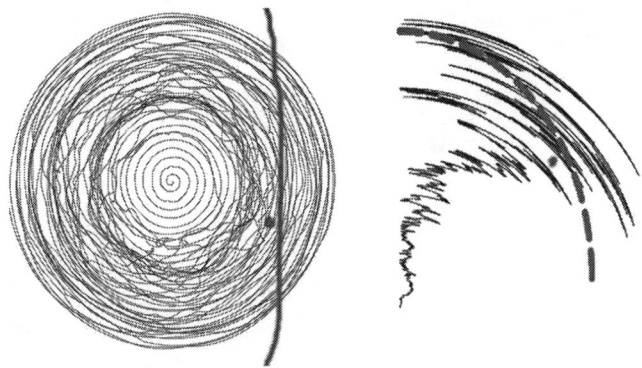

FIGURE 2. Projections of a random magnetic field line into the equatorial (left) and meridional planes (right). The expected shock shape is illustrated schematically in each panel, showing that the radial distance of the shock increases away from the direction from which the interstellar plasma flows.

SUMMARY AND CONCLUSIONS

Analysis of the acceleration of ACR at the termination shock of the solar wind, in the diffusion approximation, suggests that the general properties of the observed three-dimensional anisotropy are a natural consequence of the structure of the interplanetary magnetic field and the expected flared global shape of the termination shock.

ACKNOWLEDGMENTS

This work was supported, in part, by NASA under grants NAG5-6620, NAG5-7793, NAG5-12919,and by the NSF under grant ATM9616547 and ATM0327773. I acknowledge the essential contribution of my colleagues at the University of Arizona, Joe Giacalone and Jozsef Kóta to my understanding of the points covered in this overview.

REFERENCES

1. L. A. Fisk, e. a., *Astrophys. J.*, **190**, L35 (1974).
2. Pesses, M., Jokipii, J. R., and Eichler, D., *Astrophys. J.*, **246**, L85 (198).
3. Jokipii, J. R., *J. Geophys. Res.*, **91**, 2929 (1986).
4. Stone, E. C., *Science*, **293**, 55 (2001).
5. Barnes, A. J., *J. Geophys. Res*, **98**, 15137 (1993).
6. McDonald, F. B., Stone, E. C., Cummings, A. C., Heikkila, B., Lal, ., and Webber, W. R., *Nature*, **426**, 48 (2003).
7. Krimigis, S. M., Decker, R. B., Hill, M. E., Armstrong, T. P., Gloeckler, G., Hamilton, D. C., Lanzerotti, L. J., and Roelof, E. C., *Nature*, **426**, 45 (2003).
8. Burlaga, L. F., Ness, N. F., Stone, E. C., McDonald, F. B., Acuna, M. H., Lepping, R. P., and P.Connerney, J. E., *Geophys. Res. Lett.*, **30**, SSC9–1 (2003).
9. Gurnett, D., Kurth, W. S., and Stone, E. C., *Geophys. Res. Lett.*, **30**, 2309 (2003).
10. Jokipii, J. R., and Giacalone, J., *Astrophys. J. Lett.*, **605**, L145 (2004).

11. Parker, E. N., *Plan. Sp. Sci.*, **13**, 9 (1965).
12. Florinski, V., *Cosmic Rays in the Heliosphere*, University of Arizona Ph D thesis, Tucson, 2001.
13. Jokipii, J. R., *Astrophys. J.*, **182**, 585 (1973).
14. J. R. Jokipii, e. a., *Geophys. Res. Lett*, **22**, 3385 (1995).
15. Jokipii, J. R., Kòta, J., and Giacalone, J., *Astrophys. J. Lett.*, **611**, L141 (2004).
16. Cummings, A. C., Stone, E. C., Burlage, L. F., Ness, N., McDonald, F. B., and Webber, W. R., *Proc. 28th Int. Cosmic Ray Conf., Tsukuba, Japan*, **7**, 3777 (2003).

A Global V-Shaped Channel Structure of the Termination Shock Due to a Magnetic Pressure Effect, and Its Physical Connection to Bipolar Flow Type Planetary Nebulae

Haruichi Washimi[*], Takashi Tanaka[+], and Gary P. Zank[*]

[*]*IGPP, University of California, Riverside, CA 92521, USA*
[+] Faculty of Science, Kyushu University, Hakozaki, Fukuoka 812-8581 Japan

Abstract. It is shown that the magnetic-pressure effect in the azimuthal direction in the heliosheath region results in the formation of a global V-shaped channel structure in both the heliopause and the termination shock surfaces. The physical processes involved in the formation of the V-shaped heliosphere is closely related with that of a bipolar-flow type Planetary Nebulae.

Keywords: MHD simulation, Outer Heliosphere, Planetary Nebulae.
PACS: 96.50.Bh, 96.50.Ci, 98.58.Li.

FORMATION OF A V-SHAPED CHANNEL STRUCTURE OF THE OUTER HELIOSPHERE

The interplanetary toroidal magnetic field is thought to be enhanced in the heliosheath as the solar-wind speed decreases beyond the termination shock. The magnetic-pressure effect of this enhanced field has been studied in numerous papers (e.g., 1-6). This paper also deals this effect, but in contrast to previous works, which studied mainly magnetic-pinch effects in the radial direction, we focus on the role of magnetic pressure in the azimuthal direction. This azimuth-direction effect will be important because it plays a dominant role in forming other interesting astrophysical gas structures (e.g. 7).

The basic distribution of the toroidal magnetic field B_ϕ that we will consider here is the following: B_ϕ increases with the colatitude angle θ as $\sin\theta$ and becomes maximum at an angle, say θ_0, very near the equator. From the point where θ is θ_0, B_ϕ rapidly decreases to zero to the equator. Because the polarity of B_ϕ is opposite in the southern hemisphere, the distribution of B_ϕ corresponds to a north-south anti-symmetry. This simple distribution of $B_\phi(\theta)$ means that the magnetic-pressure works to drive plasma of middle and high latitudes poleward from the point where θ is θ_0, and at the same time drives the plasma of low latitudes towards the equator. The latter creates an equatorial neutral sheet. The above characteristics of the magnetic-pressure effectively split the heliosheath plasma into poleward and equatorward components from the point where θ is θ_0. This yields a pressure asymmetry along the heliopause surface.

The local interstellar medium (LISM), which surrounds the heliosheath, can therefore compress the heliosheath region along the line where θ is θ_0 more than at higher latitudes. In fact, when we performed a 3-D MHD computer simulation for the heliosphere, such a dynamical asymmetric compression occurs, and as a result a V-shaped channel structure along the equatorial plane is formed[8,9]. In these previous simulations, the solar wind density is prescribed to be constant along the inner boundary surface. In these cases, a weak density depression is formed in the interplanetary region around the neutral sheet which creates a disturbance in the heliosheath plasma near the termination shock. This disturbance evolves outward along the line where θ is θ_0, and finally results in the formation of the V-shaped structure.

In contrast to the above case, when the density of the sheet region on the inner-

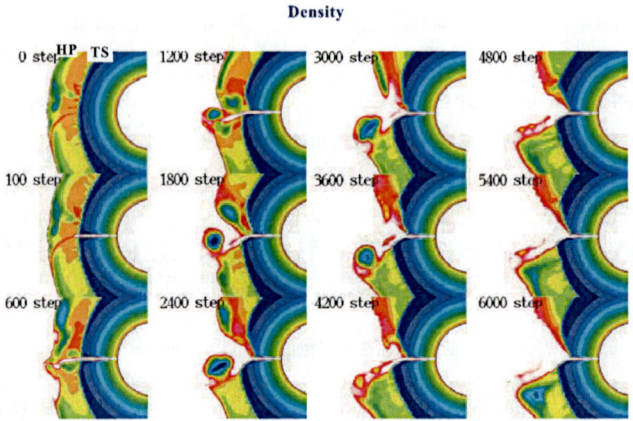

FIGURE 1. Time-development of the density structure for the inner-boundary, switching from *No Sheet* to *Sheet* conditions. The density on the Inner-Boundary is assigned to be 3 times greater in the Sheet Region.

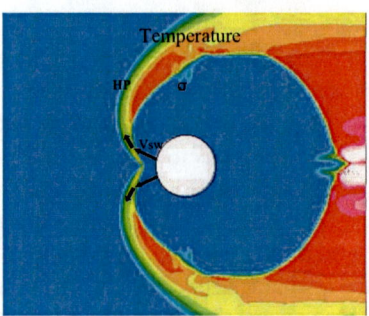

FIGURE 2. Temperature profile of the V-shaped heliosphere and schematic solar wind direction (arrows) before and after the termination shock.

boundary is sufficiently high, no evident density depression is found in the interplanetary medium. Even for this case, a V-shaped structure is formed. Figure 1 shows the case when the density of the neutral sheet region on the inner boundary is 3 times greater than that of the middle and high latitudes. The equatorial sheet is simply enhanced step by step, and reaches and pushes out the heliopause at about 600 steps, perturbing the heliopause surface. Along the surface of the sheet plasma, the disturbance evolves into the inner region of the heliosheath, and finally the V-shaped structure is formed. All these previous and present simulations indicate that the LISM can drive the heliopause deep into the heliosheath, which results in the formation of the V-shaped structure.

In our simulation, we have used a standard set of parameters both for the solar wind plasma and LISM: On the inner boundary, the solar wind is radial and isotropic, with a speed of 400 km/s. The density and the toroidal magnetic field are 5/cc and 4.2 Gamma, respectively at 1AU. The radial component of the magnetic field is neglected because its intensity is very weak in the outer heliosphere. The LISM density and magnetic field are 0.1/cc, and 0.12 micro Gauss, respectively, and the relative speed of the LISM and the sun is 25 km/s and its direction is perpendicular to the solar rotation axis. The scale of the inner and outer boundaries is 50 and 950 AU, respectively. We have performed ideal 3D MHD simulations using a TVD-Scheme.

Figure 2 shows the temperature distribution of the V-shaped heliosphere. As is shown by arrows, the solar wind velocity is distorted poleward at the oblique surface of the V-shaped termination shock. The oblique character of the termination shock causes a relatively low temperature at the outer region of the heliosheath along the heliopause surface, and the reduced heliosheath pressure (both dynamical and thermal) then allows the LISM to push the heliopause inwards thus forming the V-shaped structure.

PHYSICAL CONNECTION OF THE HELIOSPHERE AND PLANETARY NEBULAE

Planetary nebulae (PNs) are among the most beautiful objects observed in the sky, and possess a multitude of different structures. The structures are determined by a complex interplay between the stellar wind, the stellar magnetic field, and the interstellar medium. An example of a bipolar-type PN is M2-9[10], observed by Hubble Space Telescope, and it seems to be confined by a stellar toroidal magnetic field. We have a working hypothesis that the V-shaped heliosphere and the bipolar-type PNs should have some common physical characteristics. Though apparent features might not be so similar to each other, all bipolar-type PNs seem to be slim around the equatorial region, rather like the V-shaped heliosphere. This has encouraged us to investigate whether the difference between both objects may be caused by some parameter differences.

As a preliminary study, we performed 3D MHD simulations for a circumstellar gas structure for several cases by varying the relative speed between the star and the interstellar medium. For this study we assume a set of solar wind parameters for the stellar wind. For the interstellar medium, we assume a rather high temperature, 2.5×10^5 K and a negligibly weak magnetic field strength. Figure 3 shows an example of the dynamical evolution of the density and magnetic field of the circumstellar gas starting from an appropriate initial condition (0 step) when the speed of the interstellar medium relative to the star is 72.8 km/s (M=1.6) where M is the Mach number. A V-shaped structure around the equatorial region and magnetic ducts at the polar region in both northern and southern hemispheres are found. These are formed by the magnetic-pressure effect of the toroidal field in the stellar wind. The stellar wind plasma extends outwards along the bi-polar ducts. The ducts extend outwards step by step along the polar axis, and are then bent to the downstream side by the high-speed interstellar medium flow. At the initial phases of this simulation (at 1000, and 2000 steps) the

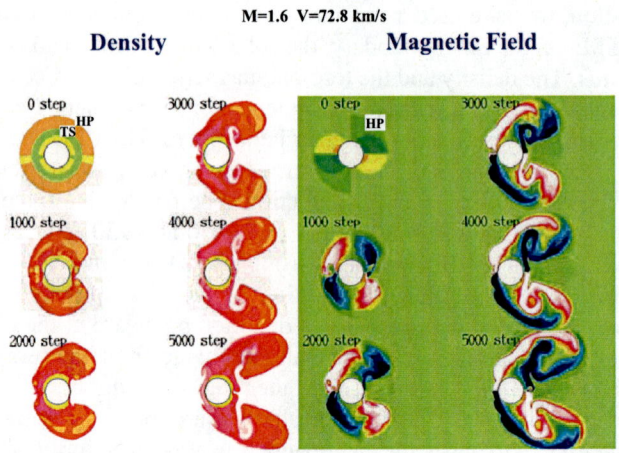

FIGURE 3. Time-development of the density and magnetic field when the interstellar medium temperature is 2.5×10^5 K and velocity is 728 km/s (M=1.6).

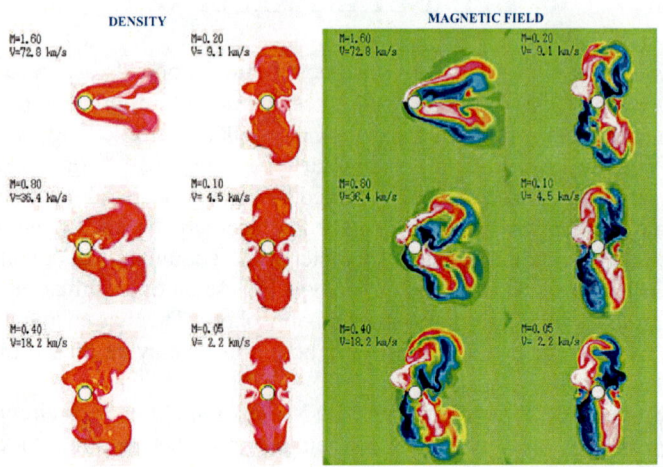

FIGURE 4. Summary of stellar-spheric structure for interstellar medium conditions M=1.6, 0.8, 0.4, 0.2, 0.1, and 0.05. The temperature of the ISM is assumed to be 2.5×10^5 K.

stellar wind plasma in low latitudes extends to higher latitudes. However, the ambient interstellar medium does not move quickly in to the low latitudes region. Hence, the stellar wind plasma which was moving to higher latitudes, turns and goes down again. Thus irregular and vortex motions of the plasma flow are found. These do not compose the eventual bipolar flow but are only an astro-debris which will eventually be removed in the course of the wind-medium interaction. Our simulations confirm that the bipolar flow plasma is formed along the magnetic ducts.

A summary of stellar-sphere structure for different values of the speed of the interstellar medium (M=1.6, 0.8, 0.4, 0.2, 0.1, and 0.05) is shown in Figure 4. Bipolar flows which emerge from the interplanetary medium are found in all cases. The inclination angle from the stellar rotation axis is reduced for smaller speeds of the interstellar medium. The inclination angle is decided by the mass-flux ratio of the interstellar medium and the bipolar flow. Bipolar flows with speeds less than a 0.02 Mach-number are found to be aligned almost along the z-axis. If the mass-flux of the simulation is greater, as often observed in PNs, the flows will be parallel to the polar axis even for high-speed interstellar winds.

CONCLUDING REMARKS

Due to a magnetic-pressure effect in the heliosheath, a V-shaped channel on the heliopause and the termination shock surfaces can form when the solar wind interacts with the LISM. We have extended our heliospheric model to conditions more appropriate to PN. In our preliminary study of PN structure, we find that we can simulate bipolar outflows as a result of the magnetic-pressure effect associated with the toroidal magnetic field of the stellar wind. The speed of relative motion between the star and the ISM and the mass flux ratio between the bipolar flow and the ISM play important roles in determining the inclination angle of the bipolar flow relative to the polar axis.

Numerical computations were carried out on the SX-6 at the National Institute of Information and Communications Technology and on the Fujitu-VPP5000 at Nagoya University. The latter was performed under the joint-research program of the Solar-Terrestrial Environment Laboratory, Nagoya University.

REFERENCES

1. Axford, W. I., *Solar Wind, NASA SP-308*, 1972, pp. 608-658.
2. Lee, M. A., *Proc. VI Solar Wind, NCAR/TN-306+Proc.*, 1988, pp. 635-650.
3. Holzer, T. E., *Annu. Rev. Astron. Astrophys.*, **27**, pp. 199-234 (1989).
4. Suess, S. T., *Rev. Geophys.*, **28**, pp. 97-115 (1990).
5. Nozawa, S. and H. Washimi, *Publ. Astron. Soc. Japan*, **49**, pp.383-388 (1997).
6. Zank, G. P., *Space Sci. Rev.*, **89**, pp. 413-688 (1999).
7. Tanaka, T. and H. Washimi, Science, 296, pp. 321-322 (2002).
8. Washimi, H. and T. Tanaka, *Adv. Space Res.* **27**, pp.509-515 (2001).
9. Washimi, H. and T. Tanaka, *Physics of the Outer Heliosphere: Third International IGPP Conf.*, pp. 87-92(2004).
10. Balick, B., http://www.astro.Washington.edu/balick/WFPC2/index.html .

The Termination Shock and Beyond: MHD Modeling

Romana Ratkiewicz*, Jolanta Grygorczuk* and Lotfi Ben-Jaffel[†]

*Space Research Center PAS, Bartycka 18a, 00-716 Warsaw, Poland
[†]Institut d'Astrophysique de Paris, 98bis Blvd Arago, 75-014 Paris, France

Abstract. The 3D MHD models of the solar wind - interstellar plasma interaction including, in a self-consistent way, interactions of various populations of plasma and neutral particles should be ready to confront their results with the forthcoming data that will be obtained from space missions. In the near future, predictions made by sophisticated theoretical models should help refine the goals and optimize the capabilities of the instruments that will explore the far heliosphere and the LISM. In this paper we are giving a short survey of the MHD models and point out the problems, which need to be solved in the near future. As the example we show our recent numerical results with the simple model of the current sheet.

Introduction

The numerical simulations of the interaction between the solar wind (SW) and the local interstellar medium (LISM) has already a history of about 30-years. The first models of this interaction were taking into account only SW and LISM plasmas. The treatment was fully gasdynamic, and as the result the heliospheric interfaces: the termination shock (TS), the heliopause (HP) and the bow shock (BS) have been created. The SW - LISM interaction is affected significantly by charge exchange of interstellar neutral particles with plasma in different regions of the heliospheric interface and in the inner heliosphere. Its importance was very early recognized. Almost simultanoeously the gasdynamic models including neutrals were built. A thorough review of the subject has been given by [1].

At present the complexity of the problem of the solar wind-local interstellar medium is very well known. Besides mentioned above this interaction encompasess first of all magnetic fields, and also pickup ions, anomalous component of cosmic rays (ACR's), energetic neutral atoms (ENA's), galactic cosmic rays (GCR's), possible time-dependent phenomena, latitudinal dependence of the solar wind, tilted heliospheric current sheet and so on.

The first attempt to include into numerical models the magnetic fields was made about 15 years ago [2]. Since then several modelers have made a big effort to model magnetohydrodynamically the heliospheric interface including interstellar magnetic field (ISMF), interplanetary magnetic field (IPMF), and both. Also neutral components of LISM were included into consideration (see Table 1.) Although none of these models is perfect, each of them gives new information or pays attention to some special aspects (see Table 1. Remarks) about this complicated interaction and deserves to be carefully discussed. The limitation of space does not however allow us to do so. Instead we would

TABLE 1. Simulation Models with Magnetic Field.

Model	Year	ISMF	N	IPMF	Remarks
Fujimoto & Matsuda [2]	1991	$\alpha = 0°$	-	-	first model
Baranov & Zaitsev [3]	1995	$\alpha = 0°$	-	-	axisymmetric
Washimi & Tanaka [4]	1996	$\alpha = 90°$	-	P	B_ϕ-current sys.
Pogorelov & Semenov [5]	1997	$\alpha = 0°$	-	-	axisymmetric
Linde et al. [6]	1998	$\alpha = 0, 90°$	GD	P	3 config.of B_{is}
Pogorelov & Matsuda [7]	1998	$\alpha = 0, 45, 90°$	-	-	nonevol. shocks
McNutt et al. [8]	1999	$\alpha = 90°$	GD	P	TS distances
Ratkiewicz et al. [9], [10]	1998,2000	$\alpha = 0, 30, 60, 90°$	-	-	asymmetries
Washimi & Tanaka [11]	1999	$\alpha = 90°$	-	P	B_ϕ-solar cycle
Aleksashov et al. [12]	2000	$\alpha = 0°$	MC	-	self-cons.
Washimi & Tanaka [13],[14]	2001,2004	$\alpha = 90°$	-	P	B_ϕ - V shape
Ratkiewicz et al. [15]	2002	$\alpha = 0, 30, 60, 90°$	CF	-	asymm-neutrals
Ratkiewicz & Ben-Jaffel [16]	2002	$\alpha = 0, 30 - 50, 90°$	CF	-	varies α & $|B_{is}|$
Ratkiewicz & Webb [17],[18]	2002,2004	$\alpha = 0°$	-/CF	-	sucking effect
Opher et al. [19], [20]	2003,2004	-	-	P	jet
Ratkiewicz & McKenzie [21]	2003	$\alpha = 0°$	CF	-	switch-on sh.
Florinski et al. [22]	2004	$\alpha = 0°$	GD	-	strong ISMF
McNutt [23]	2004	$\alpha = 90°$	PL	-	survey
Pogorelov et al. [24]	2004	$\alpha = 0, 45, 90°$ $\theta = 60°$	-	P	HCS bending
Ratkiewicz et al. [25]	2004	$\alpha = 0, 90°$ $\theta = 60°$	CF	P	numerical "reconnection"

N-neutrals
P-Parker model
GD-gasdynamic
MC-Monte Carlo Method
CF-constant flux
PL-power law

like to stress some problems which need to be solved in the near future. As the example we show our recent numerical results with the simple model of the current sheet.

Problems

The results we all have been obtaining are in general in a good agreement. We now know what kind of asymmetries caused by the interplanetary and interstellar magnetic fields one could expect, how the neutral particles influence the heliospheric interface, what may happen when the pick-up ions and ACR's are created, how GCR's may affect the heliosphere, however our knowledge about all those is in pieces. Besides it, we also have some results in which we differ from each other especially in physical interpretation. Between them are jets, bending current sheet, numerical "reconnection", V-shape. Before we build an universal code we need to understand why the differences appear. We need theory-based predictions for the current sheet, acting forces such as

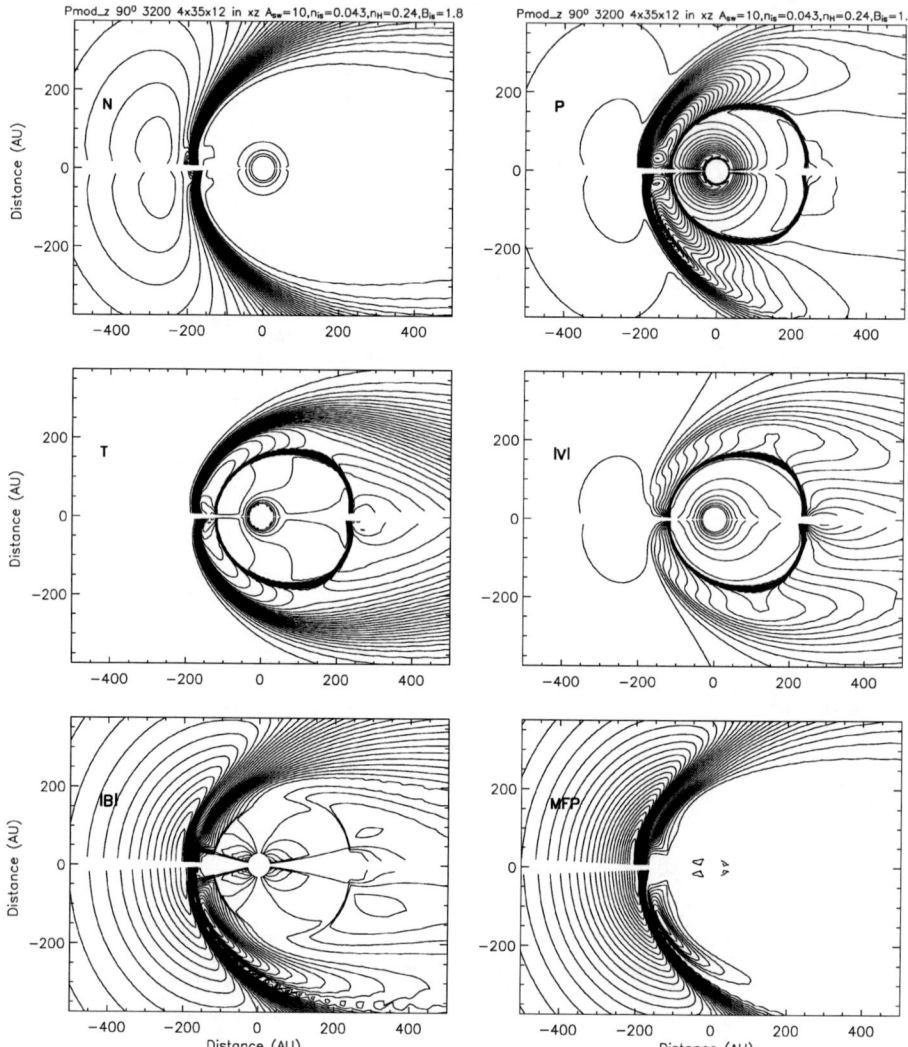

FIGURE 1. Distribution in the x-z plane of number density, thermal pressure, temperature, magnitude of velocity and magnetic field, and magnetic pressure for the perpendicular LISM magnetic field and the Parker model for interplanetary magnetic field with the current sheet, for which $\vec{B}_\phi$ increases on the inner boundary as $sin\theta$ from the pole to the grid point near the ecliptic, and decreases to zero from this grid point to the ecliptic

$J \times B$, reconnection rate, etc. And then, our numerical results should be clear, have a transparent physical meaning and interpretation, in particular, we need to understand all "discoveries" coming from our codes.

Only then we can achieve our main goal to build the self-consistent code describing fully the solar wind-interstellar medium interaction.

Our numerical results with the current sheet

As listed in Table 1. following models discuss the current sheet: [4], [11], [13],[14], [6], [19], [20], [24], [25]. Interestingly enough, in general, the heliospheric interface obtained from different models looks similar. However, they differ in details or in the interpretation of these details. In particular, similarly looking effects are considered as the bending heliospheric current sheet (HCS) [24] or numerical "reconnection" [25]. At this point it should be mentioned that a magnetic reconnection plays an important role in many space and astrophysical contexts, in particular, the physical reconnection may occur at the heliopause. On the other hand, since the solar wind and interstellar plasmas are highly conducting, they are usually treated as ideal magnetohydrodynamical (MHD) systems in which the physical magnetic reconnection cannot be reflected because of zero resistivity. However, as shown in [25] the presence of a code-dependent numerical diffusivity, even in the ideal MHD, may lead to the numerical "reconnection". The numerical "reconnection" does not give the precise reconnection rate and location. It indicates the tendencies.

Another interesting feature, V-shape has been found by Washimi and Tanaka ([13],[14]). In order to be able to compare our model with Washimi and Tanaka, we repeated the calculations using our 3D MHD model as described in [25]. The interplanetary magnetic field of solar origin (Parker model - Archimedean spiral) has opposite polarities in the northern and southern hemispheres. In the present approach, the current sheet is initially specified as a thin sheet, with finite width confined to the ecliptic x-y plane. The B_ϕ component of IPMF increases on the inner boundary as $sin\theta$ from the pole to the grid point near the ecliptic, and decreases to zero from this grid point to the ecliptic as in [14]. The interstellar magnetic field vector is perpendicular to the velocity and both vectors are in x-y plane. The interplanetary magnetic field is spiraling away from the Sun in the northern hemisphere and toward the Sun in the southern hemisphere. The results are shown in Figure 1. The numerical "reconnection" occurs in the southern hemisphere. The numerical "reconnection" effect is visible in the thermal pressure and magnetic field contours. The comparison with the results reported in [25] confirms asymmetry north-south caused by the IPMF. Also the V-shape structure appears, but in the heliopause only as shown, for example, in the velocity contours.

Conclusions

The interpretation of the results obtained from different models needs the further discussion. For this purpose all modelers should use the same conditions at the boundaries, and similar models of the current sheet to eliminate unwanted numerical effects. Our knowlegde about the solar wind is quite complete, so it is easy to fix the physical parameters for the SW. Much more unknowns we have as concerns the interstellar medium. It requires perhaps, a parametric study. But we all should use the same numbers in our calculations to be able to compare the results.

Having a complete model, observations made up to now and in the future, especially data from IBEX, we will be able to give much better estimation of LISM parameters, which still we do not know, such as, for example, an orientation and a magnitude of the interstellar magnetic field.

ACKNOWLEDGMENTS

Authors acknowledge support from CNRS/France and PAS/Poland under Program 18249 and program *ASTRO-PF*. RR and JG acknowledge the support in part from KBN Grant No.1P03D 009 27. RR acknowledges support from the IGPP at the University of California Riverside, and the hospitality of the director of the IGPP, prof. Gary Zank.

REFERENCES

1. Zank, G. P., *Space Sci. Rev.*, **89**, 413–687 (1999).
2. Fujimoto, Y., and Matsuda, T., *KUGD91-2*, **Dep. of Aeronaut Eng**, Kyoto Univ., Kyoto, Japan (1991).
3. Baranov, V. B., and Zaitsev, N. A. *Astron Astrophys*, **304**, 631–637 (1995).
4. Washimi, H., and Tanaka, T., *Space Sci. Rev.*, **78**, 85–95 (1996).
5. Pogorelov, N. V., and Semenov A. Yu., *Astron Astrophys*, **321**, 330–337 (1997).
6. Linde, T., Gombosi, T., Roe, P., Powell, K., and DeZeeuw, D., *JGR*, **103**, 1889–1904 (1998).
7. Pogorelov, N. V., and Matsuda, T., *JGR*, **103**, 237–245 (1998).
8. McNutt, R. L.,Jr., Lyon, J., Goodrich, C. C., and Wiltberger, M., *AIP*, **471**, 823–826 (1999).
9. Ratkiewicz, R., Barnes, A., Molvik, G. A., Spreiter, J. R., Stahara, S. S., Vinokur, M., and Venkateswaran, S., *Astron Astrophys*, **335**, 363–369 (1998).
10. Ratkiewicz, R., Barnes, A., and Spreiter, J. R., *JGR*, **105**, 25,021–25,031 (2000).
11. Washimi, H., and Tanaka, T., *Ad. Sp. Res.*, **23**, 551–560 (1999).
12. Aleksashov, D. B., Baranov, V. B., Barsky, E. V., and Myasnikov, A. V., *Astron. Letters*, **26**, 743–749 (2000).
13. Washimi, H., and Tanaka, T., *Ad. Sp. Res.*, **27**, 509–515 (2001).
14. Washimi, H., and Tanaka, T., *AIP*, **719**, 87–92 (2004).
15. Ratkiewicz, R., Barnes, A., Müller, H.-R., Zank, G. P., and Webb, G. M., *ASRes.*, **29**, 443–438 (2002).
16. Ratkiewicz, R., and Ben-Jaffel, L., *JGR*, **107**, SSH 2 – 1–13 (2002).
17. Ratkiewicz, R., and Webb, G. M., *JGR*, **107**, SSH 11 – 1–5 (2002).
18. Ratkiewicz, R., and Webb, G. M., *JGR*, **109**, A02111 1–4 (2004).
19. Opher, M., Liewer, P. C., Gombosi, T. I., Manchester, W., DeZeeuw, D. L., Sokolov, I., and Toth, G., *ApJ*, **591**, L61–L65 (2003).
20. Opher, M., Liewer, P. C., Velli, M., Bettarini, L., Gombosi, T. I., Manchester, W., DeZeeuw, D. L., Toth, G., and Sokolov, I., *ApJ*, **611**, 575–586 (2004).
21. Ratkiewicz, R., and McKenzie, J. F., *JGR*, **108**, SSH 6 – 1–8 (2003).
22. Florinski, V., Pogorelov, N. V., Zank, G. P., Wood, B. E., and Cox, D. P., *ApJ*, **604**, 700–706 (2004).
23. McNutt, R. L.,Jr. *AIP*, **719**, 111–116 (2004).
24. Pogorelov, N. V., Zank, G. P, and Ogino T., *ApJ*, **614**, 1007–1021 (2004).
25. Ratkiewicz, R., Grygorczuk, J., and Ben-Jaffel, L., *AIP*, **719**, 93–98 (2004).

Comparison of Voyager Shocks in Solar Cycle 23

Justin Ashmall* and John Richardson*

*M.I.T. Kavli Institute for Astrophysics and Space Research, 77 Massachusetts Ave, Rm 37-673, Cambridge, MA 02139. USA.

Abstract.
Solar cycle 23 was notable for two periods of intense solar activity (or 'events' as we shall hereafter refer to them): the 'Bastille Day Event' of 2000 and the 'Halloween Event' of 2003. In this paper we look at the signatures of the interplanetary shocks produced by these events, in particular the plasma parameters, as observed by Voyager 2 (V2) some six months after the events occurred at Sun. We compare these shocks with other large events observed by V2 during the preceding decade. We note that the plasma parameters, most notably the plasma density, are frequently not as might be expected for "typical" events.

INTRODUCTION

Voyagers 1 and 2 were launched in 1977, and as of mid-2005 were at radial distances of 96 AU and 77 AU and heliographic latitudes of 34°N and 26°S respectively. The M.I.T. plasma instruments on-board the two identical spacecraft have returned a wealth of data on the plasma environments of the outer plants and the heliosphere. Although the plasma instrument on-board V1 was damaged shortly after the Saturn encounter in 1981, the V2 instrument continues to return data daily. The Voyagers no longer have on-board data-storage and so data is only received when coverage is available from the Deep Space Network (DSN); this is currently about 50% of the time. But that we get should get any data at all from a 25 W transmitter over 11.5 billion km away is truly amazing.

A large halo CME lifted off from the sun on the 14th of July 2000 (DoY 195), subsequently becoming known as the "Bastille Day" event. It was observed at 1 AU the following day, giving measured solar wind speeds in excess of 1000 km/s. Some six months later a shock associated with this event was observed by Voyager 2 at 63 AU, 23°S. Unfortunately the shock arrived during a 17 hour data-gap (due to DSN coverage) and so the shock was not directly observed. There was a fortuitous alignment of the ACE and Wind spacecraft (at 1 AU) with V2 that allowed the shock to be tracked on its journey to the outer heliosphere.

Many large solar flares and CMEs occurred over the Halloween period in 2003, from late October to early November. This period produced the highest ever recorded solar wind speeds: in excess of 1500 km/s as measured by Wind. Roughly six months later, on the 28th of April 2004 (DoY 119), V2 observed this "Halloween Event" at 73 AU, 25°S. Unfortunately this event also arrived during a data-gap.

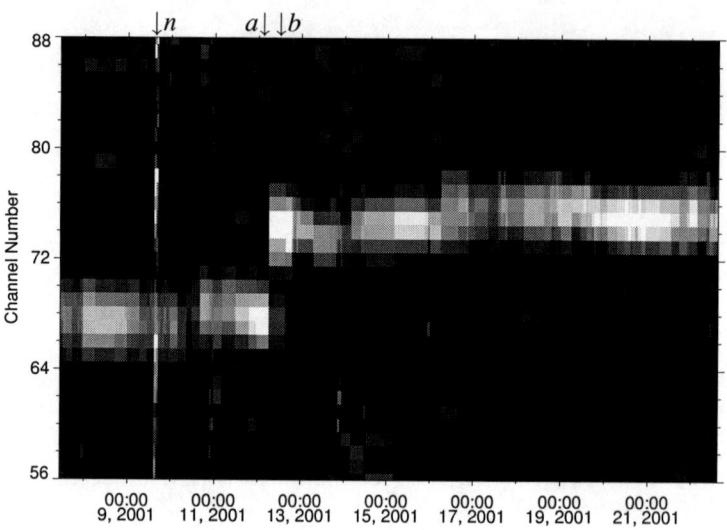

FIGURE 1. Plasma spectrogram of the Bastille Day event as measured by V2. The x-axis time-labels show hh:mm above Day-of-Year (DoY), Year. Channel counts are averaged over 20 adjacent spectra (approx one hour of data). The broad-band emission at arrowed time n is instrumental noise. Times prior to and following the shock are labelled a and b. See text for full description.

DATA AND ANALYSIS

Figure 1 shows a plasma spectrogram of the Bastille Day event as observed by Voyager 2. The light band running through the spectrogram is the signature of the solar wind. The spectrograms are particularly useful since they show, qualitatively, many features of the plasma simultaneously. The x-axis plots time and the y-axis plots energy channel, corresponding to logarithmic energy. The z-axis (i.e. the colour/shade of grey) represents the number of counts at that energy channel from low (dark) to high (light). Considering the solar wind as a convected Maxwellian enables a simple reading of the spectrogram. The bulk speed of the solar wind is given by the y-co'ordinate of the distribution (in other words the y-co'ordinate of the centre of the bright band). The temperature (or thermal speed) is represented by the width of the (Maxwellian) distribution, i.e. the broadness of the band. Finally the density can be thought of as being represented by the z-co'ordinate of the peak of the distribution, that is to say the peak brightness of the band in the spectrogram. Spectrograms can also be invaluable for identifying the feint signatures of α-particles which can be used for tracing ICMEs in the outer heliosphere.

Of course the above spectrogram description is nothing more than a broad qualitative interpretation. The calculation of actual plasma parameters is carefully performed by taking moments and fitting functions to the measured plasma distribution, taking into account the calibrated properties of the instrument. In particular the interplay between temperature and density is more complicated than described above.

The most obvious feature of the spectrogram in figure 1 is the clear jump in the bulk

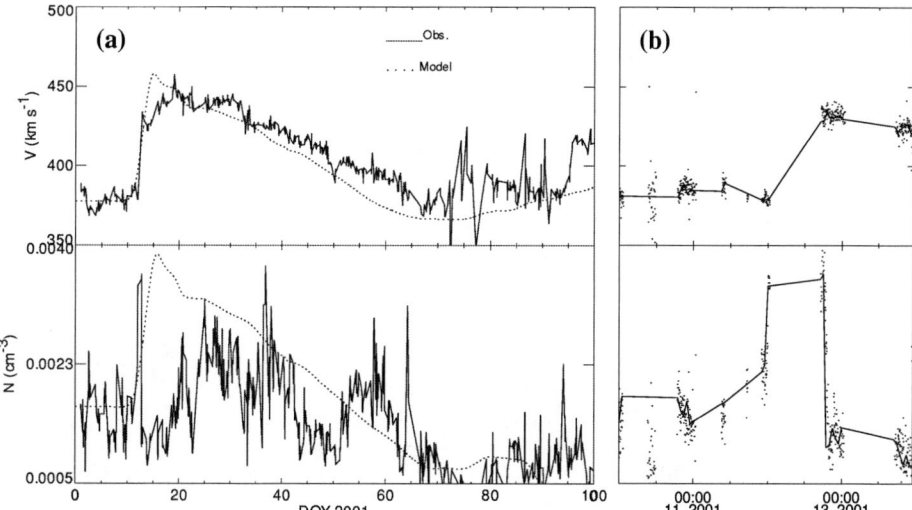

FIGURE 2. **a:** 100 days of plasma observations (solid line) and model predictions (dashed line) for the Bastille event at V2. A good example of 'typical' expectations for such an event (adapted from Wang et al. [2]). **b:** Zoomed-in section of high-resolution observations (points) and hourly averages (lines) showing four days around the shock. Note the clear increase in density that precedes the speed increase. Time labels as in figure 1.

speed of the solar wind (i.e. the discontinuity in the y-direction) occurring between labelled times *a* and *b*. Also of note is the increase in density (brightening) not only following the shock at time *b* (as might be expected) but also prior to the arrival of the shock at time *a*. Figure 2b clearly shows that the density increases before the speed jump, that is to say the density pulse straddles the shock. A more lengthy and thorough discussion of these observations has been given by Burlaga et al. [1].

This raises two questions: Firstly, is this event observed by V2 at 63 AU truly associated with the Bastille Day event seen at 1 AU, and secondly what would we expect to observe? Numerical modelling provides a convenient answer to both questions. As was mentioned previously, for this event the Wind, ACE and Voyager 2 spacecraft were fortuitously aligned with the propagation of the CME such that the Bastille Event could be observed by all three spacecraft. Wang et al. [2] used data taken at 1 AU as input into a 2.5-D MHD numerical model. Given the alignment of the spacecraft, it was reasonable to expect the output of the model would correspond to conditions observed at V2.

The model of Wang et al. [2] predicted the arrival time of the Bastille event correct to within a day or so; a truly stunning agreement between model and observation given the six months and 61 AU the event took to propagate to V2's location. This strongly suggests that the V2 event is indeed associated with the Bastille event. To answer the second question we look at the predicted plasma parameters of the Wang et al. [2] model; these are shown in figure 2a. For the speed profile the model fits well, predicting the discontinuity followed by a gradual return to pre-shock speeds. The model predicts a similar profile for the plasma density, that is a sudden increase followed by a gradual fall-

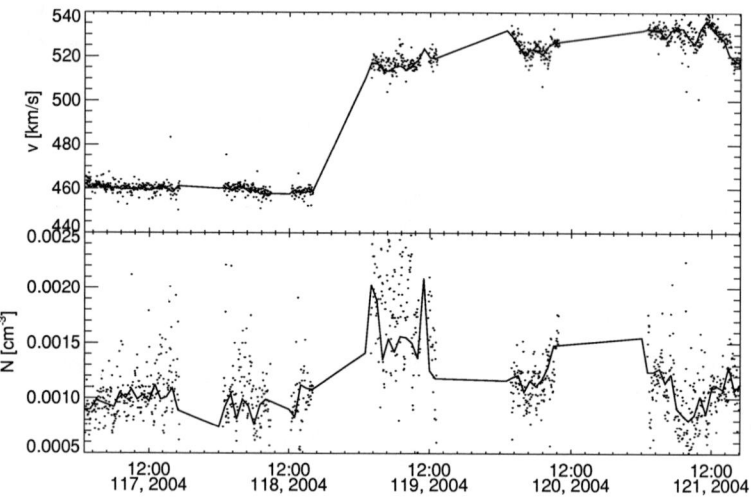

FIGURE 3. Plasma bulk speed and number density for the Halloween event from Voyager 2. Points show high-resolution measurements, and lines connect hourly averages.

off. This model prediction probably is typical of what one might (naively) expect to see. However the plasma density measurements, in particular, do not match the expectations.

Figure 3 shows a few days of the plasma parameters for the Halloween event as seen by Voyager 2 (note that figures 3 and 2b show observations only and *not* model predictions). The speed jump is quite clear, as is the data gap spanning the shock's arrival. The density measurements are considerably more scattered than the speed, and there is a brief, but distinct, increase directly following the shock.

The propagation of the Halloween event into the outer heliosphere was numerically modelled by Richardson et al. [3]. Data from 1 AU was used as input, but this time a more simple, though similar, 1-D model was employed. As can be seen in their paper [3], the model predictions for the plasma parameters were considerably more complex than the typical 'shark-fin' profiles (as predicted for the Bastille event), and the predicted timing was out by around 14 days.

The development of merged interaction regions (MIRs) from transient flows and streams in the solar wind (e.g. [4]) has been long established and much studied. The Bastille day merged interaction region seen by V2 (see [1]) was driven by a relatively short period ($\sim$ days) of activity on the Sun. The Halloween MIR, however, formed from a more extended active period ($\sim$ month). Whilst it might be expected that events in the outer heliosphere may lose 'memory' of their sources through merging and interactions, the modelling ([2] and [3]) suggests that more complex input leads to more complex output. As discussed earlier, even in the less complex case of the Bastille event, our predictions are still at odds with observations.

Unexpected observations are not restricted to such extreme events occuring once or twice per solar-cycle. In events seen by V2, in the ten years or so from the previous solar-cycle maximum to now, we can easily find a number of similarly atypical interplanetary shocks. Space prevents us from including plots here, yet we point the interested reader to the following dates (DoY-Year) in the V2 data set as examples: 170-1992, 186-1993, 24-1994, 206-1994, 82-2002, 64-2003, 4-2004. In all these cases (spanning radial distances of 38 to 72 AU) we see a discontinuity in the bulk speed accompanied with little or no signature in plasma density; again quite different to the expectation of a 'typical' event.

Although plots of the plasma temperature have been excluded for reasons of brevity, the temperature remains an important quantity to consider. For the Halloween and Bastille events at V2 there is no evidence of heating at the shocks. It must be remembered that both of these shocks arrived during data gaps, and so any brief periods of heating may have gone unobserved. Considering the further examples (whose dates were given in previous paragraph) we note that in none of the cases was there evidence of heating at the shock. However the majority of events do show periods of increased temperature following a few days after the jump in bulk speed, suggesting that the shocks have moved ahead of material they previously heated.

CONCLUSIONS

Solar cycle 23 produced two spectacular spells of solar activity: the Halloween event and the Bastille event. After travelling 60-70 AU, the signatures of these events were observed by Voyager 2. Although jumps in the solar wind speed were observed pretty much as expected for both events, in both cases the plasma densities were unusual. In the particular case of the Bastille event a strong density enhancement appeared *before* the shock and disappeared again shortly after. We have no explanation of this behaviour. Other large shocks observed by V2 in the preceding ten years show numerous examples of events with no discernible signature in the density measurements at the shock. Similarly these events show no heating at the shocks, but do show temperature increases a few days thereafter. Plasma conditions at large distances and for large events are frequently not as might typically be expected.

ACKNOWLEDGMENTS

This work was supported under NASA grant NAG5-11623 and NASA contract 959203 from the Jet Propulsion Laboratory to the Massachusetts Institute of Technology.

REFERENCES

1. Burlaga, L. F., Ness, N. F., Richardson, J. D., and Lepping, R. P., Sol. Phys., **204**, 399–411 (2001).
2. Wang, C., Richardson, J. D., and Burlaga, L., Sol. Phys., **204**, 413–423 (2001).
3. Richardson, J. D., Wang, C., Kasper, J. C., and Liu, Y., Geophys. Res. Lett., **32**, 3 (2005).
4. Burlaga, L. F., Goldstein, M. L., McDonald, F. B., and Lazarus, A. J., J. Geophys. Res., **90**, 12027 (1985).

Initial Comparison Between a 3D MHD Model and the HAFv2 Kinematic 3D Model: The October/November 2003 Events from the Sun to 6 AU

Devrie S. Intriligator[1], Thomas Detman[2], Murray Dryer[2,3], Craig D. (Ghee) Fry[3], Wei Sun[4], Charles Deehr[4], and James Intriligator[1,5]

1. Carmel Research Center, P.O. Box 1732, Santa Monica, CA 90406 USA
2. NOAA/Space Environment Center, 325 Broadway, Boulder, CO 80305 USA
3. Exploration Physics International, Inc., Huntsville, AL 35806, USA
4. Geophysical Institute, University of Alaska, Fairbanks, AK 99775, USA
5. Brigantia Building, University of Wales, Bangor, Wales, LL572AS, UK

Abstract. A first-generation 3D kinematic, space weather forecasting solar wind model (HAFv2) has been used to show the importance of solar generated disturbances in Voyager 1 and Voyager 2 observations in the outer heliosphere. We extend this work by using a 3D MHD model (HHMS) that, like HAFv2, incorporates a global, pre-event, inhomogeneous, background solar wind plasma and interplanetary magnetic field. Initial comparisons are made between the two models of the solar wind out to 6 AU and with *in-situ* observations at the ACE spacecraft before and after the October/November 2003 solar events.

INTRODUCTION

At last year's IGPP meeting, we began to discuss the possibility that effects from solar disturbances may have been in part responsible for the differences in the energetic particle measurements at Voyager 1 and Voyager 2 in August 2002. Our 3D HAFv2 results suggested that asymmetric propagation of solar events affect the dynamics of the outer heliosphere [1]. The 3D HAFv2 results suggested that in 2002, these effects - and not solely the proximity to the termination shock – contributed to the differences in the Voyager 1 and Voyager 2 energetic particle observations.

There were nineteen significant solar events between October 19, 2003 and November 20, 2003 (the "Halloween 2003" solar events). This complex system of Halloween 2003 events presents a challenge for modeling the interplanetary propagation of solar disturbances to the outer heliosphere. They also provide a special opportunity for benchmarking the interplanetary effects of these events and for determining their influence on the outer heliosphere. In Intriligator et al. [2], we used the HAFv2 model to study the propagation of these events throughout the heliosphere. We found that, while the HAFv2 model yielded many important results and insights, it would be helpful to compare its results with those of a full 3D MHD model.

In the present paper, we show some of the results from the full 3D MHD model within 6 AU. We show the time series results at ACE from both the HAFv2 model and the full 3D MHD model. In addition, we show ecliptic plane results out to 6 AU from both models for the "background" solar wind and interplanetary magnetic field (IMF) prior to the Halloween events and for the disturbed interplanetary medium after the Halloween events. Due to the limited page allowance here, in the next two sections, we discuss very briefly the HAFv2 model and then more completely the 3D MHD model. The results and conclusions are presented in the last two sections.

HAFV2 MODEL

The Hakamada – Akasofu - Fry version 2 model (Fry et al. [3, 4, 5] is a 3D kinematic simulation that inputs solar data at 2.5 Rs. The HAFv2 model is successfully used in the real time "Fearless Forecasts" from the Sun to Earth and Mars [6]. The HAFv2 model was used in real-time during the exceptional Bastille 2000 and Halloween 2003 storm intervals to predict shock arrival times at Earth (Dryer et al. [6, 7]). The model includes stream/stream interactions (Intriligator et al. [1, 2]). The details of this model are discussed in Intriligator et al. [1, 2] and references therein.

THE 3D MHD MODEL

The Hybrid Heliospheric Modeling System (HHMS) is a Sun to Earth system of coupled models designed primarily for real-time prediction of geomagnetic activity (Detman et al. [8]). The key features of the HHMS relevant to this paper are that it contains a set of empirical relationships (analogous to those in HAFv2) that translate the output of the Wang-Sheeley-Arge Source Surface (SS) model (Arge and Pizzo, [9]) into time- dependent lower boundary conditions for the IGMV (Interplanetary Global Model Vectorized) 3D MHD solar wind model ([10, 11]). The SS model is in routine daily operation at the NOAA Space Environment Center (SEC). Driven by daily solar magnetograms via the SS model, the HHMS gives the state of the slowly evolving background solar wind within the inner heliosphere from 45 degrees South latitude to 45 degrees North and for 365 degrees longitude, including stream-stream interactions and co-rotating interaction region (CIR) build-up. In addition to the time-dependent boundary condition driven by the SS model, the HHMS allows for interplanetary shock initiation at 0.1 AU based on other solar observations such as solar flares, Type II radio sweeps, and/or coronagraph observations of CMEs.

COMPARISON OF THE MODELS

The HAFv2 and HHMS models are quite different in their internal methods. The HAFv2 model is essentially a 2D (latitude, longitude) array of 1D (radial) computations; it operates by emitting pseudo particles from a grid of points fixed in Carrington coordinates. In inertial coordinates, however, the pseudo particles act like beads on a string. Their properties and interactions are designed to give conservation of mass, momentum, and magnetic flux. Also, their interaction parameters are tuned to agree with 1D MHD in the propagation of shocks. In contrast, the HHMS takes a continuum approach. It approximates the partial differential equations of MHD using the two-step Lax-Wendroff scheme [10]. It conserves mass, momentum, magnetic flux, and energy.

The HAFv2 and HHMS also have differences in their specification of lower boundary conditions. These differences, however, are not fundamental; both models have a background solar wind driven by SEC's SS maps, and both models superimpose shock inputs on their background boundary conditions based on the Fearless Forecast inputs. For HAFv2 this process uses the estimated metric Type II shock speed as a parameter in an exponential plasma speed profile that is superimposed upon the background SS model. HHMS, on the other hand, uses the Type II shock speed to compute the sonic Mach Number, hence, the Rankine-Hugoniot jump conditions of velocity, etc., at 0.1 AU for a time period suggested by the proxy piston driving time [2]. The Fearless Forecast information is shock time, flare location on the solar disk, shock speed (derived from Type II frequency sweep speed), and piston driving duration (derived from GOES X-ray flare time profile).

RESULTS

Figure 1 shows the HAFv2 results at 1 AU (Intriligator et al., [2]). The times of the simulated shock arrivals have been tuned [2] to optimize agreement with the observed (Skoug et al., [12]) shock arrivals at ACE. To begin using the HHMS for the study of shock propagation to the outer heliosphere, we focused on the October and November 2003 (Halloween) events and iteratively fine-tuned the shock inputs for agreement with ACE observations. Figure 2 shows comparisons of the HHMS simulated time series of solar wind proton speed (V), density (n), temperature (T), entropy (S), and IMF magnetic field magnitude (/B/) with corresponding ACE observations. ACE Level 2 science data with 96s resolution was combined with solar wind plasma velocity by Skoug et al. [12] for the large October 29th shock. Note the gap in density at this time due to energetic particle bombardment of the SWEPAM. These data were block averaged to 15-minute resolution, a rough match to the time step of the HHMS. ACE observations are plotted as blue symbols, the HHMS output as red, and differences are filled in green to guide the eye. The HHMS was then extended and the simulated shocks were tracked to 6 AU.

Comparison of the observed plasma speed (Figures 1 and 2) with the two models is quite good despite the unusually extreme conditions during the 10/28 to 11/05 period. The post 10/29 shock's plasma speed (HAFv2) exceeds the observed speed due to the initialization parameter used at 2.5Rs. We also note that the otherwise reasonable temperature comparison (Figure 2, panel c) of HHMS with the observations has several extreme excursions to low values. This, we believe, is due to the omission in the HHMS' ideal 3D MHD code of thermal conduction and Alfven wave damping.

Figure 3 shows the solar wind magnetic field in the ecliptic plane as calculated using the HAFv2 model out to 6 AU before (October 12, 2003) and after (November 06, 2003) the Halloween 2003 events. For comparison with the HHMS, the polarity of the field is color-coded (blue toward the Sun, red away from the Sun), and the density of the magnetic field lines is proportional to the plasma density.

The four panels in Figure 4 show solar wind plasma radial velocity (Vr) and IMF polarity before (October 12, 2003) and after (November 06, 2003) the Halloween 2003 events. In the ecliptic plots of Figure 4 the changes are evident in the configuration of the

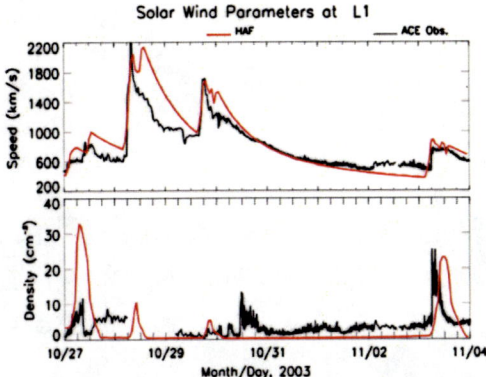

FIGURE 1. The arrival of the Halloween 2003 shocks at ACE on October 28, 29, 30, and November 4, 2003 are shown in the solar wind data and the HAFv2 results. The timings of the shock arrivals and the magnitudes of the associated speed jumps in the predictions of the HAFv2 model are very similar to the ACE speed observations at 1 AU. The Vs values in Table 1 of Intriligator et al. [2] were tuned to optimize *only* the agreement between the shock arrivals in the model results and in the ACE observations.

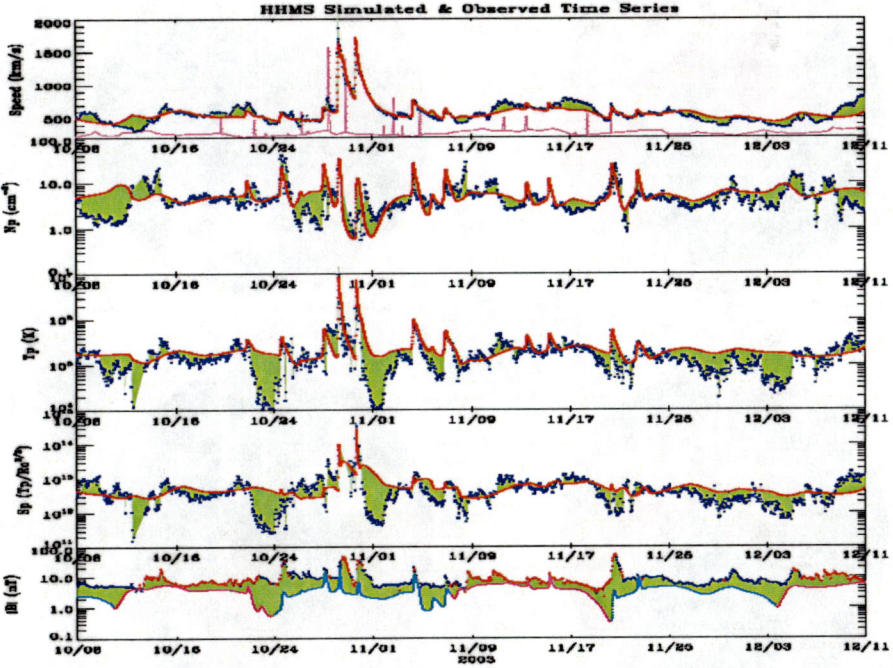

FIGURE 2. Comparison of HHMS simulated time series with ACE observations. ACE data are plotted as blue symbols, these represent 15-minute block averages as described in text. HHMS simulation output is plotted as red dots, one for each time step. Differences are filled in green to guide the eye. (a) Speed, (b) Density, (c) Temperature,(d) Entropy, (e) Magnetic field strength. Panel (a) also shows (in magenta) the model sub-Earth boundary condition speed at 0.1 AU. Panel (e) also indicates the IMF polarity changes in the red-to-blue (and vice versa) colors.

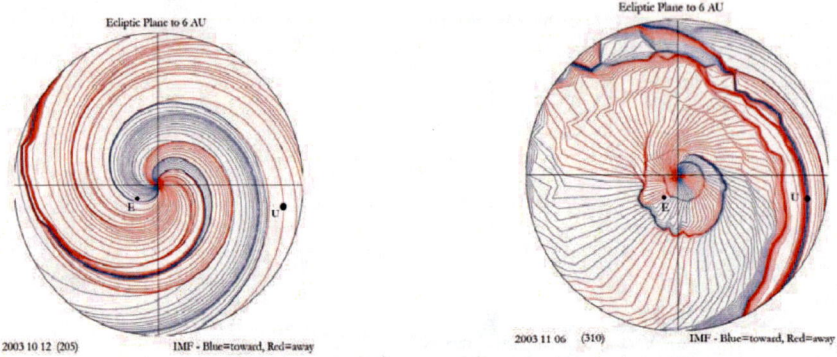

FIGURE 3. Showing the solar wind magnetic field in the ecliptic plane as calculated using the HAFv2 model out to 6 AU before (October 12, 2003) and after (November 06, 2003) the Halloween 2003 events. E and U indicate Earth's and Ulysses' locations.

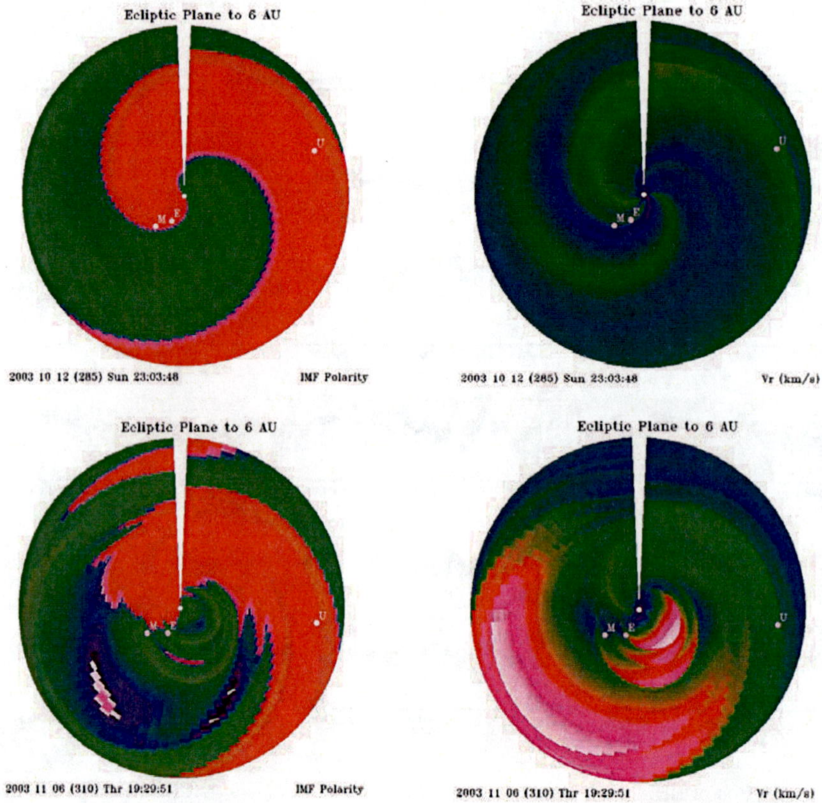

FIGURE 4. Ecliptic plane HHMS plots of the solar wind IMF polarity (left panels) and radial velocity (Vr) (right panels). Before/after the Halloween 2003 events: upper panels (October 12, 2003) and lower panels (November 06, 2003). The small circles show the locations of Earth, Mars, and Ulysses (at 5.23 AU).

interplanetary medium resulting from these Halloween events. This initial global comparison of the two models is encouraging. The HAFv2 model is shown to be an invaluable space weather tool with the HHMS model providing a more detailed examination of plasma and field conditions.

CONCLUSIONS

This work presents the first comparison of the HAFv2 3D kinematic model with a 3D MHD solar wind model. The IMF structure simulated by the HAF and HHMS models is remarkably consistent both with the ACE observations and between the simulations. In addition, the changing and asymmetric distribution of shocked solar wind is well represented by both models. The asymmetric propagation of solar events affects the dynamics of the outer heliosphere. Voyager 1 and Voyager 2 data comparisons with the 3D HAFv2 kinematic simulations and the 3D MHD simulations will enable us to achieve greater understandings of the 3D dynamics of the outer heliosphere and its interaction with the interstellar medium.

ACKNOWLEGMENTS

We thank R. Skoug and C.-C. Wu for the ACE plasma data, and the NSSDC for the available solar wind, IMF, and trajectory information. The work by DSI and JI was supported by Carmel Research Center. The work by TD, MD, CDF, WS, and CSD was supported by the DoD project, University Partnering for Operational Support (UPOS), and by NASA's Living With a Star Targeted Research and Development Program.

REFERENCES

1. Intriligator, D. S., M. Dryer, W. Sun, C.D. Fry, C. Deehr, and J. Intriligator, *Physics of the Outer Heliosphere*, AIP, V. Florinski, N. Pogorelov, and G. Zank, editors (2004).
2. Intriligator, D.S., W. Sun, M. Dryer, C. Fry, C. Deehr, and J. Intriligator, *J. Geophys. Res.*, in press (2005).
3. Fry, C.D., W. Sun, C.S. Deehr, M. Dryer, Z. Smith, S.-I. Akasofu, M. Tokumaru, and M. Kojima, *J. Geophys. Res.* **106**, 20,985-21,001 (2001).
4. Fry, C. D., W. Sun, C. S. Deehr, M. Dryer, Z. Smith, and S.-I. Akasofu, in *Solar-Terrestrial Magnetic Activity and Space Environment*, (H. Wang and R. Xu, Eds.), *COSPAR Colloquia Series on Physics and Astronomy*, 401-407 (2002).
5. Fry, C. D., M. Dryer, Z. Smith, W. Sun, C. S. Deehr, and S.-I. Akasofu, *J. Geophys. Res.* **108**, 1070, doi:10.1029/2002JA009474 (2003).
6. Dryer, M., Z. Smith, C.D. Fry, W. Sun, C. S. Deehr, and S.-I. Akasofu, *Space Weather* **2**, S09001, doi:10.1029/2004SW000087 (2004).
7. Dryer, M., C. Fry, W. Sun, C. Deehr, Z. Smith, S.-I. Akasofu, and M. Andrews, *Solar Phys.* **204**, 267-286 (2001).
8. Detman, T., C. Arge, V. Pizzo, Z. Smith, M. Dryer, and C. Fry, submitted *J. Atm. Solar Terr. Phys.* (2004).
9. Arge, C.N. and V.J. Pizzo, *J. Geophys. Res.* **105** (A5), 10465-10480 (2000).
10. Han, S.M., S.T. Wu, and M. Dryer, *Computers and Fluids* **16**(1), 81-103 (1988).
11. Detman, T.R., M. Dryer, T. Yeh, S. Han, S.T. Wu, and D.J. McComas, *J. Geophys. Res.* **96**, 9531-9540 (1991).
12. Skoug, R.M., J.T. Gosling, J. T. Steinberg, D. J. McComas, C.W. Smith, N. F. Ness, Q. Hu, and L.F. Burlaga, *J. Geophys. Res.* **109**, A09102, doi:10.1029/2004JA010494 (2004).

SESSION 6
OUTER SHOCK RELATED PHENOMENA

3-D Hybrid Simulation of Quasi-Parallel Bow Shock and Its Effects on the Magnetosphere

Y. Lin and X. Y. Wang

Physics Department, Auburn University, Auburn, AL 36849-5311, USA

Abstract. A three-dimensional (3-D) global-scale hybrid simulation is carried out for the structure of the quasi-parallel bow shock, in particular the foreshock waves and pressure pulses. The wave evolution and interaction with the dayside magnetosphere are discussed. It is shown that diamagnetic cavities are generated in the turbulent foreshock due to the ion beam plasma interaction, and these compressional pulses lead to strong surface perturbations at the magnetopause and Alfven waves/field line resonance in the magnetosphere.

Keywords: bow shock-magnetosphere coupling, pressure pulses, Alfven waves.
PACS: 52.35.Tc; 52.35.Mw; 52.35.Bj; 52.35.Hr; 52.65.Rr; 52.65.Ww.

INTRODUCTION

The Earth's bow shock is of great interest in space plasma investigation as the bow shock contains important physics ranging from kinetic to global scales. Moreover, the electromagnetic waves, often turbulent, emitted by the shock are strongly coupled to the global plasma environment of the magnetosphere. Since the shock normal angle is small at the quasi-parallel shock, the downstream thermal ions, as well as the ions reflected at the shock front, can easily backstream along the field lines and penetrate into the upstream region, leading to electromagnetic instabilities associated with ion beams, ion heating, and energetic ions at quasi-parallel shocks [1][2]. Pressure pulses are also frequently generated due to the shock intrinsic processes [3][4]. These foreshock pressure pulses and waves are carried by the solar wind to the Earth's dayside magnetopause and cause surface waves at the magnetopause [5] and ULF pulsations [6] in the magnetosphere.

Hybrid simulations have proven to be a powerful tool for the investigation of the nonlinear structure and ion heating of collisionless shocks [7][8]. Indeed, the structure of the quasi-parallel shock is found to evolve dynamically due to the wave-particle interaction. The shocks are found to go through a reformation process constantly and generate large-amplitude electromagnetic waves in the upstream and downstream regions due to reflected/backstreaming ions. Moreover, our two-dimensional (2-D) global-scale hybrid simulation for the curved bow shock-magnetosphere system [9] shows that diamagnetic cavities, with a low-density and low-magnetic field center bounded by a rim of high density and high magnetic field, are present in the foreshock region of the quasi-parallel shocks. The frequent presence of crater-like diamagnetic cavities in quasi-parallel shocks has also been observed by satellites [10][11].

The global structure the bow shock and the coupling of its waves with magnetosphere system, however, are of three-dimensional (3-D) nature. In this paper, we briefly summarize the initial results of our 3-D global-scale hybrid simulation of the curved bow shock. The focus is on the dynamic processes associated with the quasi-parallel shocks. The interaction of the pressure pulses with the magnetopause, which cannot be studied by the previous 2-D simulation, is also discussed in the present paper. The detailed simulation study will be published elsewhere.

SIMULATION RESULTS

In the hybrid code, the ions (protons) are treated as discrete particles, and the electrons are treated as a massless fluid. Quasi charge neutrality is assumed in the simulation. The description of the hybrid code can be found in *Swift* [12]. The 3-D global-scale code used here was developed by extending the 2-D model [9][12] for the dayside bow shock-magnetosheath-magnetosphere system to 3-D. The simulation domain contains the plasma regions with GSM $x > 0$, within geocentric distances $4R_E \leq r \leq 22R_E$. A spherical coordinate system is used in the simulation. A semi-cone of $20°$ polar angle around the positive and negative polar axes, which are chosen along the GSM $\pm y$ axes in the calculation, is cut out from the domain to avoid the singular coordinate line along the polar axes. For presentation, we use the convention terms of polar and azimuthal (longitudinal) directions in the geomagnetic coordinate system.

The ion gyrofrequency Ω_0 in the solar wind is chosen to be $1.0 s^{-1}$. The solar wind ion inertial length λ_0 is chosen to be $0.1R_E$. Non-uniform grid spacing Δr is used in the r direction, with a higher spatial resolution $\Delta r \approx \lambda_0$ for $r > 8R_E$. About 150 - 400 particles are used in each cell in the solar wind and the magnetosheath.

The bow shock, magnetosheath, and magnetopause form from the interaction between the superfast solar wind and the geomagnetic field. Outflow boundary conditions are used at the boundaries near $x = 0$. A perfect conducting boundary is imposed at the inner boundary at $r = 4R_E$. In the following, the magnetic field is normalized to the IMF B_0, the ion densities are normalized to the solar wind density N_0, the flow velocities are normalized to V_{A0}, and the time is normalized to Ω_0^{-1}.

For comparison with the previous 2-D simulation [9], we present the simulation results for a case in which the IMF is pointing along the Sun-Earth line, with $B_{x0} = B_0$ and $B_{y0} = B_{z0} = 0$. In such case quasi-parallel shocks occupy a large portion of the dayside domain. The solar wind Mach number $M_A = 5$. The ion $\beta_i = 0.5$.

Figure 1 shows the contour plots of the ion particle number density N_p (left column) and magnetic field strength B (right column) obtained at $t = 100$, 105, 110, and 115. The contours are shown simultaneously for a quarter of the noon-midnight meridian plane (vertical plane, for $x > 0$ and $z > 0$) and equatorial plane (horizontal plane, for $x > 0$ and $y > 0$). The average standoff distance of the bow shock, which is ~ $12-13R_E$, is reached at $t \approx 20$, while the overall shape of the curved shock front is saturated at $t \approx 30$. The shock front can be identified from the enhanced density and

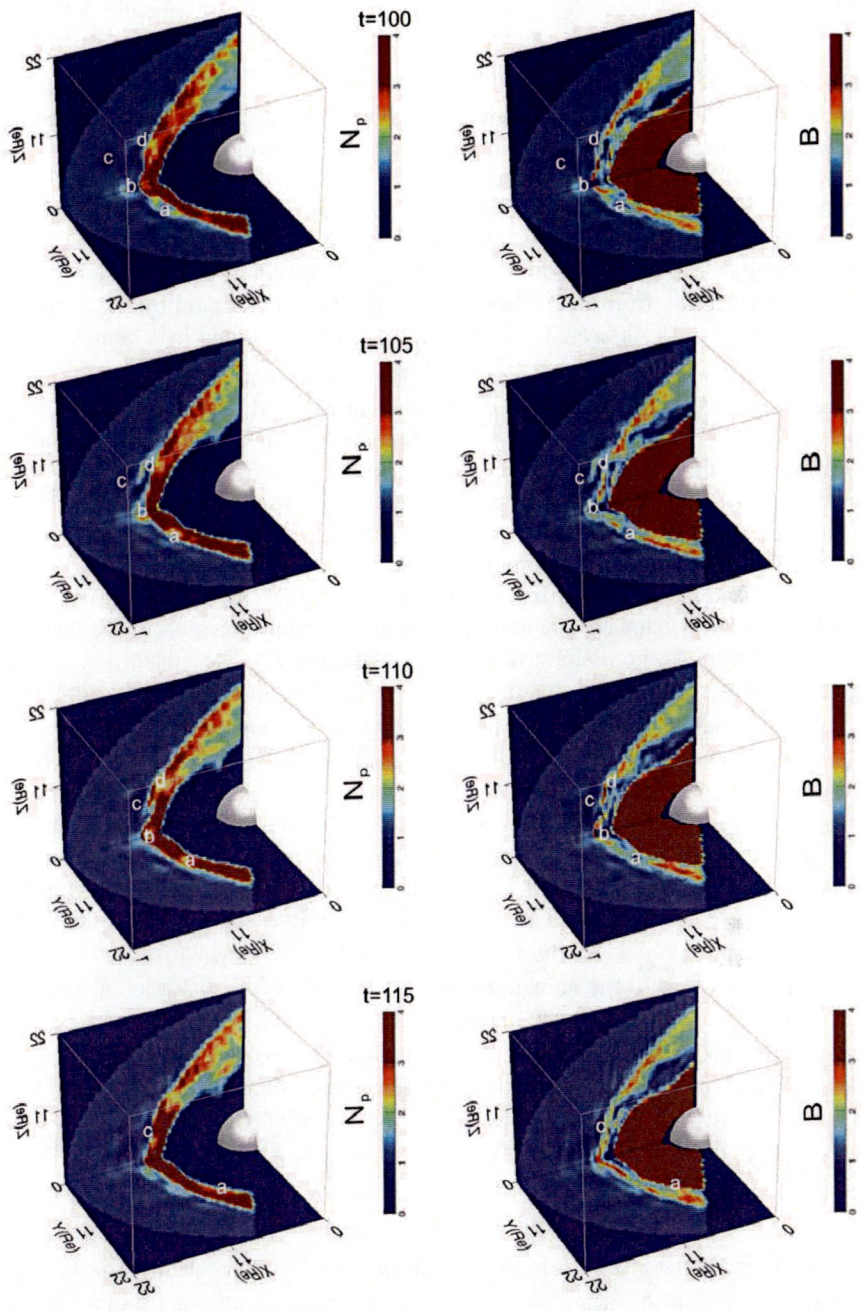

FIGURE 1. Contours of the magnetic field (B) ion particle number density (N_p) obtained at various times in the noon-midnight meridian and equatorial planes.

magnetic field strength. The magnetopause is located around $r = 10R_E$ along the Sun-Earth line, where the density N_p suddenly decreases and the B field greatly increases.

In the foreshock region of the quasi-parallel shocks ($|y|<15R_E$ and $|z|<15R_E$), the backstreaming and reflected ion beams interact with the denser and colder incoming solar wind plasma, leading to large-amplitude whistler-type electromagnetic waves (to be discussed below) with wave vector **k** nearly parallel to **B** in the solar wind frame of reference. The shock front, meanwhile, is going through a constant reformation, or steepening, process due to the transient waves, as shown in previous simulations. In addition, similar to the 2-D simulation by *Lin* [9], diamagnetic cavities are present in this foreshock region, from $t > 20$. Some of the cavities are marked by ``a'', ``b'', ``c'', and ``d'' in Figure 1. As seen for $t = 100$, the spatial variations in B and N_p are in phase and well correlated in these cavities. The cavity contains a decreased density and decreased B field (by 15–40%), and is often surrounded by enhanced density and magnetic field strength relative to the average values in the solar wind. The ion temperature, on the other hand, is enhanced inside the cavity, while the flow speed decreases. The diamagnetic cavities are essentially a pressure balance structure.

The foreshock cavities convect to the downstream magnetosheath. At $t = 105$, the cavity ``a'' has propagated into the magnetosheath. The convection of cavities ``b'', ``c'', and ``d'' are also tracked with time in Figure 1. As the cavity/rim approaches the magnetopause and is slowed down, it behaves as a dynamic pressure pulse that leads to the expansion/compression of the magnetopause. The distortion of the magnetopause is seen in the equator around $y \sim 9R_E$. The size of the distortion due to the compressional wave pulse is about $3-6R_E$ in the east-west and north-south directions. Such local pulse corresponds to a broadband wave spectrum. The arrival of the compressional waves, since much earlier times, causes the compressional and transverse waves in the magnetosphere, as to be discussed below. The foreshock diamagnetic cavities are seen to cause maximum surface perturbations among all the arrival compressional structures. The magnetopause surface perturbations mainly propagate in the east-west direction. At $t = 110$, the previous surface perturbation at the magnetopause has propagated to $y \sim 10R_E$. At $=115$, it has gained a larger tailward speed and moved toward the domain boundary. A new surface distortion also appears at the magnetopause ($y \sim 9R_E$) due to a newly arrived cavity.

Meanwhile, the foreshock cavities and their rims also evolve to structures elongated along field lines, with alternate field-aligned filaments (enhanced density) and cavity-like areas (decreased N_p), as seen at $t = 115$. These 3-D but elongated structures, with wave vector $\mathbf{k} \perp \mathbf{B}$, have a width of $1-4R_E$ in both y and z directions. The presence of the field-aligned structures has also been found in the 2-D simulation by *Lin* [9].

Figure 2 shows the spatial profiles of B, N_p, magnetic field component B_y, and ion flow component V_{py} along the Sun-Earth line, in a time sequence from $t = 20$ to 100. Large-amplitude electromagnetic waves can be seen in the shock and its vicinity. The wave particle interaction leads to the shock ion heating and diffuse foreshock ion distribution in phase space. During the re-formation cycle of the shock, the shock position oscillates within a width of $\sim 1-2R_E$. The period of the reformation cycle is

$\sim 20\Omega_0^{-1}$. The beam whistler waves generated in the foreshock, as seen in B_y and V_{py}, are right-hand polarized in magnetic field in the solar wind plasma frame of reference. Smaller amplitude perturbations in B and N_p are also seen. These waves propagate sunward with a speed $\sim 2V_{A0}$, in the solar wind frame of reference.

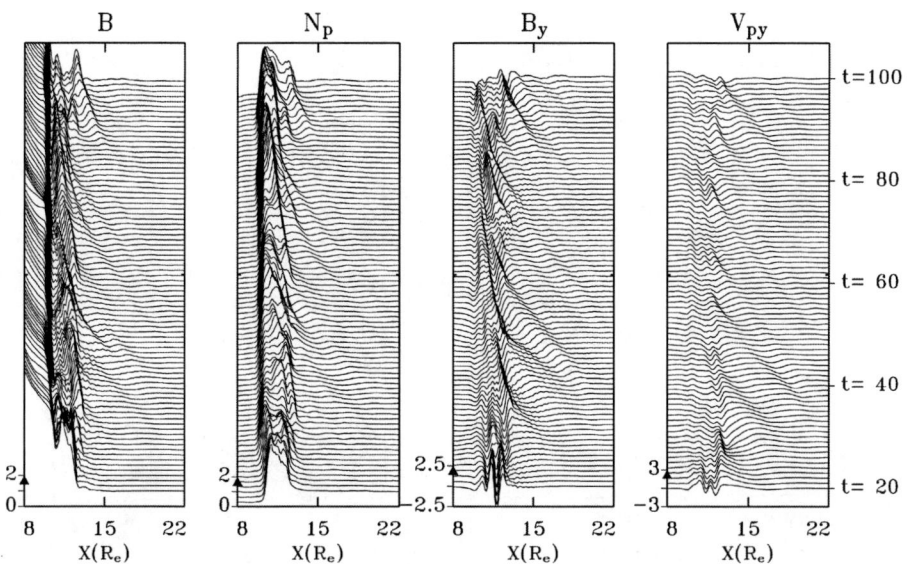

FIGURE 2. Spatial profiles of the magnetic field and plasma quantities along the Sun-Earth line, at various times.

Meanwhile, the diamagnetic cavity structures in **B** and N_p are present. They are seen more clearly within $r < 15R_E$, with amplitude $\Delta N/N \sim \Delta B/B \sim 20\%$ and parallel wavelength $\sim 3R_E$ in Figure 2. They also propagate sunward with a speed of $2V_{A0}$ in the solar wind frame, and thus move Earthward with a speed of $3V_{A0}$ in the Earth frame of reference. The waves are slowed down when they approach the shock front. The magnetopause position also oscillates in response to the foreshock pressure pulses.

As the pressure pulses approach the magnetopause, part of the compressional wave packets are transmitted into the magnetosphere. In addition, the compressional waves are coupled to the transverse Alfven mode near the magnetopause boundary and in the magnetosphere, where the magnetic field and/or density are highly nonuniform [13][14]. First, shear Alfven waves are generated at the magnetopause and in the magnetosphere due to the Alfven resonance condition. The generation of these shear Alfven waves follows the arrival of the compressional waves. As predicted by the above theories and reported by many observations [15], the process leads to the resonance of closed field lines in the dipole field, as shown in Figure 3. Second, mode conversion to short-wavelength kinetic Alfven waves (KAWs) [16][17] is seen near the magnetopause boundary, with a local enhancement of the parallel electric field $E_\parallel$

(due to electron $\nabla_\parallel p_e$). These waves appear later than the shear Alfven waves as it takes time for the energy to build up for $k_\perp \rho_i \sim 1$, where ρ_i is the ion Larmor radius. The detailed examination of the KAWs will be presented elsewhere.

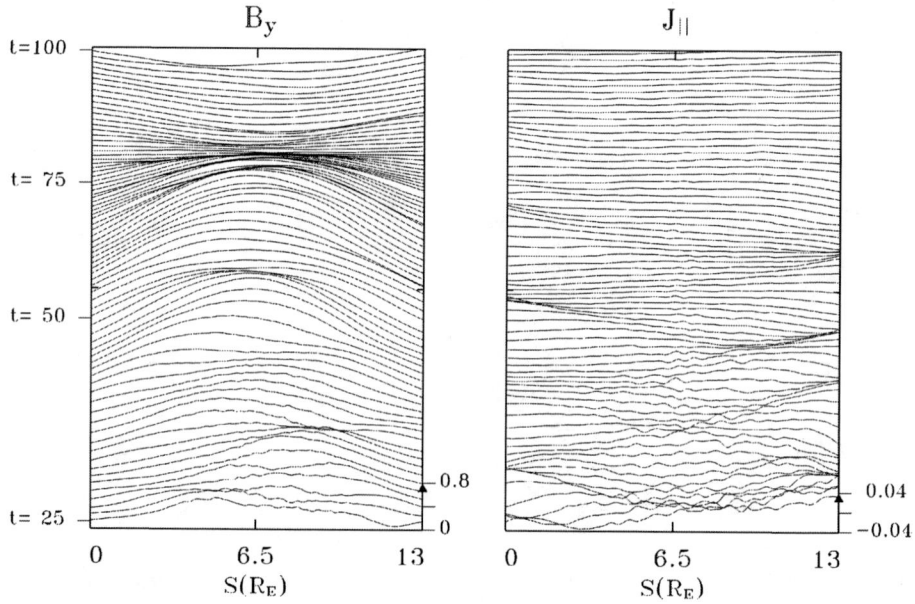

FIGURE 3. Spatial profiles of the azimuthal (east-west) magnetic field component and field-aligned current density along the field lines passing through $L = 7.5$ in the noon-midnight meridian plane, as a function of the distance S from the southern cusp to north along the field line, at various times.

Figure 3 shows time sequence of the azimuthal (longitudinal) field component, $B_y = -B_\phi$, and the parallel current density $J_\parallel$ along the field lines through approximately geocentric radius $L = 7.5$ in the noon-midnight meridian plane as a function of S, where S is the distance along the field line, with the starting point $S = 0$ ($13R_E$) approximately at the $r = 4R_E$ boundary in the south (north). Note that the tick marks on the right-side vertical axes are only for $t = 25$. During the constant arrivals of the compressional wave pulses at the magnetosphere, multiple-$k_\parallel$ perturbations in the transverse B_ϕ, as well as in $J_\parallel$ corresponding to the transverse distortion of field lines, are present and bounce between the south and north boundaries in $t < 50$. At $t > 50$, the nearly standing wave pattern is reached, and the wavelength corresponds to a 1/2 wave number between the boundary ends. Similar evolutions/patterns of the transverse field are also seen at all the other meridians, but the times to reach the stationary wave pattern are different. This smallest, fundamental resonance wave number, with $k_\parallel \Delta S \approx \pi$, may partly be due to that the source perturbations from the foreshock are mainly located near the equator, while the spectrum being broadband. Correspondingly, the resonance frequency is found to satisfy $\omega \approx V_A k_\parallel$, consistent with the resonance period $T \approx 50$ in the simulation. At $t = 100$, the field oscillation starts another cycle of the resonance. The oscillation of the magnetic field in the field line resonance is seen predominantly in the azimuthal (toroidal) direction. Different from previous MHD simulations [18], this field line resonance is driven by the compresssional wave pulses that are self-consistently and constantly generated at the foreshock of the quasi-parallel shocks. The driving sources are distributed globally, with multiple wave pulses arriving at different times and spatial locations, while each pulse has a local azimuthal extent.

In summary, our simulation shows that electromagnetic waves and diamagnetic cavities are generated in the foreshock of the quasi-parallel shock. The foreshock compressional waves/pressure pulses cause surface perturbations when they arrive at the magnetopause, and mode convert to transverse Alfven waves in the magnetosphere.

ACKNOWLEDGMENTS

This work was supported by NASA grant NAG5-12899 and NSF grant ATM-0213931 to Auburn University. Computer resources were provided by the Arctic Region Supercomputer Center. The authors thank D. W. Swift for helps in the development of the 3-D hybrid code.

REFERENCES

1. Hoppe, M. M. et al., *J. Geophys. Res.*, **86**, 4471 (1981).
2. Fuselier, S. A. et al., *J. Geophys. Res.*, **100**, 17,107 (1995).
3. Fairfield, D. H. et al. *J. Geophys. Res.*, **95**, 3773 (1990).
4. Sibeck, D. G., et al., A. , *J. Geophys. Res.*, **106**, 21,675 (2001).
5. Sibeck, D. G., et al., J., *J. Geophys. Res.*, **108**, (2003).
6. Engebretson, M. J. et al., *J. Geophys. Res.*, **96**, 3441 (1991).
7. Winske, D., Omidi, N., Quest, K. B., and Thomas, V. A., *J. Geophys. Res.*, **95**, (18,821) 1990.
8. Dubouloz, N., and Scholer, M., *J. Geophys. Res.*, **100**, 9461 (1995).
9. Lin, Y., *J. Geophys. Res.*, **108** *(A11)*, SMP 3 (2003).
10. Wibberenz, G., Zollich, F., Fischer, H. M., and Keppler, K., *J. Geophys. Res.*, **90**, 283 (1985).
11. Sibeck, D. G., et al., *J. Geophys. Res.*, **106**, 21,675 (2001).
12. Swift, D. W., *J. Comput. Phys.*, **126**, 109 (1996).
13. Southwood, D. J., *Planet. Space Sci.*, **22 (3)**, 483-491 (1974).
14. Chen, L., and Hasegawa, A., *J. Geophys. Res.*, **79**, 1024 (1974).
15. Walker, A. D. M., and Greenwald, R. A., *J. Geomagn. Geoelectr. (Japan)*, **32 (suppl. 2)**, 111 (1980).
16. Hasegawa, A., and Chen, L., *Phys. Fluids,* **19**, 1924 (1976).
17. Johnson, J. R., and Cheng, C. Z., *Geophys. Res. Lett.*, **24**, 1423 (1997).
18. Lee, D-H., and Lysak, R. L., *J. Geophys. Res.*, **96**, 3479 (1991).

3D Global Simulation of the Interaction of Interplanetary Shocks with the Magnetosphere

Chi Wang*, Zhaohui Huang*, Youqiu Hu† and Xiaocheng Guo†

*Key Laboratory of Space Weather, Center for Space Science and Applied Research, Chinese Academy of Sciences
†School of Space and Earth Sciences, University of Science and Technology of China

Abstract.
Using the recently developed PPMLR-MHD code, we present 3D global simulation of the interaction of the interplanetary(IP) shock with the magnetosphere with emphasis on the the effect of shock orientation. We investigate its impact on the sudden commencement(SC) of geomagnetic storms and the magnetospheric responses. Generally speaking, a highly oblique shock causes asymmetric compression of the magnetosphere with respect to the noon-midnight meridian, and requires more time to compress the forward part of the magnetosphere, which producing longer rise time of SC. Even if the solar wind dynamic pressure changed similarly at a solar wind monitor, the responses of the magnetosphere could vary depending on the orientation of the causative IP shock. The effect of the shock orientation is of great help to understand space weather processes.

Keywords: Shock interaction, Numerical simulation
PACS: 96.50.Fm

INTRODUCTION

It is well known that the sudden commencement (SC) of geomagnetic storms may be caused by the global compression of the magnetosphere as a result of interplanetary disturbances such as shocks, dynamic pressure pulses etc. [1, 2]. *Chao and Lepping* [3] used a large experimental data base to show that a significant number of SC's are caused by shock waves and only a small number are caused by tangential discontinuities. *Takeuchi et al.* [4] found that the orientation of an interplanetary (IP) shock plays an important role in determining the SC field observed on the ground, and concluded that knowing the orientation is important for the space weather prediction.

The interaction of IP shocks with the bow shock and their transmission through the magnetosheath to the magnetopause, have been studied using multiple spacecraft observations [5, 6] and modeled mainly by gas dynamic modeling [7, 8]. However, to our knowledge, no systematic 3D global MHD simulation of the interaction between the magnetosphere and IP shocks with different orientations has been conducted yet. In this paper, much attention has been paid to the effect of the shock orientation on the interaction processes by use of a 3D global MHD approach.

NUMERICAL MODEL

Numerical method

Recently, *Hu et al.* [9] developed a so-called PPMLR-MHD code to implement the global simulation of the solar wind-magnetosphere-ionosphere system. The Lagrangian version of the piecewise parabolic method (PPMLR) developed by *Collela and Woodward* [10] is extended to MHD for solving 3D ideal MHD equations. A characteristic method similar to that proposed by *Dai and Woodward* [11] is employed to calculate fluxes at interfaces between numerical zones. The scheme has a formal accuracy of the third order in space and the second order in time. It has a negligible numerical dissipation, and can capture shocks/discontinuities within 1 or 2 mesh points.

We take a Cartesian coordinate system with Earth centered at the origin and the x-, y-, and z-axes pointing to the Sun, the dawn-dusk direction, and the north, respectively. The numerical box extends from -300R_E (where R_E is the Earth radius) to 30 R_E in x-direction, and from (-150,150)R_E in both y and z directions. The numerical box is divided into $160 \times 162 \times 162$ grid points. The grid spacing is constant (0.4 R_E) within a radial distance of 10R_E, and increases outside as a numerical series with a common ratio of 1.05 along each axis. As usual, the inner boundary is taken at 3R_E for the magnetosphere to avoid the complexity associated with plasmasphere and strong geomagnetic field.

Initial-Boundary conditions

The Earth's magnetic field is approximated by a dipole field with a dipole moment of $8.06 \times 10^{22} A\ m^{-1}$ in magnitude. The solar wind parameters are 400 km s^{-1} in speed, 5 cm^{-3} in number density, and 0.91×10^5 K in temperature, corresponding to a Mach number of 8. The interplanetary magnetic field (IMF) is taken to be a purely spiral field, i.e. **B** = (-3, 3, 0) nT. Such a quasi-steady magnetosphere is obtained in two steps. First, a quasi-steady magnetosphere solution is obtained for the due northward IMF case, which is symmetrical with respect to the noon-midnight meridional and equatorial planes. Then, a tangential discontinuity is introduced to interact with the magnetosphere, resulting in a new quasi-steady solution that is associated with the required spiral IMF.

At the inner boundary (3 R_E), the normal component of the deviation field **B**$'$ (which is defined as $\mathbf{B} - \mathbf{B}_d$, here $\mathbf{B}, \mathbf{B}_d$ are the total magnetic field and the dipole field, respectively) is 0, the tangential components are determined by equivalent extrapolation; the normal component of the flow velocity $v_n = 0$, and the tangential components are determined from the convection velocity $\mathbf{u} = \mathbf{E} \times \mathbf{B}_d / B_d^2$ in such a way that $\mathbf{v} = \mathbf{u} - u_n \mathbf{B}_d / B_{dn}$. Here **E** is the electric field mapped from the ionosphere to the inner boundary. A number density of 370 cm^{-3} and a temperature of 505 K are fixed at the inner boundary. At the inflow boundary($x = 30R_E$), all quantities are specified according to interplanetary conditions. At three free boundaries ($y, z = \pm 150 R_E$ and $x = -300 R_E$) all quantities are evaluated by equivalent extrapolation of except **B**$'$, which is evaluated from equivalent extrapolation of the total magnetic field **B**.

MODEL RESULTS

In an attempt to investigate the effect of the shock orientation, all IP shocks in this study are launched from the inflow solar wind region and have the same solar wind dynamic pressure of 2.98 nPa and a vanishing B_z downstream. However, they have different shock normal orientations: one parallel and the others oblique to the x-axis. The orientation angles (the angle between the shock normal and the x-axis) are $\lambda = 180^o$ (for the parallel case), $\lambda = 150^o, 135^o$, and 120°, respectively. All the shock normals are parallel to the x-y plane. Figure 1 shows the snapshots of the thermal pressure contours in the equatorial plane at different times ($t = 200, 400$ and $600\tau_A$, where τ_A is the Alfven time unit with a value of 0.94s) for each case (a-d represent cases in which the impingement IP shock has an orientation angle of $180^o, 150^o, 135^o$, and 120°, respectively).

The arrival of IP shocks causes the abrupt displacement of the whole bow shock-magnetopause system toward Earth, as would be expected. After colliding with the bow shock, the transmission shock propagating in the magnetosheath keeps its original planar configuration, and keep pace with its counterpart in the interplanetary space, implying that shocks propagate with about the same speed in two regions. This result is consistent with that obtained from the gasdynamic model of *Spriter and stahara* [8]. The impingement of an IP shock on the magnetosphere causes a complicated disturbances in the magnetosphere, with the wave pressure front projecting tailward within the magnetosphere because of a higher Alfven speed there.

In order to illustrate the effect of the shock orientation on the SC rise time, *Takeuchi* [4] introduced the concept "geoeffective magnetopause", and conjectured that there would be a border somewhere within which the geoeffective magnetopause is confined. The time for an IP shock to seep by the geoeffective magnetopause predominantly determines the SC rise time. The model results shown in Figure 1 demonstrate that a highly oblique shock require more time (in the order of minutes) to compress the forward part of the magnetosphere, producing longer rise time of SC. Thus, the shock orientation plays an important role in determining the SC rise time.

If an IP shock is inclined toward dusk (or dawn) side, compression of the magnetosphere would start at a point away from the subsolar point. The compression wavefront therefore propagates in the magnetosphere asymmetrically with respect to the noon-midnight meridian. Figure 2 shows the response of the magnetic field in the geostationary orbit ($\sim 6.6R_E$) to the magnetospheric compression due to the shock interaction. The magnetic field strength is plotted against local time (LT) at two different times ($t = 200$ and $400\tau_A$). The solid lines represents the case with a shock orientation angle of $\lambda = 180^o$ (case a), and the dotted lines indicates the case with $\lambda = 135^o$ (case b). In both cases, the magnetic field strength in the geostationary orbit at the Sun-Earth line jumps from 121 to 133 nT after the arrival of the IP shock. The magnetic field strength in case b increases more in the dusk side and less in the dawn side than those in case a, resulting from the asymmetric compression of the magnetosphere. Eventually, after the passage of the wave front to the magetotail, the dayside magnetosphere will be in a similar compressed state determined by the enhanced dynamic pressure behind the shock. Therefore, the effect of orientation is detectable only during the early part of the geomagnetic disturbance.

The model results show clearly the important role of the shock orientation on the

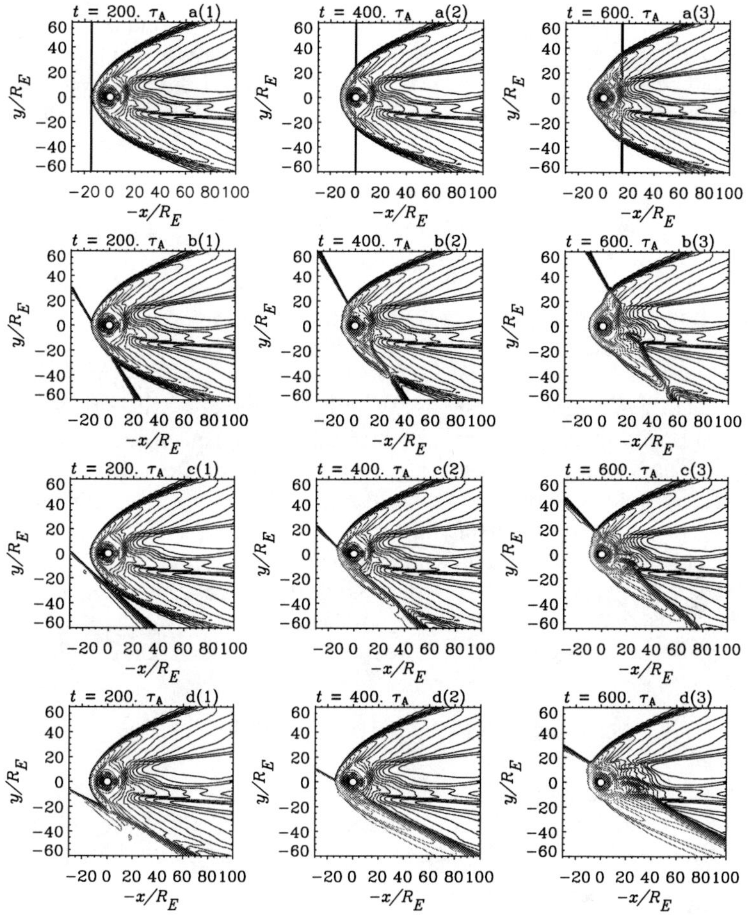

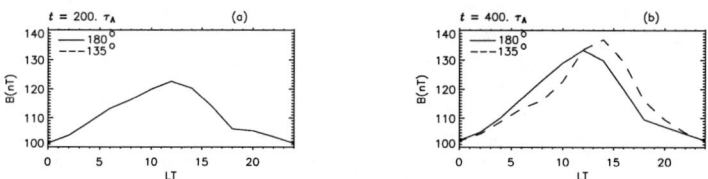

FIGURE 1. The interaction of IP shocks with different orientations between the magnetosphere. Shown are thermal pressure contours in the equatorial plane at different times.

FIGURE 2. Magnetic field strength in the geostationary orbit versus local time. In figure 2a, the solid and dashed lines are in coincide.

response of the magnetosphere. Even if the change in the solar wind dynamic pressure is abrupt at a single point, the magnetospheric response may be quite different. We can not predict magnetospheric response precisely based on the real time solar wind observations without information of the shock orientation. All these numerical results support the findings by *Takeuchi et al.* [4].

SUMMARY

Using the recently developed PPMLR-MHD code, we present 3D global simulations of the interaction of the interplanetary shock with the magnetosphere. Much attention has been paid to the effect of the shock orientation on the dynamical processes. The initial magnetosphere is in a quasi-steady state, embedded in a uniform solar wind and a spiral interplanetary magnetic field (IMF). As expected, the arrival of an IP shock causes a compression of the magnetosphere, and therefore an enhancement of the magnetic field strength. The shock propagating in the magnetosheath keeps its original planar surface. A highly oblique shock requires more time to compress the forward part of the magnetosphere, producing longer rise time of SC. Even if the solar wind dynamic pressure changed similarly at a solar wind monitor, the corresponding responses of the magnetosphere could vary significantly depending on the orientation of the causative interplanetary shock. Any study concerning the response of the magnetosphere to the interplanetary shocks cannot neglect the effect of the shock orientation.

ACKNOWLEDGMENTS

This work was supported by grants NNSFC 40325010 and 40474053. We are grateful to the Supercomputing Environment Construction and Applications Project of Chinese Academy of Sciences (INF105-SCE-02-15).

REFERENCES

1. B. T. Tsurutani, *J. Geophys. Res.*, **100**, 21717–21733 (1995).
2. T. Takeuchi, T. Araki, A. Viljanen, and J. Watermann, *J. Geophys. Res.*, **107**, 1096–1100 (2002).
3. J. K. Chao, and R. P. Lepping, *J. Geophys. Res.*, **79**, 1799–1807 (1974).
4. T. Takeuchi, C. T. Russell, and T. Araki, *J. Geophys. Res.*, **107**, 1423–1433 (2002).
5. A. Szabo, C. W. Smith, and R. M. Skoug, "The transition of Interplanetary Shocks through the Magnetosheath," in *Proceedings of the Tenth International Solar Wind Conference*, edited by M. Velli et al., American Institute of Physics, New York, 2003, pp. 782–785.
6. C. T. Russell et al. *J. Geophys. Res.*, **105**, 25143–25154 (2000).
7. W. W. Shen, and M. Dryer, *J. Geophys. Res.*, **77**, 4627–4644 (1972).
8. J. R. Spriter, and S. S. Stahara, "Computer modeling of solar wind interaction with Venus and Mars," in *The comperative Study of Venus and Mars: Atmospheres, Ionospheres and Solar Wind Interactions*, AGU monograph 66, Washington, D.C., 1992, pp. 345-383.
9. Hu et al., "Osillation of the Quasi-steady Earth's magnetosphere", to be submitted, 2005
10. P. Collela, and P. R. Woodard, *J. Comput. Phys.*, **54**, 174–201 (1984).
11. W. Dai, and P. R. Woodard, *J. Comput. Phys.*, **121**, 51–65 (1995).

Spiral shocks in astrophysical disks

W. K. M. Rice[*], G. Lodato[†] and P. J. Armitage[**,‡]

[*]*Institute of Geophysics and Planetary Physics and Department of Earth Sciences, University of California, Riverside, CA, 92521*
[†]*Institute of Astronomy, Madingley Road, Cambridge, CB3 0HA, UK*
[**]*JILA, Campus Box 440, University of Colorado, Boulder, CO 80309-0440*
[‡]*Department of Astrophysical and Planetary Sciences, University of Colorado, Boulder, CO 80309-0391*

Abstract. Spiral shocks waves are present in many astrophysical systems, including galactic disks, binary systems such as cataclysmic variables, AGN disks, and are probably present in disks around newly forming stars. In this paper we will discuss, in particular, spiral shocks resulting from the growth of a gravitational instability. We investigate how these spiral waves can transport angular momentum outwards and mass inwards - an important aspect of star formation - and a process that may play a role in the secular evolution of disk galaxies, leading to the formation of bulges. In some cases the instability can be sufficiently violent for the disk to fragment into gravitationally bound objects. This may explain the origin of the stellar population orbiting the galactic center, and has also been suggested as a mechanism for forming gaseous planets similar to Jupiter and Saturn. We consider the conditions required for fragmentation and whether such a process could indeed produce gaseous planets.

PACS: 97.10.Bt, 97.10.Gz

INTRODUCTION

Spiral shock waves are present in many astrophysical systems including galactic disks, disks present in binary systems such as cataclysmic variables, in disks at the center of active galaxies, and are probably present in disks around newly forming stars. In some cases these spiral waves are a consequence of a tidal interaction with some kind of companion. If, however, the disk is sufficiently massive, its own self-gravity may be important, and spiral waves may result from the growth of a gravitational instability [1, 2, 3].

The condition for a disk to be gravitationally unstable is that the Q parameter, defined as

$$Q = \frac{c_s \kappa}{\pi G \Sigma}, \quad (1)$$

be of order unity. In equation 1, c_s is the sound speed, κ is the epicyclic frequency (which equals the angular frequency in a Keplerian disk), and Σ is the disk surface density. The gravitational instability can therefore generally only operate in cold, massive disks (measured relative to the mass of the central object). When self-gravity is important, the disk will either evolve into a quasi-steady state or, if sufficiently unstable, will fragment into gravitationally bound objects [4, 5].

The ultimate evolution of a self-gravitating disk is largely determined by the relationship between the rate at which the disk cools, and the rate at which it heats up through dissipation of the turbulence generated by the instability. If cooling is extremely fast, the disk will fragment. Gammie [4], using a local, two-dimensional model showed that fragmentation occurs if $t_{cool} \leq 3\Omega^{-1}$ where Ω is the angular frequency, a result that was largely confirmed by Rice et al. [5] using global, three-dimensional models. If $t_{cool} > 3\Omega^{-1}$, the disk then evolves into a quasi-steady state in which gravitational instabilities lead to the outward transport of angular momentum.

In this paper we consider the dynamical evolution of self-gravitating disks. In §2 we consider how quasi-steady, self-gravitating disks transport angular momentum, in §3 we discuss the conditions required for disk fragmentation, and in §4 we consider the evolution of solid particles in self-gravitating disks and suggest a way in which disk self-gravity may aid in the formation of planets in disks around young stars.

ANGULAR MOMENTUM TRANSPORT

It is generally accepted that during star formation, a large fraction of the final stellar mass is accreted from a circumstellar disk. The mechanism by which the angular momentum is transported outwards, allowing mass to accrete onto the central star, is still somewhat unclear. In the absence of a general theory, it is generally assumed that the viscosity, ν, can be parameterized as $\nu = \alpha c_s H$ [6], where c_s is the local sound speed, H is the disk scaleheight, and α is the viscous parameter that will be discussed in more detail below. In disks that are sufficiently ionized, the magnetorotational instability [7] may provide the required viscosity. Disks around young protostars are, however, thought to be magnetically dead (especially near the disk midplane), in which case such a mechanism will not work. If the disk is sufficiently massive, however, its self-gravity may provide the means to transport angular momentum. Such a mechanism may also play a role in the secular evolution of disk galaxies, leading to the formation of bulges [8].

To investigate the transport of angular momentum in self-gravitating disks we use three-dimensional Smoothed Particle Hydrodynamics (SPH)[9], a lagrangian hydrodynamics code. The gas is assumed to have an adiabatic equation of state, with $\gamma = 5/3$, and is allowed to heat up through both PdV work and viscous dissipation. Cooling is implemented by adding a simple cooling term to the energy equation, $du_i/dt = -u_i/t_{cool}$, with a cooling time given by $t_{cool} = \beta\Omega^{-1}$ [4]. Since we wish to consider angular momentum transport in disks that attain quasi-steady states and do not fragment, we use $\beta = 7.5$. The left panel in figure 1 shows the surface density structure at the end of a simulation in which the disk mass was one-tenth that of the central object. It can be shown [10, 4] that, in thermal equilibrium, the cooling time and the viscous parameter α [6] are related through

$$\alpha = \frac{4}{9\gamma(\gamma-1)} \frac{1}{t_{cool}\Omega}. \qquad (2)$$

The viscous parameter α can be computed from the Reynolds, $<T_{R\phi}^{grav}>$, and gravitational, $<T_{R\phi}^{Reyn}>$, stresses in the disk through

$$\alpha = \frac{2}{3} \frac{T_{R\phi}^{grav} + T_{R\phi}^{Reyn}}{\Sigma c_s^2}. \qquad (3)$$

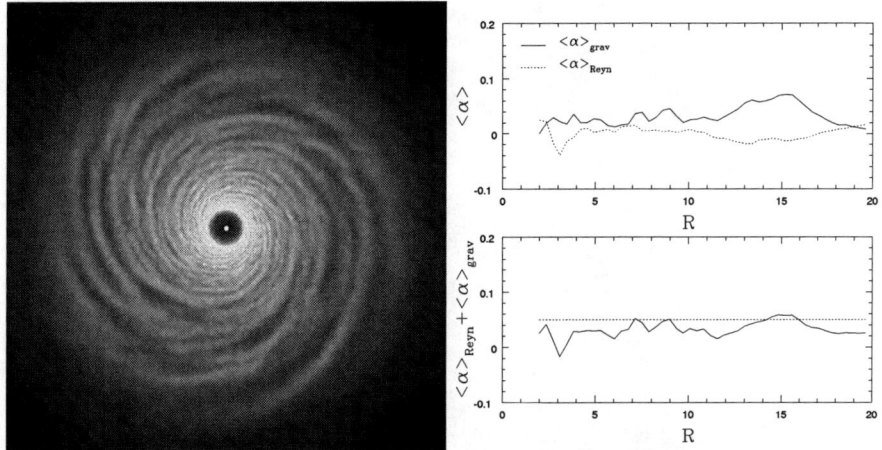

FIGURE 1. Left planel : Disk surface density structure at the end of a simulation in which the disk mass was one-tenth that of the central object. The spiral structures are due to the growth of the gravitational instability. Right panel: Top figure shows the Reynolds and gravitational stress contributions to the viscous α, while the bottom figure shows the total α compared to that expected based on the imposed cooling (dashed line).

Since we can use our simulation results to calcuate the Reynolds and gravitational stresses, we can compare the expected value of α (equation 2) with the value we get from our simulations. The right panel of Figure 1 shows the viscous parameter α in our simulations. The top figure shows the separate contributions of the Reynolds and gravitational stresses, while the bottom figure compares the α computed using the Reynolds and gravitational stresses, with that expected based on our imposed cooling (dashed line).

In almost all of our simulations the α value determined using the stresses compared well with the expected value [11]. In general it appears that self-gravitating disks with masses less than that of the central object are able to evolve into a quasi-steady state in which the transport of angular momentum occurs in a manner that agrees well with standard viscous accretion disk theory [11]. The disk then evolves on a viscous timescale, suggesting that the self-gravitating phase may be reasonably long-lived [12].

DISK FRAGMENTATION

In the above simulations, the cooling time was chosen such that the disk should settle into a quasi-steady state, rather than fragment into bound objects. It has been shown [4, 5], that fragmentation occurs if $t_{cool} \leq 3\Omega^{-1}$. Figure 2 shows a disk in which the cooling time is such that fragmentation occurs. It is this fragmentation that has been suggested as a possible mechanism for forming stars in disks around active galaxies [13], and for forming gaseous planets in protostellar disks [14]. Although this is an attractive mechanism for forming gaseous planets, since it is extremely rapid and overcomes all

FIGURE 2. Surface density structure of a disk in which the cooling time is such that the disk fragments into gravitationally bound objects.

the timescale issues related to the standard core accretion model [15], it appears unlikely that the cooling rate in protostellar disks is fast enough to satisfy the fragmentation condition [16]. The standard fragmentation condition ($t_{cool} \leq 3\Omega^{-1}$) was, however, determined using equations of state with $\gamma = 2$, and $\gamma = 5/3$. In protostellar disks, the specific heat ratio may be as small as $\gamma = 7/5$. It has been suggested that in a quasi-steady, self-gravitating disk, there is a maximum stress, as measured by α, above which the disk undergoes fragmentation [12]. Equation 2 would then suggest that the fragmentation boundary should increase with decreasing γ. This could lead to fragmentation for $t_{cool} \leq 10\Omega^{-1}$ in disks with $\gamma = 7/5$ [12]. There may then be regions of protostellar disks around very young protostars where such cooling times could be achieved.

GRAIN GROWTH IN SELF-GRAVITATING DISKS

The implication that disk fragmentation requires rapid cooling has lead to the view that gaseous planets do not form in this way. The more commonly accepted formation mechanism is that dust grains grow to form kilometer sized planetesimals that then coagulate to form a planetary core. Once sufficiently massive ($> 10 M_{Earth}$), this planetary core then accretes a gaseous envelope [15]. A problem with this process is that dust grains experience a drag force due to the disk gas that tends to cause them to migrate in towards the central star. The inward migration rate depends on the grain size, and there is a particular size range ($\sim 10 - 100$cm) for which the migration rate may be extremely rapid, removing these particles before they become large enough to decouple from the disk gas [17].

In a self-gravitating disk, the migration process is not as simple. The drag force tends to cause dust grains and small particles to migrate towards the centers of the spiral arms. The most significant effect is for those particles with the largest migration rates. Figure 3 shows the results of two simulations, one with 50 cm particles and the other with 1000 cm particles. The surface density structure of the 1000 cm particles matches that of the disk gas (not shown). The 50 cm particles are, however, strongly concentrated in the centers of the spiral arms and may achieve densities comparable to that of the disk gas. Such a process may enhance grain growth and may allow particles to grow sufficiently large that gas drag is unimportant. A self-gravitating phase in a protostellar disk may play an important role in the planet formation process [18], despite disk fragmentation in protostellar disks being unlikely [16].

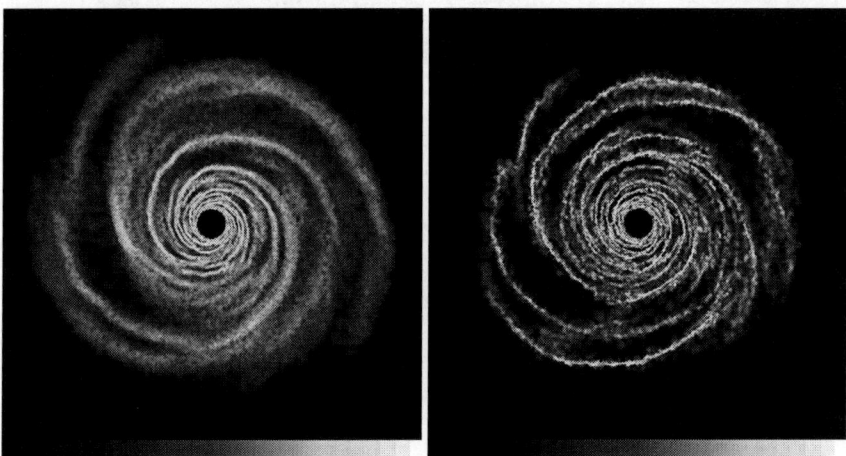

FIGURE 3. Surface density structure of 1000 cm particles (left panel) and 50 cm particles (right panel) embedded in a self-gravitating gas disk. The 50 cm particles are strongly concentrated in the centers of the spiral arms leading to density enhancements that may accelerate grain growth.

CONCLUSION

We have considered the dynamical evolution of self-gravitating accretion disks. For cooling times that are sufficiently long, the disk evolves into a quasi-steady state in which angular momentum is transported outwards and mass inwards. Despite this being a non-viscous process, the evolution of a self-gravitating disk is comparable to that of a viscously evolving disk, at least for disk masses less than that of the central object [11, 12]. For extremely rapid cooling, the disk may fragment into bound objects, a process that has been suggested as a mechanism for forming gaseous planets similar to Jupiter and Saturn. The rapid cooling required for this process makes it unlikely in protostellar disks [16], although there is a suggestion that the required cooling time may be longer in disks with specific heat ratios of $\gamma = 7/5$. Although disk fragmentation may not produce gaseous planets, a self-gravitating phase in protostellar disks may play an important role in grain growth. Particles that would migrate rapidly into the central star in a non-self-gravitating disk, may collect at the center of the spiral arms where they could grow large enough to decouple from the disk gas [18].

REFERENCES

1. V. S. Safronov, *Ann. d'Astrophys.*, **23**, 979 (1960).
2. A. Toomre, *Astrophys. J.*, **139**, 1217 (1964).
3. C. C. Lin, and F. Shu, *Astrophys. J.*, **140**, 646 (1964).
4. C. F. Gammie, *Astrophys. J.*, **553**, 174 (2001).
5. W. K. M. Rice, P. J. Armitage, M. R. Bate, and I. A. Bonnell, *Mon. Not. Royal Astron. Soc.*, **338**, 227 (2003).
6. N. I. Shakura, and R. A. Sunyaev, *Astron. Astrophys.*, **24**, 337 (1973).
7. S. A. Balbus, and J. F. Hawley, *Astrophys. J.*, **376**, 214 (1991).
8. J. Kormendy, and R. C. Kennicutt, *Ann. Rev. Astron. Astrophys.*, **42**, 603 (2003).
9. J. J. Monaghan, *Ann. Rev. Astron. Astrophys.*, **30**, 543 (1992).
10. J. E. Pringle, *Ann. Rev. Astron. Astrophys.*, **19**, 137 (1981).
11. G. Lodato, and W. K. M. Rice, *Mon. Not. Royal Astron. Soc.*, **351**, 630 (2004).
12. G. Lodato, and W. K. M. Rice, *Mon. Not. Royal Astron. Soc.*, **358**, 1489 (2005).
13. J. Goodman, *Mon. Not. Royal Astron. Soc.*, **339**, 937 (2003).
14. A. P. Boss, *Nature*, **393**, 141 (1998).
15. J. B. Pollack, et al., *Icarus*, **124**, 62 (1996).
16. R. Rafikov, *Astrophys. J. Lett.*, **621**, 69 (2005).
17. S. Weidenschilling, *Mon. Not. Royal Astron. Soc.*, **180**, 57 (1977).
18. W. K. M. Rice, G. Lodato, J. E. Pringle, P. J. Armitage, and I. A. Bonnell, *Mon. Not. Royal Astron. Soc.*, **355**, 543 (2004).

On the Fitting of Ion-Ion Drifting Plasma

E. Kh. Kaghashvili*, G. P. Zank* and B. J. Vasquez[†]

*Institute of Geophysics and Planetary Physics, University of California, Riverside
Riverside, California 92521
[†]Institute for the Study of Earth, Oceans and Space, University of New Hampshire
Durham, New Hampshire 03824

Abstract. It has been known for almost three decades that the solar wind proton population can often be represented as a sum of two proton distribution components of nearly equal temperatures with the diluted component streaming relative to the denser component. The differential streaming speed is often near the local Alfvén speed. The usual way for deriving such important quantities as a mean speed of individual proton distributions and its characteristic thermal temperature is based on separation of observed distribution into two parts and fitting them independently with a given well-defined distribution function. As we will show below this kind of analysis can be misleading when the thermal spread of the particle distributions is wide enough to make these distributions overlap. In the summary, we argue where this kind of analysis can be important.

Keywords: ion-ion beams, observation fitting procedure
PACS: 95.75Pq

INTRODUCTION

Observations showing that a proton-proton drift that is always present in the fast solar wind at high latitudes has stimulated considerable interest recently. Observations show the presence of two clearly identifiable proton components. For a given streaming speed and parameter regime the drifting proton beams enter the macroinstability regime.

Linear microinstabilities in a uniform plasma and magnetic field can be driven by the free energy of streaming and are influenced significantly by the anisotropy of the components and other plasma parameters. Such instabilities occur above some threshold speed and lead to deceleration and saturation at some new speed below the range of instability [e.g., Winske and Omidi, 1992; Daughton and Gary, 1998; Daughton et al., 1999; Hellinger et al., 2003]. A difficulty with the straightforward application of linear beam instabilities concerns the observed distribution of average differential streaming speed. The mean speed can be significantly below the expected threshold within 1 AU [Tu et al., 2004] and especially beyond 1 AU where plasma $\beta \geq 2$ [e.g., Goldstein et al., 2000; Reisenfeld et al., 2001].

The usual way for deriving important quantities such as the mean speed of individual proton distributions and the characteristic thermal temperature is based on the separation of the observed distribution into two parts and fitting them independently with a given well-defined distribution function [e.g., Goldstein et al., 2000; Tu et al., 2004]. As we show below, this kind of analysis can be misleading when the thermal spread of the particle distributions is wide enough to ensure the distributions overlap. Here, we consider the problem from the perspective of hybrid simulations and draw some inferences about the temporal behavior of field-aligned beams with drifted velocities.

RESULTS OF HYBRID SIMULATIONS

In a recent paper, we examined the evolution of the proton-proton distribution using two and a half dimensional $\left(2\frac{1}{2}-D\right)$ hybrid simulation for cases with and without initial Alfvén waves [*Kaghashvili et al.*, 2004]. The background plasma consisted of two proton components - a relatively dense "core" component and a beam component. Protons in the simulation were represented as particles. The positively charged proton population was embedded in a quasineutralizing electron fluid. We chose common parameters for all our simulation runs. The time step was $0.05\Omega_p^{-1}$, where the proton cyclotron frequency, Ω_p^{-1}, was used as a time normalization unit. Positions in our simulation box were a normalized to the inertial length, c/ω_p, where c is the speed of light and ω_p is the proton plasma frequency. Simulations were performed on a 128 by 128 grid in the xy plane with cell sizes of $1.96\ c/\omega_p$. The major results relevant to this study were the following:

- Without waves, significant linear beam instability occurs as expected at super-Alfvénic speeds, and at greater rates for larger beam densities and initial speeds. Cyclotron and Landau resonances contribute to the excitation of oblique proton-proton cyclotron waves when the beams are unstable. Beam speeds decelerate to below V_A and saturate well below V_A, where V_A is the local Alfvén speed. Where linear instability occurs, the corresponding beam deceleration rates with initial waves are greater than without waves. During the beam evolution, beam protons maintain a Maxwellian distribution while core protons with a parallel velocity develop a tail extending in the direction of the beam.
- For highly unstable cases, the initial shape of the proton distributions (i.e., Gaussian-shaped) is strongly affected. For the beam proton population, the instability results in a decrease of the bulk velocity of beam protons, but the shape of the distribution remains symmetric around the bulk speed and nearly Maxwellian. However, the bulk proton distribution develops a significant tail in the direction of the beam. The tail forms because the slower bulk protons are accelerated by the growing waves which travel in the direction of the beam. Because of this asymmetry, the bulk proton distribution cannot be fitted by a Maxwellian or Kappa distribution. For other strongly unstable cases with $U_{pp} \geq V_A$ and with or without initial waves, proton distributions follow the same trend as above but to a lesser extent, which can leave the bulk and beam components distinct at later times. (U_{pp} is the absolute value of the streaming speed of the proton beam.) The evolution of the main proton tail in these cases is found to be proportional to the deceleration rate. A large deceleration rate, corresponds to a rapidly developing, well populated tail stretching over a large range of $V_\parallel$, whereas cases with smaller rates show less developed tails at the same time. For cases with $U_{pp} \leq 0.5V_A$ and initial waves, where the deceleration rate is small, little change with time from the Maxwellian shape is seen for both beam and bulk protons.

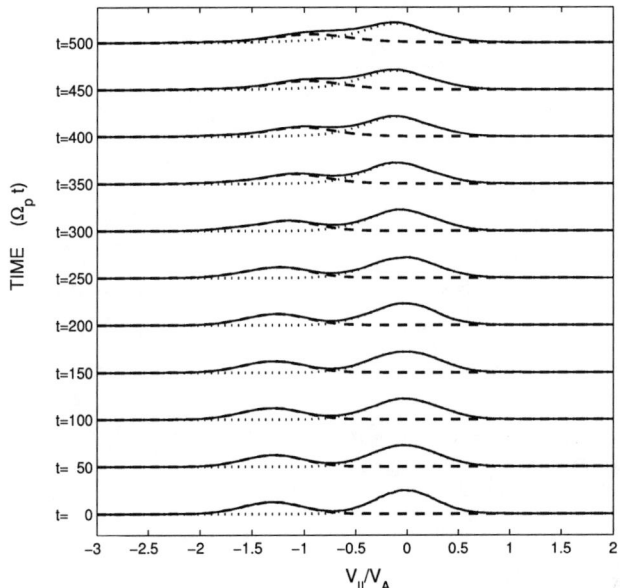

FIGURE 1. Evolution of proton parallel velocity distribution function $f(V_\parallel)$ for the most unstable case starting from $U_{pp} = 1.57$, $n_b = 0.5$, and with initial waves. Plots show proton relative number normalized to the maximum value of the main protons at $t = 0$ as a function of $V_\parallel$ for different values of t. The solid lines correspond to separate beam and main proton distributions. The thick solid line corresponds to the sum of the beam and main protons.

- Due to the presence of initial Alfvén waves in the system, the parameter space for which deceleration occurs is expanded and the deceleration rates are amplified. A reduction of the threshold streaming speed can be an important factor in accounting for observed mean differential speed distributions of solar-wind ions. Our simulation results apply only to low β plasma, appropriate to conditions within 1 AU. For moderate proton beam densities, there are linear instabilities which have a threshold near V_A and peak growth times of times of 1000. Deceleration of these beams due to a linear instability is consistent with observed protons maintaining a mean speed near local V_A without invoking finite-amplitude wave effects. Weak proton beams can generate much slower growing linear instabilities when speeds are near V_A, and these instabilities might not be effective if they compete with faster processes. For these beams, as with minor ions, wave-amplitude wave effects could be an important factor in providing an effective threshold near V_A.

ANALYSIS OF THE ION-ION BEAM DRIFT

The most unstable case of our simulations is demonstrated in Figure 1 (which corresponds to a case given by *Kaghashvili et al.*, 2004; their figure 10). Figure 1 captures

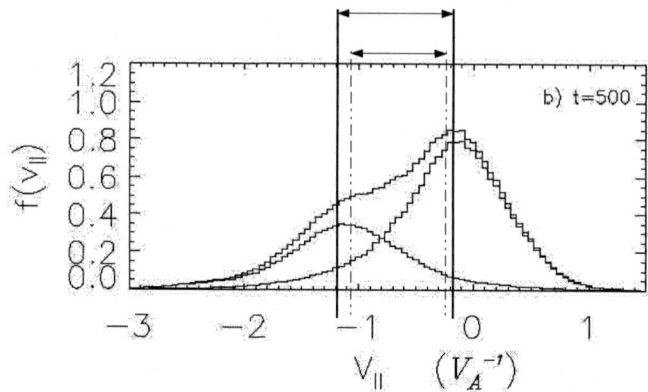

FIGURE 2. Analysis of the proton core and beam distribution functions at t = 500 Ω^{-1} of Figure 1. The figure shows the distribution of the core and beam protons; and the composite distribution (observed quantity) of two. A least square fit is done for two cases: the total distribution is separated into two parts, which are then fitted independently (dotted lines); and the total distribution is fitted as a sum of two Maxwellian (solid lines).

the earlier stage of the development of the core proton and beam-proton distributions. It is seen that at $t = 400$, the peaks of the summarized curve (observational data curve) deviates from individual peaks of both the core and beam distributions.

The usual way to fit the proton-proton distribution is to split them at a lowest point between two distinctive peaks. Afterwards, one tries to fit the separate parts with a predefined distribution function. In Figure 2, this kind of analysis corresponds to the dot-dashed lines. We suggest that when an instability is expected in the system and the thermal spread of the particle distributions is wide enough to make these distributions overlap, more accurate plasma characteristics can be obtained if one fits the observational data with the sum of two predefined distribution functions. In Figure 2 (which corresponds to Figure 1 when t = 500 Ω^{-1}), we show the result of this kind of analysis (solid lines). As one can see the absolute value of the streaming speed is always higher in this case than in the previous analysis.

The results of the two analyses are given in Table 1. As expected the absolute value of the streaming speed is somewhat higher in the case when the observational data is fitted with the sum of two Maxwellians. The absolute difference between the two is $\sim 0.11 V_A$, which is roughly 10% of the value obtained by usual fitting. It has to be mentioned that this is the most unstable case in our simulations. Table 1 shows that other important parameters also deviate for the latter case.

SUMMARY

We have shown that when the thermal spread of the ion-ion particle distributions is wide enough to make these distributions overlap, the usual way of splitting the total

TABLE 1. Proton-Proton Distribution Fit Parameters. The fitting function used is given as $A \exp[-B(v - v_0)^2]$.

	A	B	v_0
Separately			
Core	0.8836	5.5833	-0.1215
Beam	0.4500	3.6051	-0.9689
As Sum			
Core	0.8632	6.0288	-0.1106
Beam	0.3992	5.5285	-1.0252

distribution into two parts and fitting them independently can be misleading. For a particular hybrid simulation example, the fitting of the total distribution as a sum of two predefined distribution functions can produce a more accurate result. This kind of analysis is relevant to proton-proton streaming in the solar wind, for pick-up ion populations as well as reflected particles at the quasi-perpendicular shocks [e.g., Zank et al. 2001] where a strong ion-ion instability is expected.

ACKNOWLEDGMENTS

E.Kh.K. and G.P.Z. are supported by NSF grants ATM-0296113 and ATM-0317509 to the University of California, Riverside. B.J.V. is supported by the NASA Sun-Earth Connection Theory Program under grant NAG5-11797 to the University of New Hampshire and NASA SR&T grant NAG5-10988 to the University of New Hampshire.

REFERENCES

1. W. Daughton, and S. P. Gary, 1998, Electromagnetic proton/proton instabilities in the solar wind, *J. Geophys. Res.*, *103*, 20613.
2. W. Daughton, S. P. Gary, and D. Winske, 1999, Electromagnetic proton/proton instabilities in the solar wind, Simulations, *J. Geophys. Res.*, *104*, 4657.
3. B. E., Goldstein, M. Neugebauer, L. D. Zhang, and S. P. Gary, 2000, Observed constraint on proton-proton relative velocities in the solar wind, *J. Geophys. Res.*, *27*, 53.
4. P. Hellinger, P. Trávníček, A. Mangeney, and R. Grappin, 2003, Hybrid simulations of the expanding solar wind: Temperatures and drift velocities, *Geophys. Res. Lett.*, *30*, 1211, doi:10.1029/2002GL01409.
5. E. Kh. Kaghashvili, B. J. Vasquez, G. P. Zank and J. V. Hollweg, 2004, Deceleration of relative streaming between proton components among nonlinear low-frequency Alfvén waves, *J. Geophys. Res.*, *109*, A12101, doi:10.1029/2004JA010382.
6. D. B. Reisenfeld et al., 2001, Helium energetics in the high-latitude solar wind: Ulysses observations, *J. Geophys. Res.*, *106*, 5693.
7. C.-Y. Tu, E. Marsch, and Z.-R. Qin, 2004, Dependence of the proton beam drift velocity of the proton core plasma beta in the solar wind, *J. Geophys. Res.*, *109*, A05101, doi:10.1029/2004JA010391.
8. D. Winske, and N. Omidi, 1992, Electromagnetic ion/ion cyclotron instability: Theory and simulations, *J. Geophys. Res.*, *97*, 14779.
9. G. P. Zank et al., 2001, The "injection problem" for quasiparallel shocks, *Physics of Plasmas, 8*, 10, 4560.

Coronal shock waves observed in images

H. S. Hudson

Space Sciences Lab, UC Berkeley

Abstract. The large-scale coronal shock waves observed from radio type II bursts and from Moreton waves have proven surprisingly difficult to detect in coronal images. I review the evidence for such waves in radio, optical, EUV, and soft X-ray images. The data generally support the conclusion that the metric type II bursts can be identified with weak fast-mode shock waves launched at the impulsive phase of the associated flares. Other coronal waves, well seen by EIT, are more closely related to CMEs.

Keywords: Corona, Shock Waves, Flares, Coronal Mass Ejections
PACS: 96.60.Pb, 96.60.Rd, 96.60.Wh

INTRODUCTION

Flares and coronal mass ejections (CMEs) involve restructuring of significant volumes of the coronal magnetic field, and thus launch large-scale waves. The discovery of radio type II bursts showed immediately that these waves became shocks, following the now well-established theoretical picture that non-linear effects at the shock create Langmuir waves at the plasma frequency, and these then scatter and emit electromagnetic radiation (e.g., [1]). The further evidence from the phenomenon of the "Moreton wave" as observed in the chromosphere, via Uchida's unified theoretical interpretation [2], convincingly demonstrated that large-scale shock waves commonly occur in the solar corona. In the heliosphere one has direct *in-situ* and geomagnetic observations of shocks driven by CMEs, but the relationship between these shocks and the ones sensed remotely in the solar corona still remain somewhat controversial.

New observational facilities have become available in the last decade of the twentieth century; these have made it possible to glimpse coronal shock waves at new wavelengths, including most directly soft X-rays (which show the direct thermal emission of the million-degree plasma). The soft X-ray images can be interpreted in terms of density and temperature jumps at a shock front. Because of observational limitations we can never actually resolve the structure of a coronal shock front; only the non-thermal signatures such as radio waves or particle acceleration can confirm that a given disturbance or wave has actually approached the shock condition. We present these briefly and then comment on the implications for our understanding of coronal dynamics, the relationship between the coronal phenomena and those observed in the heliosphere, and the physical parameters of the shocked material in the corona.

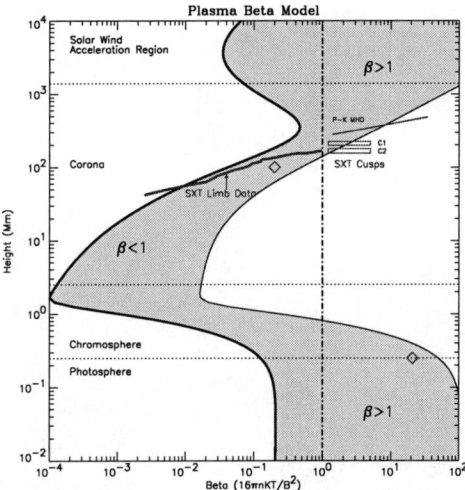

FIGURE 1. Distribution of plasma beta (ratio of gas to magnetic pressure), from [3]; for coronal conditions the Alfvén speed is roughly $200\,\beta^{-1/2}$ km s^{-1}. Note that filaments, not shown on this diagram, represent high-beta inclusions in the low corona.

WAVE DRIVERS AND IMAGES

We still do not have much knowledge of large-scale wave generation and propagation in the corona (see discussion below), even with a long history of MHD model development (e.g., [4]) for flares and CMEs This reflects our ignorance of the actual structural changes during the flare impulsive phase and the acceleration of the CME. To create a large-amplitude wave requires a motion perpendicular to the field that is rapid on the Alfvén time scale ([5]; [6]). We believe these initial motions to be in regions of low plasma beta, and the wave disturbance will therefore not reach the shock condition until it or the disturbance itself has propagated some distance (see Figure 1). Interplanetary shock waves observed in the heliosphere should have the character of bow waves; if the disturbance propagates into a region of reduced Alfvén speed, a high Mach number may result [7].

The imaging observations give us a means for assessing the Mach number of any given disturbance via the Rankine-Hugoniot jump conditions (see [8] for a discussion in this context). The first step here is to estimate the local plasma density (or temperature) from the observations. This is tricky because the observations only give line-of-sight integrations of a 3D structure of unknown dimensions (see below on modeling).

The relative roles of solar flares and CMEs in launching and driving large-scale waves remains controversial; for simplicity some would prefer to have CMEs be responsible for all large-scale waves, since the association of CME-induced driver gas and heliospheric shock waves is well established in the interplanetary medium. On the other hand the time development of the metric type II burst and Moreton wave point quite clearly to the flare itself. The observational situation is often confused because a well-studied major

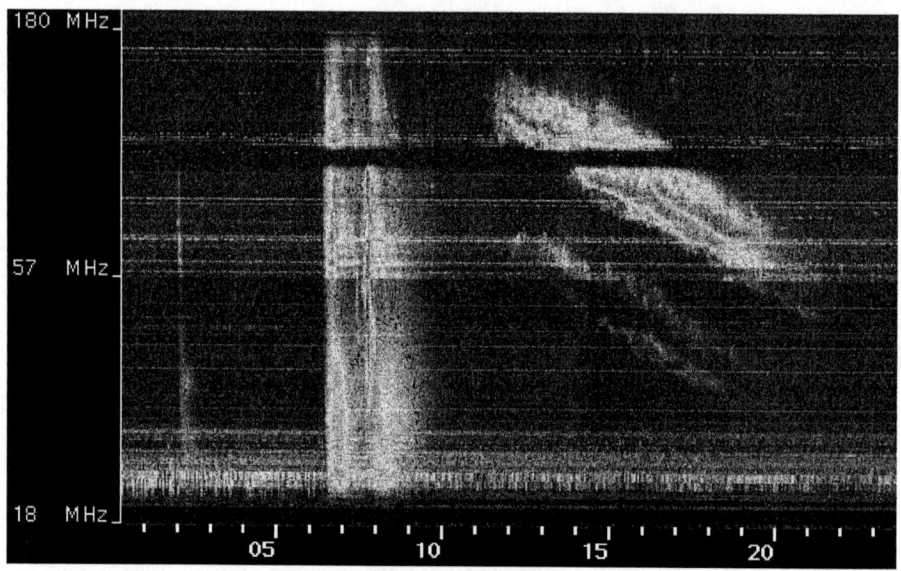

FIGURE 2. Dynamic radio spectrograph of a protypical event: impulsive type III bursts plus a very regular type II burst, fundamental and harmonic. Time scale is in minutes. Note the gap in time between the type III and type II, interpreted as the interval between the wave origination and the onset of the shock condition (e.g., [6]).

event will usually have examples of all phenomena (Kahler's "big flare syndrome"). Theoretically we do not know much about the sources of the waves in the lower corona, which are usually outside the field of view of a coronagraph; worse yet, the usual PFSS[1] modeling framework deliberately introduces geometrical artifacts between the corona and the solar wind.

Figure 2 illustrates the possible ambiguities. In this example, an observation from the Culgoora spectrograph, the type III emission presumably mark the impulsive phase of the solar flare. The type II emission begins about 5 minutes afterwards, with its plasma-frequency fundamental band at about 60 MHz. At an exciter speed of $\sim$2000 km s^{-1} (the Alfvén speed in an active region at $B = 300$ g, $n = 10^9$ cm^{-3}, this would correspond to a travel distance on the order of 1 R$_\odot$ to a location at a density of some 5×10^7 cm^{-3}. How does this information fit into the real corona at the time of the observation? This density is much higher than that of a spherically-symmetric model corona at a height matching this travel distance (e.g., [9]). On the other hand [10] note a good agreement between UV coronagraphic observations of density and the directly observed type II burst frequency for a flare event on June 11, 1998.

---

[1] Potential Field Source Surface

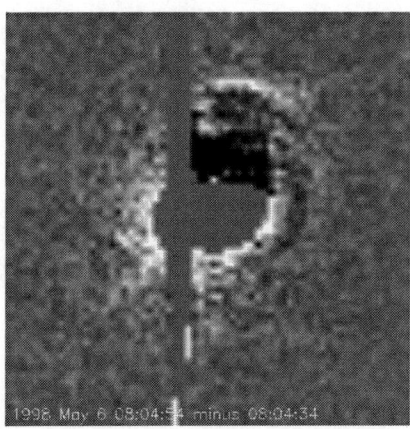

FIGURE 3. Soft X-ray observations of a large-scale wave from a W limb flare on May 6, 1998 ([14]). The image is about 0.2 $R_\odot$ across and the limb runs vertically through the (saturated) flare core; the wave thus extends into the corona in projection.

NEW OBSERVATIONS

There are as many as five new observational windows on large-scale coronal shock waves: the EUV observations of "EIT waves" ([11]; [12]); soft X-ray observations from *Yohkoh*/SXT ([13]; [14]; Figure 3); microwave observations at 17 GHz from Nobeyama ([15]); meter-wave observations from Nançay ([16]); and HeI 10830Å observations from Mauna Loa ([17]). All of these except for the Nançay data reflect thermal emissions; at metric wavelengths one presumably sees synchrotron emission from relativistic electrons trapped in expanding CME structures (e.g., [18]).

The passage of a compressive wave through the corona will produce a front of temperature and density increase. The thermal response at radio wavelengths will be free-free emission (bremsstrahlung), with the intensity increasing directly with temperature according to the Rayleigh-Jeans law. At short wavelengths the wave will have more complicated effects, depending upon the passband of the detector. A broad-band observation, such as soft X-rays, will generally see an increase at the shock front, but a narrow-band observation could actually show a decrease in intensity if the temperature change results in the disappearance of a strong emission line from the passband. The sensitivity in the EUV or in a coronal emission line would therefore depend not only on the geometry of the wave structure, but also the ambient temperature.

The new observations generally are consistent with a "two-wave" scenario: a weak fast-mode blast wave in the lower corona, plus a CME-driven disturbance. The extensive EIT observations appear to show both kinds of wave activity ([19]; [20]). Figure 4 illustrates a good association between an EIT wave (seen in EUV) and a Moreton wave (seen in bbHα).

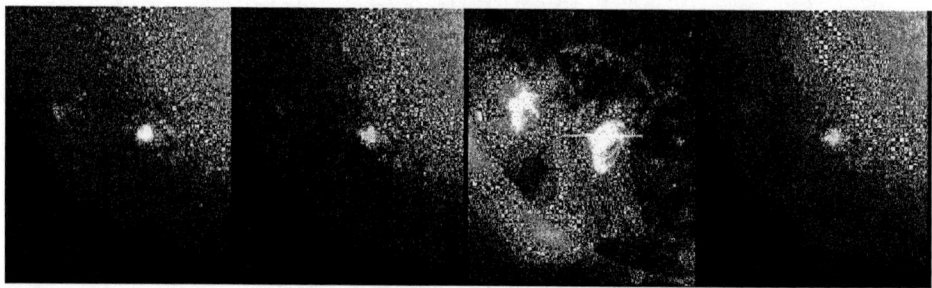

FIGURE 4. Moreton wave and EIT wave ([21]); Hα images except for the EUV image second from right. This particular EIT wave is one of the sharply-defined ones best associated with a flare-launched blast wave ([19]).

MODELING IDEAS AND CONCLUSIONS

Blast waves running freely away from their launch site propagate through unperturbed coronal field. A model corona consisting of a realistic extrapolation of the photospheric field, plus a mass loading consistent with the X-ray observations, could in principle be used to study blast-wave propagation. The refraction of these waves around magnetic obstacles is known observationally [14] and also expected theoretically [2] as the wave normals bend towards regions of lower Alfvén speed. Ideally the mass loading would be computed self-consistently along with the field structure, but an easier approach would be to make use of a PFSS (potential-field source surface) model, as for example in the recent simulations of the global corona by [22]. This model makes allowances for electric currents flowing in the body of the solar corona by using a fictitious current system on a solar-wind source surface, typically taken to be at 2.5 $R_\odot$. This approach provides suprisingly good fidelity in many applications, for example in tracing the locations of open field lines, but it clearly distorts the field geometry near the source surface itself. Nevertheless, until a self-consistent model based on better knowledge of the physics appears (e.g., [20]), the PFSS framework should be explored as a guide to interpreting images in terms of large-scale waves.

The imaging of large-scale wave disturbances in the solar corona, together with models of ambient (and transient) coronal structure can help us to understand the formation of the waves. It may also ultimately help us to understand the basic mechanisms of coronal restructuring by flares and CMEs.

ACKNOWLEDGMENTS

Corona solaris in undas divisa est. This work has been supported by NASA under grant NAG5-12878. For more graphics related to the conference presentation, please see **http://sprg.ssl.berkeley.edu:80/ ~hhudson/presentations/palmsprings.050303/**.

REFERENCES

1. T. S. Bastian, A. O. Benz, and D. E. Gary, *ARAA*, **36**, 131 (1998).
2. Y. Uchida, *Solar Phys*, **4**, 30(1968).
3. G. A. Gary, *Solar Phys*, **203**, 71 (2001).
4. R. S. Steinolfson, *Solar Phys*, **94**, 193–202 (1984).
5. R. S. Steinolfson, *Washington DC American Geophysical Union Geophysical Monograph Series*, **35**, 1 (1985).
6. B. Vršnak, and S. Lulić, *Solar Phys*, **196**, 157–180 (2000).
7. G. Mann, A. Klassen, H. Aurass, and H.-T. Classen, *A&A*, **400**, 329 (2003).
8. N. Narukage, H. S. Hudson, T. Morimoto, S. Akiyama, R. Kitai, H. Kurokawa, and K. Shibata, *ApJL*, **572**, L109 (2002).
9. G. L. Withbroe, *ApJ*, **325**, 442 (1988).
10. J. C. Raymond, B. J. Thompson, O. C. St. Cyr, N. Gopalswamy, S. Kahler, M. Kaiser, A. Lara, A. Ciaravella, M. Romoli, and R. O'Neal, *GRL*, **27**, 1439 (2000).
11. D. Moses, et al., *Solar Phys*, **175**, 571–599 (1997).
12. B. J. Thompson, S. P. Plunkett, J. B. Gurman, J. S. Newmark, O. C. St. Cyr, and D. J. Michels, *GRL*, **25**, 2465(1998).
13. J. I. Khan, and H. Aurass, *A&A*, **383**, 1018 (2002).
14. H. S. Hudson, J. I. Khan, J. R. Lemen, N. V. Nitta, and Y. Uchida, *Solar Phys*, **212**, 121 (2003).
15. S. M. White, and B. J. Thompson, *ApJL*, **620**, L63–L66 (2005).
16. D. Maia, M. Pick, A. Vourlidas, and R. Howard, *ApJL*, **528**, L49 (2000).
17. H. R. Gilbert, and T. E. Holzer, *ApJ*, **610**, 572 (2004).
18. T. S. Bastian, M. Pick, A. Kerdraon, D. Maia, and A. Vourlidas, *ApJL*, **558**, L65 (2001).
19. D. A. Biesecker, D. C. Myers, B. J. Thompson, D. M. Hammer, and A. Vourlidas, *ApJ*, **569**, 1009 (2002).
20. P. F. Chen, C. Fang, and K. Shibata, *ApJ*, **622**, 1202 (2005).
21. B. J. Thompson, B. Reynolds, H. Aurass, N. Gopalswamy, J. B. Gurman, H. S. Hudson, S. F. Martin, and O. C. St. Cyr, *Solar Phys*, **193**, 161 (2000).
22. C. J. Schrijver, A. W. Sandman, M. J. Aschwanden, and M. L. DeRosa, *ApJ*, **615**, 512 (2004).

Proton, Electron and Ion Temperatures in Fast Shocks

John C. Raymond* and Kelly E. Korreck[†]

*Center for Astrophysics, 60 Garden St., Cambridge, MA 02138
[†]Department of Atmospheric, Oceanic, and Space Sciences, University of Michigan, 2455 Hayward, Ann Arbor, MI 48109

Abstract. The Coulomb equilibration time scale among various particle species behind a fast collisionless shock can be much larger than the dynamical time scale in a supernova remnant or CME. Ultraviolet and optical emission line profiles can be used to measure proton, electron and ion temperatures. Particles are fairly close to thermal equilibrium behind a relatively slow (350 km/s) shock, but very far from equilibrium in faster (2000-3000 km/s) shocks.

Keywords: Collisionless Shock Waves, Electron Heating
PACS: 52.35.Tc, 95.85.Kr, 95.85.Mt

INTRODUCTION

Because collisions drive a plasma toward thermal equilibrium, the proton, electron and ion temperatures should be equal behind a collisional shock front. In a collisionless shock, however, one might expect the different particle species to separately thermalize the energies they dissipate, so that for a strong shock

$$T_s = 2.4 \times 10^5 \frac{m_s}{m_p} V_{100}^2 \qquad (1)$$

where V_{100} is the shock speed in units of 100 km/s [18]. The real situation is more complex, of course, in that plasma turbulence may bring the particles more quickly into equilibrium, or it may preferentially heat some particle species depending on charge to mass ratio. Another complication is that particles may interact in different ways with the shock precursor or ramp, again leading to preferential heating (e.g. [14]). In addition, a significant fraction of the energy dissipated by the shock may go into non-thermal particles or magnetic fields.

For many years astrophysicists generally assumed complete equilibration behind shock fronts, mostly for lack of a better assumption. We got away with that either because Coulomb collisions brought the plasma into equilibrium on a time scale shorter than the other relevant time scales, or because the available data provided only weak constraints. In cases where the Coulomb collisions are fast compared with the time scales required to excite the radiation that makes the shocked gas observable, thermal equilibrium is an adequate approximation. However, the Coulomb time scale in a young supernova remnant is generally much longer than the age of the remnant, and one makes a serious error in identifying T_e determined from the X-ray spectrum with the equilibrated shock temperature. This paper summarizes recent results for the electron-ion equilibration in fast shocks. A paper in this volume by T. Zurbuchen et al. discusses ion-ion equilibration measurements.

Astrophysical shocks can be classified as radiative, in which case the heat generated in the shock is converted to radiation as the gas cools, or non-radiative, in which case the cooling time scale is long compared to the dynamical time [18]. In the radiative shocks, Coulomb collisions usually equilibrate T_e with T_p before most of the radiation is emitted, so that the signatures of collisionless shock physics are erased before the observable radiation is produced. In non-radiative shocks, however, optical and UV radiation arises in a thin ionizing layer just behind the shock, and the collisionless shock physics can be studied. The non-radiative shock emission is faint, but it presents unique diagnostics that complement *in situ* measurements of shocks in the heliosphere, in particular extending to very high Mach numbers.

NON-RADIATIVE SHOCK DIAGNOSTICS

When a neutral atom encounters a collisionless shock, it does not feel the electromagnetic fields or turbulence, so it passes through undisturbed. Finding itself immersed in hot plasma, it will soon be ionized. Before that happens, however, it may be collisionally excited by an electron, or for shocks faster than 1000 km/s by a proton. The line profile of the resulting emission will correspond to the pre-shock velocity distribution. It is also possible for the neutral to undergo charge transfer before it is excited, in which case the newly formed neutral has the velocity of the incident proton. In that case the line profile width corresponds to the post-shock proton temperature. Thus a hydrogen line from a non-radiative shock has two components of roughly equal intensity; the narrow component width is a measure of the pre-shock kinetic temperature, while the broad component width is comparable to the shock speed, and it measures the post-shock proton temperature [3]. Furthermore, the intensity ratio of the broad and narrow components depends on the ratio of charge transfer and ionization rates, and therefore on the electron temperature [13, 5].

The optical emission from a fast non-radiative shock is an almost pure hydrogen Balmer line spectrum with extremely faint helium lines. Ultraviolet lines of He II, C IV, N V and O VI are also observed [23, 24]. Their line widths directly measure the post-shock ion temperatures of those elements, and their intensity ratios provide consistency checks on the electron temperature. X-ray spectra provide another means of measuring electron temperatures provided that one has high enough spatial resolution to measure the spectrum just behind the shock before Coulomb collisions have much effect. Several papers have compared electron temperatures derived from X-ray spectra with proton temperatures derived from Hα line widths or from shock speeds [9, 16, 20]. It is also possible to compare line widths with Doppler shifts or known shock speeds to constrain the ratios of proton, electron and ion heating in coronal shocks [25, 26, 17]. Several more shock candidates have been recently found in UVCS observations of Coronal Mass Ejections [2], but they have not yet been fully analyzed.

TABLE 1. Electron, Proton and Oxygen Temperature Ratios

SHOCK	V_s	T_e/T_p	T_O/T_p	Reference
1E102-7219	6000	<0.02		9
SN1987A	3500	0.05-0.15		19
Tycho(X-ray)	3000	~0.15		21
SN1006(opt)	2890	<0.07		6
SN1006(UV)	2890		8	16
Tycho(opt)	2100	<0.20		5
CMEs(UV)	1200	<0.50	>8	26,17
DEML71(opt,X-ray)	1000	<0.40		7,20
DEML71(opt,X-ray)	800	0.4-0.8		21
Comet(UV)	700	~0.2		25
RCW86(opt)	600	0.4-0.5		5
DEML71(opt,X-ray)	550	>0.85		7,20
Cygnus(opt)	350	>0.70		5
Cygnus(UV)	350		<2.7	27

RESULTS

Rakowski [21] presents the ratios T_e/T_p for eight SNRs measured so far. Table 1 adds the the heliospheric remote sensing measurements and the ratios T_O/T_p of the kinetic temperatures of oxygen and protons.

There is an obvious tendency for T_e/T_p to decline as the shock speed increases and for T_O/T_p to increase toward the mass ratio. Ideally one would select the one shock model that matches these trends and call it a day. Unfortunately, such a model does not seem to exist. Electron heating in shocks has been modeled as resulting from a combination of Buneman and ion-acoustic instabilities [1], lower hybrid waves (e.g., [31, 12]) or the shock electric field [29]. Cargill & Papadopoulos [1] predicted $T_e/T_i \sim 0.2$, and the lower hybrid heating produces a hard tail on the electron distribution. As long as the electrons are not too hot, Coulomb equilibration may convert the tail produced by lower hybrid waves into general electron heating. Measurements of shocks in the solar wind indicate electron heating of the order of 20% of the proton heating, but with large scatter [28]. Two major difficulties in connecting the astrophysical T_e/T_p determinations with theory are that the Alfvén speed is only roughly known and the angle between the shock normal and the magnetic field is essentially unknown.

Finally, it is worthwhile to mention some caveats. One is that the line widths may contain contributions from the bulk motion of the shocked gas if the shock front is curved. This can be assessed from the curvature of the shock front inferred from the size of the remnant or from the scale of ripples in the observed Hα filaments (e.g. [27, 22]), and is generally small. Another contribution to the line widths could arise from turbulence in the shock. This turbulence will die away very quickly if the turbulent speeds are large enough to significantly broaden the lines, so it is unlikely to be important [10].

A second caveat is that there could be other contributions to the narrow component of the Hα line, and that could lead to erroneous values of T_e/T_p. There is reason to believe that this happens in Kepler's SNR, where the broad to narrow intensity ratio in several filaments lies below any value predicted by the models. Other non-radiative shocks show broad to narrow ratios near the lower end of the allowed range. The additional narrow component contributions might come from radiative shocks (particularly in Kepler) or a photoionized precursor such as that around Tycho's SNR [4]. It is also possible that Hα could be excited in a a shock precursor heated by cosmic rays and turbulence or by fast neutrals that overtake the shock [30, 8]. The cosmic ray precursor is required by diffusive shock acceleration models, and the diffusion coefficient can be constrained by Hα profiles. Lim & Raga [15] modeled the precursor due to fast neutrals and found the heating to be small. However, they did not consider wave generation by the protons formed when those neutrals are ionized. Korreck [11] investigates that process. Fully self-consistent models of the shock and precursor are needed for more complete analysis of the non-radiative shock observations.

ACKNOWLEDGMENTS

This work was partially supported by NASA grants NAG5-12446 and NAG5-12827 to the Smithsonian Astrophysical Observatory.

REFERENCES

1. Cargill, P.J. & Papadopoulos, K. 1988, ApJL, 329, L29
2. Ciaravella, A., Raymond, J.C. & Kahler, S.W. 2005, submitted to ApJ
3. Chevalier, R.A., & Raymond, J.C. 1978, ApJL, 255, L27
4. Ghavamian, P., Raymond, J.C., Hartigan, P. & Blair, W.P. 2000, ApJ, 535, 266
5. Ghavamian, P., Raymond, J.C., Smith, R.C., & Hartigan, P. 2001, ApJ, 547, 995
6. Ghavamian, P., Winkler, P.F., Raymond, J.C., & Long, K.S. 2002, ApJ, 572, 888
7. Ghavamian, P., Rakowski, C.E., & Hughes, J.E. 2003, ApJ, 590, 833
8. Hester, J.J., Raymond, J.C., & Blair, W.P. 1994, ApJ, 420, 721
9. Hughes, J.P., Rakowski, C.E., & Decourchelle, A. 2000, ApJL, 543, L61
10. Korreck, K.E. 2005, PhD Thesis, University of Michgan
11. Korreck, K.E., Raymond, J.C., Zurbuchen. T.H., & Ghavamian, P. 2004, ApJ, 615, 280
12. Laming, J.M. 2001, ApJ, 546, 1149
13. Laming, J.M., Raymond, J.C., McLaughlin, B.M., & Blair, W.P. 1996, ApJ, 472, 267
14. Lee, L.C., & Wu, B.H. 2000, ApJ, 535, 1014
15. Lim, A.J. & Raga, A.C. 1996, MNRAS, 280, 103
16. Long, K.S., Reynolds, S.P., Raymond, J.C., Winkler, P.F., Dyer, K.K. & Petre, R. 2003, ApJ, 586. 1162
17. Mancuso, S., Raymond, J.C., Kohl, J.L., Ko, Y.-K., & Wu, R. 2002, A&A 383, 267
18. McKee, C.F., & Hollenbach, D.J. 1980, ARAA, 18, 219
19. Michael, E., et al. 2002, ApJ, 574, 166
20. Rakowski, C.E., Ghavamian, P., & Hughes, J.P. 2003, ApJ, 590, 846
21. Rakowski, C.E. 2005, submitted to Adv. Sp. Res. (astro-ph/0503636)
22. Raymond, J.C. 2003, RevMexAA (Serie de Conf.) 15, 258
23. Raymond, J.C., Blair, W.P., Fesen, R.A., & Gull, T.R. 1983, ApJ, 275, 636
24. Raymond, J.C., Blair, W.P. & Long, K.S. 1995, ApJL, 454, L31
25. Raymond, J.C., et al. 1998, ApJ, 410, 417
26. Raymond, J.C., et al. 2000, GRL, 27, 1439
27. Raymond, J.C., Ghavamian, P., Sankrit, R., Blair, W.P., & Curiel, S. 2003, ApJ, 584, 770
28. Schwartz, S.J., Thomsen, M.F., Bame, S.J. & Stansberry, J. 1988, JGR, 93, 12923
29. Scudder, J.D. 1995, Adv.Sp.Res. 15, no. 8/9, 181
30. Smith, R.C., Raymond, J.C., & Laming, J.M. 1994, ApJ, 420, 286
31. Shapiro, V.D., Bingham, R., Dawson, J.M. et al. 1998, Phys. Scr. T75, 39

Author Index

A

Achilleos, N., 109
Acuña, M. H., 267
André, M., 79
Ao, X., 170
Armitage, P. J., 325
Ashmall, J., 299

B

Bale, S. D., 79
Balikhin, M. A., 123
Bamert, K., 129
Baring, M. G., 207
Bedros, R., 64
Behlke, R., 79
Ben-Jaffel, L., 294
Bertucci, C., 109
Blanco-Cano, X., 27
Burgess, D., 17
Burlaga, L. F., 261, 267
Burton, M., 109

C

Cohen, C. M. S., 227
Cummings, A. C., 227, 261, 273

D

Dandouras, I., 146
Dasgupta, B., 64
Decker, R. B., 180
Deehr, C., 304
Desai, M. I., 180, 219, 227
Detman, T., 304
de Zeeuw, D. L., 201
Dougherty, M. K., 109
Driscoll, J., 56
Dryer, M., 304
Dubinin, E., 141, 146
Dwyer, J. R., 219

F

Fisk, L. A., 252
Florinski, V., 170
Fränz, M., 146
Fry, C. D., 304
Fujimoto, R., 56
Fukazawa, K., 95

G

Giacalone, J., 213
Glaßmeier, K.-H., 146
Gloeckler, G., 252
Gombosi, T. I., 201
Grygorczuk, J., 294
Guo, X., 320

H

Haggerty, D. E., 227
Hamza, A. M., 116
Heikkila, B. C., 261
Hilchenbach, M., 129
Ho, G. C., 180
Hu, Q., 180, 233
Hu, Y., 320
Huang, Z., 320
Hudson, H. S., 336

I

Intriligator, D. S., 304
Intriligator, J., 304

J

Jokipii, J. R., 201, 283
Jones, F. C., 165
Joy, S. P., 95

K

Kaghashvili, E. K., 331
Kallenbach, R., 129
Karimabadi, H., 56
Khurana, K., 95
Kivelson, M. G., 95
Korreck, K. E., 342
Korth, A., 146
Kóta, J., 201
Krauss-Varban, D., 56
Kucharek, H., 32, 79, 151

L

Labrador, A. W., 227
Lal, N., 261
Lanzerotti, L. J., 252
Lario, D., 180
Lee, M. A., 151, 240
Le Quéau, D., 89
le Roux, J. A., 170, 191, 196
Leske, R. A., 227
Li, G., 170, 191, 227, 233
Lin, R. P., 246
Lin, Y., 313
Liu, Y. C.-M., 151
Lodato, G., 325
Looper, M. L., 227
Lucek, E. A., 79

M

Maclennan, C. G., 227
Manchester, W. B., 201
Mason, G. M., 219, 227
Matsukiyo, S., 22
Mazelle, C., 89, 109, 116, 146
Mazur, J. E., 219, 227
McDonald, F. B., 261, 267
McKenzie, J. F., 141
Melrose, D., 135
Merka, J., 84
Mewaldt, R. A., 227
Meziane, K., 89, 116
Möbius, E., 32

N

Ness, N. F., 261, 267

O

Ogilvie, K. W., 72
Ogino, T., 95
Omelchenko, Y., 56
Omidi, N., 27

P

Parks, G. K., 116
Perumalla, K., 56
Pogorelov, N. V., 42

R

Ratkiewicz, R., 294
Raymond, J. C., 342
Rème, H., 146
Rice, W. K. M., 170, 325
Richardson, J. D., 261, 278, 299
Russell, C. T., 3, 27, 109

S

Sauer, K., 141, 146
Scholer, M., 22
Shapiro, V. D., 159
Smith, C. W., 129, 180
Smith, E. J., 109
Stone, E. C., 227, 261, 267, 273
Summerlin, E. J., 207
Sun, W., 304
Szabo, A., 37

T

Tanaka, T., 289
Tsurutani, B. T., 109
Tylka, A. J., 185

U

Üçer, D., 159

V

Vasquez, B. J., 331
Verkhoglyadova, O. P., 196
Viñas, A.-F., 180

W

Walker, R. J., 95
Walker, S. N., 123
Wang, C., 278, 320

Wang, X. Y., 313
Washimi, H., 289
Webb, G. M., 64, 141, 170, 191
Webber, W. R., 261
Wiedenbeck, M. E., 227
Wilber, M., 89, 116
Winske, D., 50

Y

Yin, L., 50

Z

Zank, G. P., 64, 170, 191, 233, 289, 331